T0301863

Particle Therapy Technology for Safe Treatment

Particle Therapy Technology for Safe Treatment

Jay Flanz

CRC Press
Taylor & Francis Group
Boca Raton London New York

CRC Press is an imprint of the
Taylor & Francis Group, an **informa** business

Cover Art: Heaven's Charged Particle Scanning System

First Edition published 2022
by CRC Press
6000 Broken Sound Parkway NW, Suite 300, Boca Raton, FL 33487-2742

and by CRC Press
2 Park Square, Milton Park, Abingdon, Oxon, OX14 4RN

ISBN: 978-0-367-64014-9 (hbk)
ISBN: 978-0-367-64311-9 (pbk)
ISBN: 978-1-003-12388-0 (ebk)

DOI: 10.1201/9781003123880

Typeset in Times LT Std
by KnowledgeWorks Global Ltd.

Dedicated to Nancy, Adam and Scott

Thank you for your patience while putting up with it all.

Fun with Family helps to keep Physics Fun!

Cover Art: Heaven's Charged Particle Scanning System

Contents

1 Introduction

The path from clinical requirements to technical implementation is filtered by the translation of the modality to the technology. For example, the word 'safe' is defined in the context of the application. The appropriate interpretation of that context is essential. It's reasonable for there to be a deterministic flow from requirements to implementation in any application. For particle therapy, it helps to understand what clinical parameters affect the safety of a treatment, and then the filter is constructed when determining how the technology can affect those parameters. Unsafe scenarios are to be deduced or inferred and mitigated. This all begins as it ends by asking if you would feel comfortable having a family member receive treatment.

Particle therapy is a multidisciplinary application that benefits from the insights of experts in these disciplines. It is the intention to weave many of the disciplines together to construct the framework for the application. This includes the context for the meanings and the relevance of the parameters as well as limits imposed by the physics and engineering. Well, in fact, it is difficult to be exhaustive in a book this size, so a sampling of this is offered, based on personal experience. The book builds with a discussion of some tools, an introduction to the application and the technology of the components that are needed. At almost every step the question, what could go wrong, is asked. The statement that the system must be safe is insufficient. A look into the clinical and technological considerations to achieve a realizable set of preconditions that has safety at its core could be a sufficient start. This may not be done explicitly in every section and for every subject, but it is hoped that there will be sufficient clarification for a safety filter to be internalized.

Somewhat underlying the above is the approach of obtaining the necessary information about the application before considering how to create the technical components. The information necessary to learn about the applications to be presented can be acquired from many sources. The corollary to this is that many good sources exist and it is not the author's intent to reinvent the wheel. Yet, there are many assumptions in these different sources, so it is hoped that fewer assumptions are made herein. Some of the why is included with the what. If the reader finds the topics interesting and needs additional perspective, that information probably exists.

The approach used is one that leads toward the identification of sensitivities and tolerances. The technology is interpreted while applying a filter that includes the clinical application and safety. Much of this filter has to do with considering what the tolerances are and what could go wrong. It is not intended to fully design a system or even identify all the final specifications, but it is hoped there is enough for many back of the envelope estimates. It's probably best to get a new box of envelopes now.

To accomplish this, it will be necessary to review some of the fundamental principles. It is assumed that the audience for this book could be from various multidisciplinary fields. The times that the author taught the 'Medical Applications of Accelerators' course at the US Particle Accelerator School, the class usually included participants from both the medical and accelerator communities. A foundation is offered that includes the clinical application, physical principles related to the interaction of particles with matter, charged particle beam acceleration and transport and even some engineering considerations. These are described at a level sufficient to build on in the text. Almost everything that is introduced is used, although it may not always be obvious. Some of it will be new for some and other topics will be review for others.

Putting all this together in one place at a level to understand how these can be used together is the challenge. Much that is discussed in this book is placed in a context of safety, even if the word 'safety' is not used on every page. The word 'safety' is used several times in this introduction, to make

DOI: 10.1201/9781003123880-1

up for that. Asking questions about how a parameter might affect a treatment and how a component might affect a parameter will help the reader to get what's intended from what's been written. In the end, the reader is rewarded with this. Sections entitled 'what could go wrong' are not meant to cause concern, but only to heighten the thought process.

Part of the reason this book has been written after the author's 45+ years in these fields is the wish that a book like this existed when I started. I sincerely hope that you will find this contribution helpful and perhaps enjoy a few smiles along the journey.

2 Evolution of Medical Particles

The discovery of particles at the atomic level and smaller and their interactions with matter has been the subject of considerable interest for over 100 years. As the properties of the particles and their interactions became better understood either quantitatively or qualitatively, applications to use these particles were developed. To better study particles and their interactions, accelerators have been constructed and the field of beam physics was developed. It was only recently (in the last few decades) that the study of the physics of beams has become a legitimate field of investigation. Generally, it was justified by the applications it served such as nuclear and high-energy physics. Furthermore, it may be felt by some that nuclear and high-energy physics themselves are only a stepping stone to develop useful applications for humanity. The ultimate application for society may be medicine. The fundamental search for knowledge has always paid off whether it has been the main goal of research or not. The key to applying the results of research is to have in mind an overview of the relevant phenomena. Beginning with an understanding of what is required for a particular application, the requirements can be followed to determine the details of the implementation.

The applications of particles in medicine have been recognized from the earliest time that particles were discovered. The difficulty involves the preparation and delivery of these beams (from naturally occurring beams [e.g. radioisotopes] to accelerator-produced beams). From the time, in 1895, when Roentgen discovered X-rays, and in 1913, when Coolidge developed the vacuum X-ray tube, it has been shown that energetic particles can be useful for medical applications. It is clear that there is a wide range of applications and, therefore, also a wide range of requirements for the particles that must be considered.

Almost in parallel with the discovery of physical phenomena and the development of physical devices, people started using these discoveries for treating those who were ill. Some of these discoveries include:

- Magnetic fields
- High frequencies
- X-rays
- Electrons
- Protons
- Lasers

Most of the accelerators existing today are used for applications in the field of medicine. While accelerators were originally introduced for research, the use of energetic particles for medicine is natural. Medical applications require energetic beams (particles with energies higher than thermal energies). Table 2.1 highlights some of the firsts in the chronology of accelerators used for medicine.

Some of the early history is quite interesting and highlights the synergy and interactivity of physics and medicine and money. In the 1930s, early trials at CalTech following the work of Milliken and Lauritsen gave rise to a 750 keV X-ray generator and this was used for patient treatment. Lauritsen's students operated the machine during the day for patient treatments. W.W. Kellogg (of Cornflakes) donated money to establish the Kellogg Labs at CalTech in 1931. W.W. Kellogg was the brother of John Kellogg who ran the Battle Creek Sanitarium in Michigan where James Case had installed the first 200 kV generator that was used for patient treatment.

DOI: 10.1201/9781003123880-2

TABLE 2.1

Highlights of Technology for Medicine

Year	Energy	Particle	Event
1895	5 kV	X-ray	K. Roentgen discovers X-rays
1913	keV range	X	W.E. Coolidge develops vacuum X-ray tube
1931	80 keV	H	Ernest Lawrence develops a cyclotron (1939 Nobel Prize)
1937	1 MV	X	Air-insulated Van de Graaff installed at Huntington Memorial Hospital (Boston)
1939	8 and 21 MeV	d,n	First medical cyclotron, Crocker Lab, Berkeley (USA) (R. Stone)
1940	1.25 MV	X	Pressure Van de Graaff accelerator at Massachusetts General Hospital
1946	Any MeV	p, Ion	Robert Wilson suggests medical use of protons and heavy ions
1949	20 MV	X	First patient treated with 20 MV X-rays from betatron at the University of Illinois
1950	8 MeV	e, X	First medical rf linear accelerator at Hammersmith Hospital in London
1954	100 MeV	p, He	First proton beam treatments at Berkeley
1954–1958	4 MeV	e, X	First 140° linac gantries in United Kingdom, New Zealand, Australia
1956	6 MeV	E, X	First United States 6 MeV linac in operation at Stanford Lane
1957	200 MeV	p	Proton synchrocyclotron beam treatments begin at Uppsala, Sweden
1959	50 MeV	E	First scanning electron beam (5–50 MeV) from linac for cancer (Chicago, USA)
1961	160 MeV	p	Proton beam treatments begin at the Harvard Cyclotron
1962	6 MeV	e, X	Isocentric (360° gantry) linear accelerator installed in the United States
1976	50 MeV	Π	First irradiation with a pion beam in an rf linear accelerator LAMPF (USA)
1982	30 MeV	N	First cyclotron used for neutron rotation therapy in the United States
1990	30 MeV	N	First superconducting cyclotron for neutron therapy (Harper, USA)
1990	250 MeV	p	First hospital-based proton synchrotron used for radiotherapy (Loma Linda, USA)
2001	230 MeV	p	First commercially built proton therapy system for a hospital (MGH, USA)

During that time, it was said that 'such formidable installations would be prohibitive for the average radiologist to consider and would be limited to those institutions with engineering and physics skills available and should be centralized'. So was planted the seed for the first generation of particle therapy (PT). In the same time period, 1–2 MV generators were developed by General Electric. In 1932, Van de Graaff traveled from Princeton to MIT. He heard Dr. George W. Holmes, a radiologist from MGH speak. He realized that his high-energy accelerators could be of use. With a grant of $25,000, the first 1 MV Van de Graaff generator was built for therapy. The first patient was treated on March 1, 1937. Later in the 1930s, Donald Kerst developed the first betatron at the University of Illinois. While he was a teaching assistant at the University of Wisconsin, he took part in seminars at the University of Wisconsin General Hospital and recognized the utility of high-energy electron accelerators in medicine. Later he worked for GE and developed a 20 MeV medical betatron in 1942. In the 1950s, the first linear accelerator for therapy was installed in Hammersmith Hospital in London and patients were treated from 1952 through 1969. Shortly thereafter one was installed in Stanford. In 1946, Robert Wilson published an article recognizing that the physics of the Bragg peak could be useful for medical therapy. Table 2.2 traces some of the history of charged PT.

All in all, this is quite an interesting history. While it's true that personal connections were certainly involved, the fact that physicists went to listen to medical talks, informally and vice-versa, is

TABLE 2.2
Charged Particle Therapy Milestones

Time Since the Start of the Universe	Event	Time Since the Last Development
$t = 0$ second	*The universe was created*	∞
$t \sim 0.2$ second	*Protons were created*	$\Delta = 0.2$ second
$t \sim 10 \times 10^9$ years	*First charge particle scanning – Northern and Southern Lights (see book cover)*	$\Delta = 10^{10}$ years
$t = 1780$	*Coulomb developed Coulomb's laws*	$\Delta = 10^9$ years
$t = 1892$	*Lorentz introduced the Lorentz force*	$\Delta = 100$ years
$t = 1946$	*Wilson foresaw a clinical use*	$\Delta = 50$ years
$t = 1957$	*Beginning of proton therapy in Sweden*	$\Delta = 10$ years
$t = 1975$	*Koehler used double scattering w/protons*	$\Delta = 30$ years
$t = 1980s$	*PTCOG formed to develop Particle Therapy*	$\Delta = 5$ years
$t = 1985$	*First hospital-based facility (Loma Linda)*	$\Delta = 5$ years
$t = 2001$	*Second hospital based (MGH), (1ˢᵗ Vendor IBA)*	$\Delta \approx 5$ years
$t > 2002$	*Proton facilities multiply*	$\Delta \approx 1$–2 years
$t = 2008$	*First hospital-based scanned beam treatment*	$\Delta \approx 6$ years
$t = 2021$	*I wrote this book*	$\Delta \approx 13$ years

something that may be lost in today's society. Granted there were new innovations and the young inventors were certainly looking for opportunities to utilize their systems. There is perhaps a not-so-fine line between 'Have Accelerator, Will Travel' (an accelerator exists, can you use it?), and identifying the requirements of a modality prior to developing a system for that. Still today, there are people who want to offer different accelerator ideas for use in PT. So far, no accelerators, other than the cyclotron and synchrotron, have proved successful in this field.

Accelerators of various types are used to generate these beams. The accelerator is one of the key components of this equipment, and its continued future will be dictated by its ability to accommodate evolving clinical requirements. It is not the only, or possibly even the most important technical component. Beam delivery systems tailor the beam for treatment. The field of PT is quickly growing and yet its more widespread adoption is limited by size and cost. In order to fully realize the benefits of this modality, the equipment used to generate and deliver the beam is evolving.

The growth of the number of PT facilities in the world is fascinating to follow. While it's clear that, at the start, people with forward-thinking minds, usually at academic medical centers, would be part of the inception and the early adoption of the modality, it's perhaps not as obvious what would be the next step. One can look at a map of the United States to partially answer that question. Figure 2.1 is a population density map of the United States with the regions of high population density denoted by darker shades.

Presently, the location of PT centers in the United States is shown in Figure 2.2. Note that all these centers are located in regions of high population density. There are still high density population centers lacking PT nearby, particularly in the center of the country. However, considering the distribution of the high population centers in the United States, there are wide regions between high density populations and it is likely that the distance to travel for many will be prohibitive, so there is likely to be more located in lower density areas. The same logic holds for Europe and one can imagine the equivalent

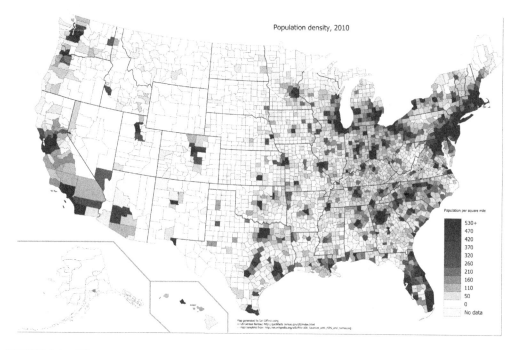

FIGURE 2.1   US population density map.

FIGURE 2.2   North American particle therapy facilities.

logic for China, which has vast regions with population densities higher than that in the United States. A recent accounting of the proton centers in the world is shown in Figure 2.3. Of note is the concentration between 20°N and 66°N latitude.

Some years ago, treatments were suspended in the one PT treatment center in South Africa. A table with more information about these facilities is in Appendix A. The growth over time has been almost exponential as shown in Figure 2.4. Projections forward (after the break in the curve)

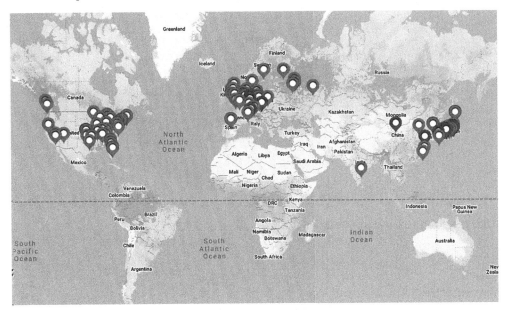

FIGURE 2.3   World particle therapy facilities.

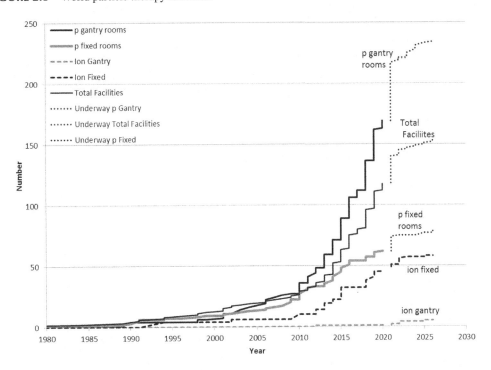

FIGURE 2.4   Particle therapy number history.

seem flatter, but that has always been the case (for projections), and even more so now in the middle (end?) of the coronavirus pandemic. The distribution across the world is shown in the pie chart of Figure 2.5. If a region is not included, PT does not yet exist there. Those same regions are plotted in Figure 2.6(left) for megavoltage (e.g. Linac)-based therapy (Japan is part of Asia). They are remarkably similar. However, Figure 2.6(right) shows all the MV therapy in the world and the picture changes with the numbers in Africa and South America.

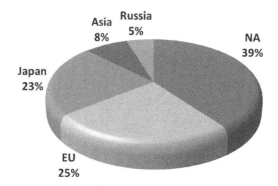

FIGURE 2.5   Distribution of particle therapy.

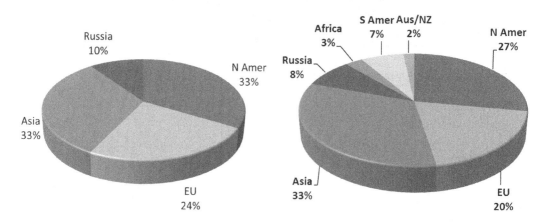

FIGURE 2.6   MV therapy world distribution (left) and MV therapy world distribution (full) (right).

It is instructive to examine some of these regions further by socioeconomic factors. Table 2.3 includes information about the main geographic regions in PT. The population is given in millions and the number of PT rooms (not just facilities) is listed, including currently operating, under construction and in planning stages. This total is compared to the population in that region. The rightmost columns are millions of people per PT room.

Figure 2.7 contains the same data sorted in different ways. The gross domestic product (GDP) and GDP/Capita (GDP/Cap) are plotted with the millions of patients per room total from Table 2.3. The graph in Figure 2.7(top) shows that the countries with the lowest population per room are also in the higher GDP category, so there is a direct relation between the resources of a country and the number of PT facilities constructed. This results from both government and industry factors in those countries. Figure 2.7(bottom) shows the same data sorted with increasing GDP. This shows a couple of things. The GDP/Cap while overall showing the same trend as the GDP does show

TABLE 2.3
Regional Statistics

Region	Population Millions	Rooms Operating	Under Construction	Sub Total	In Planning	Total	Millions Per Room	Mill/Rm Incl Planning
Asia	4,336	28	62	**90**	20	**110**	**48.18**	39.42
EU	508	82	11	**93**	12	**105**	**5.46**	4.84
Japan	126.8	51	9	**60**	6	**66**	**2.11**	1.92
United States	327.2	94	14	**108**	0	**108**	**3.03**	3.03
Russia	144.5	7	5	**12**	2	**14**	**12.04**	10.32
		262	101	**363**	40	**403**		

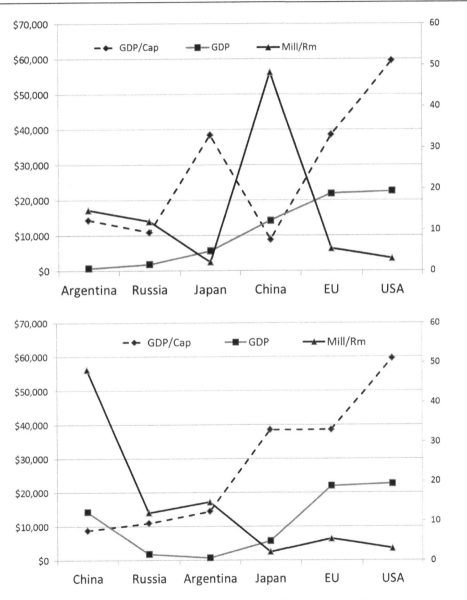

FIGURE 2.7 Plot of various socioeconomic data by region (top) and plot of various socioeconomic data by region resorted (bottom).

some fluctuations. The correlation, however, between GDP/Cap and millions/room is not smooth (the GDP/Cap is more or less flat for the first two points, but the Mill/Rm is not). Perhaps the main reason is the situation of China. Without China, the GDP/Cap would have a smooth correlation with the inverse of the millions/room. This is likely to start changing fairly soon. Thus, the correlation of PT that is available is closer to the GDP/Cap than to the GDP. China has a higher GDP but with a huge population.

All of this points to the fact that the costs of PT seem to be an impediment to even faster growth. Given the similar distribution between MV and PT therapy, it isn't about the modality. Differences include a larger proportion of electron accelerators in Asia and a larger proportion of particle facilities in Japan. Therefore, there seems to be cultural aspects to the distribution also.

It can be noted that some of this has to do with the capital cost of a particle accelerator. Many who read this sentence could believe that this is a sarcastic statement of the obvious. However, there are some misconceptions. It is important to compare apples and oranges when comparing the costs of a PT system with an electron linac system for photon therapy. Some factors for proton therapy facilities are listed in Table 2.4.

TABLE 2.4
Proton Therapy Capital Costs

Factor	Proton	Photon
Building	$20,000,000[1]	$0? (within existing building)
Equipment	$50,000,000[2]	$6,000,000
No. of rooms	3	1
Lifetime	20 (maybe 30)	10[3]

[1] Many systems are quoted as costing $100M or double of that. Most of the time, this includes the construction of an entirely new cancer center with diagnostic equipment and other services. This cost should not be directly compared. The figure used is that of a modest new building, but it can be possible to install proton equipment in an existing building (e.g. as was done at MGH).

[2] Most proton centers include multiple treatment rooms. The cost shown is for a 3-room system.

[3] Medical linacs are not built for a long lifetime and sometimes used to have gantry bearing issues, for example. But really the issue is that new features are introduced and medical centers would like to improve their patient treatment capabilities. Depreciation is a standard part of owning a linac. Depreciation is not generally taken for PT.

With these considerations, the total cost of a proton facility/room/10 years is $8.3M (or $10M depending on how well one negotiates), which is more directly comparable to the $6M linac.

Beyond this, an article by Martin Jermann and Michael Goitein analyzed various other cost considerations, including operations, business expenses, building, infrastructure, preventive maintenance etc. The conclusion was that the cost per treatment of a proton system is 2.39 times the cost of the X-ray system. However, a critical challenge to these numbers can identify some specific questions. A subjective, devil's advocate response to these questions can include:

- The project management (PM) costs of proton equipment should not be significantly higher than X-ray equipment. Do not include the PM costs of a brand new cancer center.
- The building cost has been mentioned in Table 2.4; however considering a new building for X-ray equipment and a new building for a single-room proton, equipment could be comparable, especially with modular approaches being developed.
- Equipment operations should be within a factor of 2 of each other.

- It's unclear why the business costs should be much different, now that proton therapy installation is more mature.
- The treatment time per fraction is closer and closer to X-ray beam setup and delivery.

All in all, the author has analyzed the various factors and the spreadsheet shows that it should be possible to achieve similar patient throughput and the cost/treatment can be closer than a factor of 1.2. New development of systems must take into account these factors to ensure that the actual costs are known, and therefore the perceived cost differential is minimized. Then the impediment to faster growth can be reduced. Perception is generally taken as reality. It remains for reality to influence the perception by example. MGH showed that it's possible to install a proton beam inside an existing and operating radiotherapy department. Single-room systems are becoming more available. Community health centers are now offering proton therapy. The field is changing and growing.

3 A Personal Historical Perspective

The evolution of particle therapy is now in its fourth generation. In 1946, Robert Wilson published the famous observation that the behavior of charged particles in matter could be beneficial for the treatment of cancer. At that time, production of charged particles with the energy and beam properties necessary for particle therapy was only possible in national accelerator laboratories and yet treatments began there over 50 years ago. In 1957, patient treatments began in Sweden and early adopters included Berkeley National Laboratory, the Harvard Cyclotron Laboratory, UC Davis, Fermi National Accelerator Laboratory (*uncharged* neutrons). The development actually paralleled, in time, the start of linac-based electron and photon therapy, but that modality was possible to construct by private industry vendors, and it spread quickly.

In the 1980s–1990s, it was imagined that proton therapy could be possible in a hospital environment. The second generation of particle therapy began when Loma Linda built the first hospital-based system, with the help of Fermilab and commercial firms. A wonderful collaboration among interested parties to specify the requirements of that facility saw the formation of the Particle Therapy Co-Operative Group (PTCOG), started by Jim Slater, Michael Goitein and Herman Suit. Next, the Massachusetts General Hospital constructed the second hospital-based system, but the first one entirely (well almost) constructed by a commercial vendor, IBA. Around that time, NCC East in Japan also constructed a similar facility with the help of Sumitomo Heavy Industries.

I joined MGH in 1993 after spending 15 years at MIT involved in nuclear and accelerator physics. I was hired at MIT by an individual who may have had the most influence in my professional life. I was fortunate to be at a program advisory committee meeting where, as a student, I was to present an experimental proposal (which was accepted). It was not typical for a student to present a proposal, so I was nervous. I was waiting my turn in the back of the room when a person walked in wearing a torn sweater with stained fingers, unkempt hair and a scruffy beard. He sat in a chair, crossed his legs in a lotus position and started to roll something that would be smoked. I couldn't believe he had the audacity to enter this high-level meeting and I lost focus on what was happening at the front of the room. Well, what was happening was that several individuals got into a disagreement about physics principles and experimental techniques. This went on for some minutes without any resolution. Then, this individual, who I thought shouldn't have been in the room, got up from his chair, began to speak and everyone went silent. He categorically clarified every point of confusion, sat down and lit his smoke. I told myself that I wanted to work with him one day. It was quite a happy day when I was talking to Phil Sargent a few years later about my employment offers (there were no open positions at the time at MIT), when he said – no don't go there, I'll hire you. At MIT, I designed a beam recirculation system for a 200 m long high precision linac. I also designed a one GeV electron pulse stretcher/storage/synchrotron for nuclear physics. (We built both and they operated well.) Through it all I owe much to Phil, Bill Turchinetz and Peter Demos.

During an external review for the ring, a reviewer, Jose Alonso from Berkeley, told us about the proton therapy initiative at MGH. I had some peripheral familiarity with the medical field at that time, having had the high pleasure of interacting with Jake Haimson, who was the designer of the MIT-Bates accelerator (and Linacs in Saclay and Saskatoon among many other things). It turned out he was engaged for the MIT-Bates design after working with Varian on the design and construction of one of the first 360-degree rotatable Linac gantries as head of accelerator research. And before that was responsible for hospital Linac installations for MVEC. That company installed the first medical Linac in Hammersmith Hospitals in 1953. The patents in the early 1960s for a linac on a

DOI: 10.1201/9781003123880-3

gantry all bear his name. He also invented the technology for the first fast CT scanner, the Imatron, using plasma-focused electron beams, among other innovations in medical technology. Imagine my surprise a decade or two later when I met his friend Sarah Donaldson and found out who she was. During my work on the electron storage/stretcher ring, I had another honor to interact with Mikael Eriksson of Lund Sweden, famous for among other things the Max-Lab Lund synchrotrons. Earlier in his career, he contributed (may be too weak a word) to the MM50 electron microtron, an accelerator for medical treatment. Other groups were interested in high duty factor pulse-stretching synchrotrons, such as the team in Saskatoon. I interacted with so many talented individuals, including Dennis Skopik, but special mention is for one who became a longtime mentor of mine (and other friends of mine). He was Roger Servranckx coming to Saskatchewan by way of Belgium and the Belgium Congo. Beyond all the technical aspects of ring design, I learned that there were more beers than pilsners. I do want to mention here my interactions with Klaus Halbach and Harald Enge, both fathers of magnetics and optics. I, myself, benefitted from a look into Klaus's famous black book (which later was published as an aqua green book)! Harald, who used to use a draftspersons table to trace rays, before computers, was the optics designer of the Loma Linda proton therapy corkscrew gantry, and the first compact gantry for PSI. Eventually, I joined MGH during the proposal stage and met Yves Jongen of IBA. I learned much of what I know about cyclotrons from him. Yves tried to convince me that two people can disagree on a technical solution while both are right. I'm not sure that I ever fully bought into that. The rest of my cyclotron education comes from my Sensei Yukio Kumata whose deep experience he openly shared. Figure 3.1 summarizes some of this. This isn't a full acknowledgment list (there are so many more friends and colleagues), but a short digression from the narrative to which I now return.

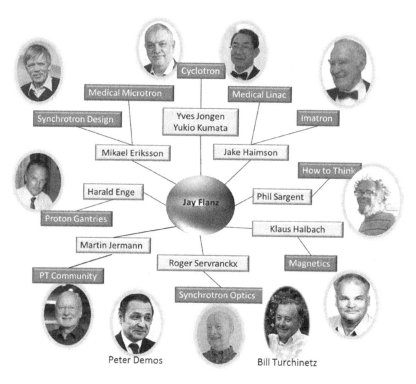

FIGURE 3.1 Author's mentor tree.

Considering the unfortunate dwindling funding situation for nuclear physics in the United States, at the time and the extremely interesting sounding health-care project, I inquired about the opportunities at MGH and joined the project in the proposal stages. There were cultural differences to which I had to adapt. Physicians used 35 mm slides, not 'dirty' overhead transparencies with 'fingerprints'. Some say that there were political machinations, of which I was unaware, but eventually MGH was awarded funds for the construction of a proton facility. The history of this award is not without some interest. In the 1980s, the promise of neutron therapy was courted. In fact, the National Cancer Institute (NCI) awarded funds for the production of several neutron therapy systems that were cyclotron based. Not all these facilities were completed and a primary company had some issues. This was the last time that the NCI provided funding for the construction of medical therapy equipment of this magnitude. The funds awarded to MGH were for the construction of the building, not the technical equipment.

Initially, we had enough funds for only one gantry. However, some crafty contractual terms (which I suggested) led to the possibility to consider two gantries when I redesigned the proposed beamline, considerably reducing the cost of that portion of the project. Finally, with the help of my colleagues from MIT and the MGH technical team, we constructed the fixed beamlines that were able to accommodate the eye treatment station and the stereotactic radiosurgery station that were in use at Harvard. I was told by Michael Goitein and Al Smith, at the start of my tenure at MGH, that it would be a boring job in that everything would be done by the outside company and the system would run without any intervention needed. I'm not sure if it was good or bad, but it was never actually boring.

At the time, in this second generation of particle therapy, it was thought by wise people in the field that the need for proton therapy in the United States would be three facilities. This is much like the feeling in the 1930s about the 'formidable installations' of the early accelerators. This would include the Loma Linda facility, the MGH facility and one in the Midwest. Some years later, a facility was constructed as an outgrowth of the Indiana University Cyclotron Facility (IUCF), called the Midwest Proton Radiotherapy Institute (MPRI). I was happy to see this. IUCF was the proton sister lab to the MIT Bates lab in which I had worked and the community was already connected and growing. However, the original predictions fell far short of reality. With particle therapy offering equal or better efficacy (and note here that I am including the word equal), one can argue that this modality can serve a large fraction of the existing cohorts of radiotherapy patients. With thousands of linac-based therapy systems, this translated to some more than three particle facilities in the United States.

As particle therapy expanded, the capabilities of photon therapy also grew, with IMRT, for example. In addition, the resources within the particle therapy community grew as well as the desire to improve the beam delivery capabilities. In the early electron beam days, Anders Brahme explored the use of scattered beam delivery and scanned beam delivery. (Well, truth be told, if someone is thinking about something in the field today, Brahme had probably already considered or tested it much earlier.) Some decades later, the scanning beam for protons found its way in Japan, Berkeley, and for dedicated clinical use at PSI. These were still in a laboratory environment. In 2008, Hitachi installed a scanning system in MD Anderson, and MGH did the same. I was very proud to have worked with a talented team, including MGH, IBA and Pyramid Technical Consultants to create the scanning beam modality at MGH in a hospital environment. Since then many other centers have benefited from the same technology. This began the third generation of particle therapy.

The first adopters of this 'new' modality were academic medical centers. Owing to their larger patient population, multiroom centers were constructed. This is reminiscent of the MM50 electron microtron that was built for electron/photon therapy serving two rooms. The multiroom centers, however, come with a higher building cost. Sometimes, complete cancer centers were constructed around them. This greatly swayed the cost of the proton therapy facility when compared with that of a linac in a bunker in an existing radiotherapy department. A system with three rooms might last two or three times as long as a linac. In addition, a new building structure is not always needed and is not comparable with a single room linac cost. If one properly includes these numbers, it turns out that the capital cost may be within a factor of 2 or less per room. The academic centers did need multiple rooms given their patient population. As the number of facilities grew, many of the academic centers had their own systems and the industry focused more on the needs of smaller hospitals. In addition, some companies were formed with the idea of providing freestanding systems to serve community-based environments. In addition, industry began to build single-room systems and reduce the cost of the components. This contributed to the start of what has been called 'the democratization of proton therapy'. This began the fourth generation of facilities. Much of this last generation is focused on protons since heavier particles continue to be a larger, more expensive system. However, superconducting technology and good engineering have contributed to heavier ion gantries (pun intended) at such places as Heidelberg and HIMAC.

Scanning has indeed improved the conformity of particle radiotherapy. However, there is still more to be gained. There are margins that may be possible to reduce, if clinically warranted and mitigation of organ motion to address. The integral dose can be significantly reduced with available technology. There are more papers being published about adaptive therapy, and work is ongoing toward the development of proton tomography. This can all lead toward the onset of the fifth generation of particle therapy.

As in the past, what has been thought about before can come back to the present. There is now renewed interest in exploring the effects of high dose rate and striped beams. Very high dose rate with high doses seems to spare healthy tissue but control diseased tissue, and beams that are spatially separated in that region seem to spare tissue and only affect the tissue when the beam spreads out filling the spaces in between. This, inspired by the reintroduction of biology in particle therapy, instead of simply assuming proton therapy has an RBE of 1.1, is giving rise to new possibilities in the application of particle therapy. This may also be the beginning of the sixth generation – perhaps more medicine and biology than physics and technology.

To date, there are about 105 particle therapy facilities in the world, with strong growth continuing. The growth rate does follow the economy, and the effect of COVID has been noticeable. As President of PTCOG with the help and support of Martin Jermann, I have personally seen a rapid growth in the availability of particle therapy to the patient population that needs it. In addition, the people involved in this growth are motivated, energetic and contribute significantly to the community. It is and has been very rewarding to be a part of this.

4 Flow of Requirements

4.1 DIRECT REQUIREMENTS

At the highest level, the goals of radiotherapy are to:

- Deliver the required dose
- Deliver that dose with the prescribed dose distribution and
- Deliver that dose in the right place
- Do it safely
- Keep the patients and staff safe and happy

These goals are achieved through the careful identification and cross referencing of requirements. The requirements of the systems that enable particle therapy are interrelated. Above all, the overall system design and design of the specific components must be safe and must satisfy the desired clinical specifications. As part of this, the clinical beam requirements and sensitivities should be defined and then it should be determined which ones are related to the technology. The chart in Figure 4.1 indicates the flow of requirements, in some semblance of order from requirements to technology.

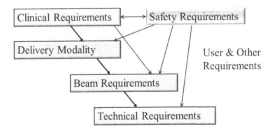

FIGURE 4.1 Flow of requirements.

Starting from any position in the chart other than the top will likely result in compromised treatment parameters. Note that the clinical requirements are buffered, in part, by the delivery modality. This is so, because the delivery modality will affect the type of beam that is needed to achieve a desired clinical goal, and therefore what the beam requirements will be. For some parameters, the characteristics of the accelerator are critical to the beam delivery process, and for other parameters they are almost irrelevant Furthermore, at each step of the analysis, the safety requirements must be evaluated in order to be sure that they are all captured and well translated before designing the component, such as the accelerator. Note how safety and user requirements flow into every level.

Clinical beam parameters, such as dose, dose rate, range, distal falloff, penumbra and dose conformity, among others, will be associated with beam parameters such as beam time structure, beam current, beam energy, beam shape, size and position. When discussing the beam production and delivery technology, it is always important to remember and associate the beam parameters with the clinical parameters; however, the association can be one to many or many to one. In addition, and perhaps even more importantly, the tolerances associated with each of these parameters are critical. A change in the beam range, or position, or shape could deposit dose outside the target, especially in the case of beam scanning.

DOI: 10.1201/9781003123880-4

The beam delivery system, which is in between the accelerator and the patient, will play a role in how the safe delivery of clinical beam parameters is related to the accelerator parameters. It is necessary to first understand these dependencies and to identify potential safety concerns. Figure 4.2 shows an example of what a high level dependency analysis might yield. Most of these terms will be discussed in the book.

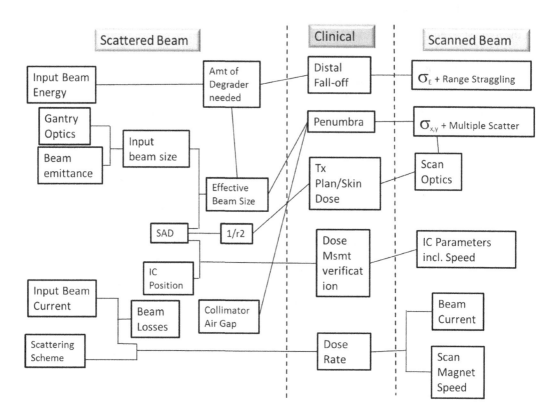

FIGURE 4.2   Parameter dependencies.

There are many links, but just to pick one, consider the penumbra. If the beam spreading modality under consideration is scattering, a partial list of some of the parameters that will affect the beam penumbra includes the following:

- Gantry optics
- Beam emittance
- SAD
- Collimator
- Air gap

In the case where the beam spreading modality is scanning, the beam size and scanning optics will affect the penumbra. These dependencies have real consequences.

As an example of this, Table 4.1 shows a few possible parameters and the flow of values from the clinical values to the beam parameters and then to an accelerator parameter that is involved for the case of a beam scanning delivery system.

Many of the terms used in this section will be defined in subsequent chapters. Keep these considerations in mind while studying this book.

TABLE 4.1

Sample of Flow from Clinical Values to Accelerator Parameters

Clinical Parameter	Sample Clinical Value	Beam Parameter	Accelerator Parameter
Dose rate	1 Gy/L min	$\sim100 \times 10^9$ protons/min	Beam current
Range	32 cm (in water)	226.2 MeV protons	Beam energy
Scanned-beam penumbra	80–20% falloff = 3.4 mm (in air)	4 mm sigma ($e^{-1/2}$ for a Gaussian beam)	Beam size, beam emittance

4.2 DEVELOPMENTAL REQUIREMENTS

It may also be useful to consider the evolution of beam delivery in the near future. The following are some of the recent themes that have been driving the development of particle therapy. Evolution to achieve improved treatment parameters is also a requirement.

- *Beam scanning* ('pencil' or 'crayon'). The method of choice for spreading the beam is beam scanning: more particularly, using magnetic fields to move the beam across the target, thus 'painting' the desired area. The size of the 'brush' is the beam size, which is strongly related to the properties of the largely unperturbed beam emerging from the accelerator (and the subsequent focusing systems). The depth of penetration of the beam is primarily determined by the beam energy. The desire is to do this quickly.
- *Image-guided radiation therapy* (IGRT). The beam position is determined by the use of imaging technology of some sort. For moving targets, the beam properties may require adjustment by feedback during the motion. With charged particles it is possible to image anatomy and directly determine the effective stopping power along the path to the target. Particle radiography and tomography depend upon the ability of the beam to penetrate the patient, and thus require appropriate beam energy.
- *Adaptive radiotherapy.* Imaging techniques and treatment planning must evolve to a point where a target today that has a different geometry from yesterday (or a minute ago) can be effectively treated. The treatment parameters need to be modified almost on-the-fly. This has implications not only for beam delivery but also for quality assurance.
- *End of range.* Currently, there is some uncertainty in the range of the particles in the patient. This uncertainty results from errors in conversion from X-ray-based imaging and from organ motion or redistribution. Such range information can potentially be obtained more accurately using particle-based imaging or other on-line detection methods, which would then require adjustment of the delivered beam energy during delivery.
- *Ions.* It has been suggested that the treatment of a single tumor could benefit from the use of multiple particles with different values of linear energy transfer, delivered during a single irradiation.
- *Effective cost.* It is a continuing concern that the capital investment is higher for particle facilities than for some other modalities. The basis of that conclusion may be from inappropriate comparisons. In any case, the goal must be to achieve a cost balance in terms of capital investment, patient throughput and treatment accuracy and efficacy so as to be competitive with other modalities.

Consideration of the above goals of particle therapy, as well as the specific clinical requirements placed on the beam parameters, could and probably should be factored into the requirements for the technology.

5 External Beam Systems

Therapeutic treatment using external beams involves the use of energetic beams to ameliorate abnormal conditions in patients. The beams come from a source external to the patient which are generated and controlled by various systems. They are used either to kill abnormal cells or prevent their reproduction or multiplication. The energetic particles in the external beams lose energy when entering the body (one way or another) and this energy in turn leads to damage of the cells. Examples of some types of external beam therapy systems include

1. X-ray therapy
2. Electron beam therapy
3. Particle beam therapy

The key goals of radiotherapy are to

- Deliver the required dose (including the required overall dose)
- Deliver that dose in the right locations

An extension of these two points is to deliver the prescribed dose distribution within the three dimensional volume of the target at the appropriate time.

That seems simple enough. However, understanding how each and every component that is to be used, affects the clinical beam and how the parameters of the clinical beam can affect the patient is the reason why it took some 50 years to develop modern radiotherapy systems.

Any and all of these systems include a fairly large number of components that serve to produce and deliver the therapeutic beam. One must clearly define the clinical beam requirements and determine how these are related to the system design. The 'system design' can be defined to include three primary components.

- Beam production – Produce the beam at the desired energy level (using an accelerator) and direct the beam to the room in which treatment takes place. This could include a particle accelerator and a beam transport system.
- Beam delivery – Take the beam from the beam production system and deliver it to a desired point in space with the prescribed clinical parameters. This includes the components that will tailor the beam to match the needs of the patient and components to orient the beam relative to the patient, such as a gantry.
- Positioning – Identify the location in the patient which is to be the target of the beam and position the patient in such a way that this target is coincident with the beam from the beam delivery system. This includes components that image the target and immobilize and position the patient.
- Each of these components has controls and systems of safety included.

These components can be organized as in Figure 5.1.

For some parameters, the characteristics of the systems are critical to the beam delivery process, and for other parameters, they are almost irrelevant. As much as possible, one must design the system to achieve all the desired clinical goals. While figuring out how to apply an existing system to some clinical goals can work temporarily, designing a system with the primary purpose of safe

DOI: 10.1201/9781003123880-5

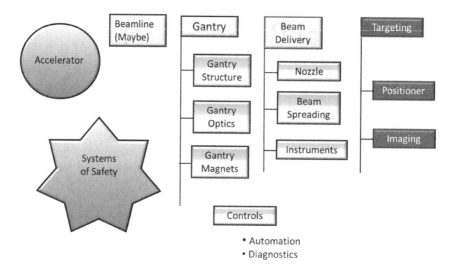

FIGURE 5.1 Parts of a particle therapy system.

therapy is optimal. The components used will depend on the choices. Figure 5.2 is an example layout of the components of a cyclotron-based particle therapy system which includes the beam production components composed of the accelerator (cyclotron) and energy selection system; the beam transport components, a gantry and the nozzle/beam delivery components.

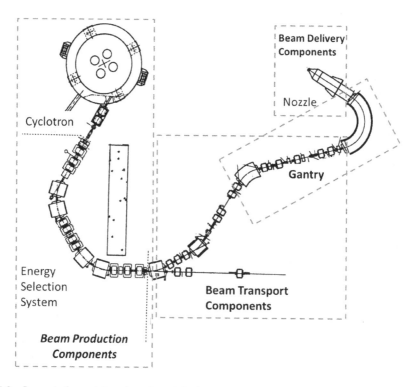

FIGURE 5.2 Layout of a cyclotron-based particle therapy system.

The modality discussed in this book is that of an external charged particle beam that is directed to the target. In Figure 5.3, with the beam coming in from the left, slices of a spheroidal target, are in the transverse plane. With respect to that beam, one can define four coordinates in the target.

- The longitudinal direction (z) also known as the depth direction. It is the direction of the traveling beam in the patient.
- The transverse directions (x and y), or the plane perpendicular to the depth.
- The time (t), of which the other three directions are a function, i.e. $x(t)$, $y(t)$ and $z(t)$.

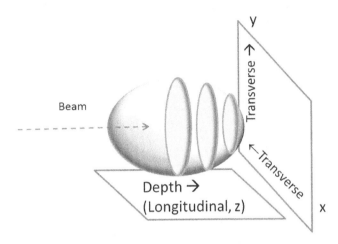

FIGURE 5.3   Target planes.

These directions, or parameters that define some of the beam characteristics, are important in describing not only the desired action of the external beam, but also the technology required to control the beam and the systems of safety that must be implemented.

All of these systems are integrated to some extent, depending upon how modularized the designers have made them. One subjective opinion about how these systems could interact is that they should be modular. There are various levels of this modularity. At the highest level, the beam production system defined earlier controls the trajectory of the beam to pass through a specific point in space. This is handed off to the beam delivery system which tailors the beam for patient treatment and directs it to the patient treatment location. The patient targeting system positions the patient, so that the target is centered at that point in space. Variations of this are possible. The patient targetry can be implemented quite independently. Interfaces that are physical and control related must, of course, be carefully integrated. Of these systems, this book does not cover the controls architecture, patient targeting and positioning systems.

There is a long list of 'ilities' that are important to include in system design including, but not limited to

- Useability
- Maintainability
- Operability
- Upgradeability
- Diagnosability

If any one component is too tightly integrated with all the others, it may be hard to upgrade the system when needed. Similarly, when, in the inevitable eventuality, it is necessary to fix things, too tightly coupling components can make it very difficult to diagnose where an issue might lie or how best to repair it. Repairability is directly related to forms of safety.

Right from the start, the system design has to take possible failure modes and errors into account. Strategies of mitigation of potential errors are critical to integrate at the outset. Having auxiliary systems that are redundant is great, but not a substitute for good system design. Questions like how many layers of redundancy and should two out of three logic be used, must be answered. A system that does not work is not safe, since the patient is likely not getting the needed treatment.

At every step along the way, in this book, systems are torn apart to their base purpose. The question, what can go wrong is addressed at the appropriate intervals. While it might not be referred to in every chapter and section, this information will eventually be used to help define tolerances and determine the sensitivities of the clinical treatment and equipment error. When reading much of this book, ask what does this do? Why does it do this? How does it do this? What should be considered to implement it safely?

6 How to Damage Unwanted Cells

It is desired to eliminate unwanted cells in living tissue. There are a number of considerations when identifying the optimal approach. Many references discuss the medicine, radiobiology, chemistry and physics of these considerations. The goal in this chapter is to identify some parameters that would provide insight to the technology and safety considerations applied to particle therapy. This short section does not include the complexity and depth of the relevant biology underlying the modality.

Eradicating unwanted cells can be done by killing the cell outright or by inhibiting its ability to reproduce. This can be done by depositing sufficient energy into the cell or introducing chemicals that react with the cell components to destroy the cell. One can, alternatively, be more controlled about the amount of energy or chemicals that are introduced, and use them to affect the DNA of the cell in such a way that the cell cannot reproduce. One of the most important cellular functions is the ability of the cell to divide. If a cell retains this ability following irradiation, it is said to have survived.

Biological systems are sensitive to radiation. Take for granted that depositing 5 Joules(J) in 1 kilogram(kg), which is written as 5J/kg, in a body or cell is sufficient to disrupt its life cycle. A cell is mostly made up of water.

- Therefore, converting this energy into calories, 5 J/kg = 5/4.18 cal/kg.
- The specific heat of water is 10^3 cal/(kg °C)
- And, $\Delta T = 5/4.18$ cal/kg $\times 10^{-3}$ kg/cal C $= 1.195 \times 10^{-3}$ °C

Thus, this amount of energy raised the temperature of the cell by about a millidegree, or not much of a temperature rise! So the damage must come from other factors. This energy, once released, gives rise to a variety of interactions that may damage molecules of biological importance such as DNA and so lead to cell death or the inability to divide. Basically there are direct effects and indirect effects. Figure 6.1 illustrates the difference between direct and indirect actions of ionizing radiation that can affect the DNA.

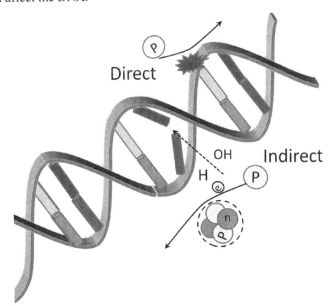

FIGURE 6.1 Types of DNA damage.

DOI: 10.1201/9781003123880-6

6.1 DIRECT EFFECTS

A direct action is that of the ionizations produced by energy released directly to break the bonds of a specific structural component of an organism, such as

Molecular:
- Protein
- Enzyme
- DNA

Complex cell structure:
- Chromosome
- Ribosome
- Mitochondria

6.2 INDIRECT EFFECTS

Since biological material is primarily composed of water, most of the damage resulting from ionizing radiation comes from chemistry. Here exists a remarkable confluence of physics, chemistry and biology. The viability of a cell can be affected indirectly through radiochemical effects. Cells are 70% water (H_2O) and the energy released from radiation is absorbed in the water. From this, the following reactions can occur:

$$H_2O + energy \rightarrow H_2O^+ + e^- \rightarrow H_2O + e^-_{aqueous} \tag{6.1}$$

$$\text{or } H_2O^+ \rightarrow \cdot OH + H^+ \tag{6.2}$$

$$\text{or } H_2O + energy \rightarrow H_2O^* \rightarrow H \cdot + \cdot OH, \tag{6.3}$$

where the •OH is a hydroxyl radical, a molecule with an unpaired electron and an excited state is marked by (*). The products are very reactive acid and base radicals. A radical is a molecule that has a relatively high energy content due to the presence of an unpaired electron. (Lack of spin pairing decreases the stability of the molecule.) This results in reactions like oxidation (loss of the electron) or reduction (gaining of an electron).

This was first recognized by Fricke in 1927, followed by a clearer understanding in the 1960s.

The reactions can cause damage to the DNA – as depicted in Figure 6.1. The types of DNA damage that can occur include

1. Base changes
2. Breaks
3. Cross links

Molecular masses obtained from radiation studies are different than those obtained without radiation due to these chemical reactions that are initiated by the released energy. In the presence of oxygen, the ionized molecule combines with oxygen forming a peroxyl radical that wreaks havoc with the target. When oxygen is not present, any ionized molecule might just be reconstituted. Therefore, there is an advantage to having oxygen to help the reaction. This is called the oxygen enhancement effect. The ratio of the dose required to achieve a level of cell inactivation without oxygen (anaerobic) to that required to achieve the same level of inactivation with oxygen (aerobic) is called the oxygen enhancement ratio (OER).

Cancer cells may develop faster than the vascular systems needed to properly support them, but they can survive, and hence, some cancer cells are anaerobic. Thus, it could be that cancer cells are less radiosensitive than the surrounding healthy tissue. However, different radiations differ in their ability to cause a specific biological effect. The radiobiological efficiency (RBE) is defined as the ratio of the dose of X or γ rays needed to produce a specific biological effect compared with to the dose of the radiation medium in question needed to produce the same effect. Therefore, the dose equivalent of a particular charged particle = RBE × dose (X).

It is important to minimize collateral damage, in the sense that it is important to protect the normal cells. So it would be best to localize the effects and consider the concept of a target. Charged particles produce excited and ionized atoms and molecules. The distance between these events depends upon the particle and the energy it releases per unit distance. The linear energy transfer (LET), in general, is the energy transferred per unit distance. The range of the particle will be related to the incident energy of the particle divided by the average LET (which will change as it loses energy). There are various categories of this having to do with where the energy is lost. The stopping power is related to the LET and will be discussed in Chapter 9. The LET is proportional to the charge and inversely proportional to the velocity (to some power) of the charged particle traveling in the target. The energy transferred ionizes the material in the medium. The specific ionization (SI) is the number of ion pairs produced per unit length as indicated in Figure 6.2 by the star like (darker gray) explosions along the particle path. Higher LET particles produce ion pairs closer together. The energy transferred, the LET will be related to SI × W (energy required/ion pair). A characteristic distance of the DNA is on the order of nanometers. Indirect effects make up about 80% of the action for lower LET particles.

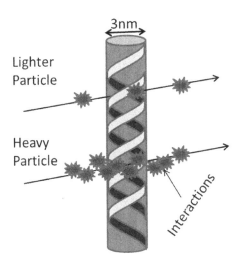

FIGURE 6.2 Interactions in and near DNA.

Some typical values of LET are

Particle	LET
^{60}Co	0.25 keV/μm
Alpha	250 keV/μm
Proton	1–30 keV/μm
Neutron	30 keV/μm
Carbon	40–100 keV/μm

After these effects to the cell, there are a number of pathways to cell inactivation. The cell could basically enter a programmed death called apoptosis. It could die from necrosis or from a nutrient deficiency called autophagy.

To damage unwanted cells using ionizing radiation, it is necessary to deposit the energy locally near the critical structures of the unwanted cells. Choosing an appropriate particle that will create a sufficient number of ionization pairs near the DNA will enhance the probability of cell death. However, the cell can repair itself, depending upon the type of damage. Elements of this are referred to as four Rs of radiotherapy and they include

- *R*epair of sublethal damage. Single strand breaks can be repaired.
- *R*epopulation by surviving cells. The surviving bad cells repopulate.
- *R*edistribution of cells throughout the division cycle. The sensitivity of a cell to radiation depends on the biological phase of the cell's survival cycle. Different cells are in different phases of the cycle at any given time, so not all are affected in the same way at the same time. One of the factors contributing to the sensitivity of cells to radiotherapy is the position of a cell in its proliferation cycle:
  - Mitosis – division takes place
  - Synthesis – DNA is synthesized
- *R*eoxygenation of hypoxic cells. Oxygen is important for the chemical reactions. Proliferation of cancerous cells takes place faster than the ability to generate blood vessels for oxygen transport. Soon some cells get buried and don't get enough oxygen to be radio-sensitive to the indirect mechanisms.

7 Exponentials

7.1 e

The fields of beam physics and medical physics are steeped in the properties of exponentials. It is useful to review some of the high-level properties. The exponential or Euler number (e) is given by,

$$e = \lim_{n \to \infty}\left(1 + \frac{1}{n}\right)^n \tag{7.1}$$

Or it can be represented as $e = \sum_{n=0}^{\infty}(1/n!) = 2.718281\ldots$ The functions e^x and e^{-x} are plotted in Figure 7.1. e^1 equals the value of $e = 2.718281\ldots$ and $1/e^1 = 0.3678$. This $1/e$ value is used frequently, as frequently as $1/10$ might be used. e^x is sometimes written as $\exp(x)$. The plot of e^x grows and its rate of increase grows, well, exponentially. It has an interesting property in that

$$\frac{d}{dx}e^x = e^x \tag{7.2}$$

So, its derivative is equal to itself. This means that its slope at any given point is equal to its value at that point. Yes this seems trivial, but it means that if a function increases or decreases at a rate proportional to its present value, the function can be an exponential. Money might be the first thing that comes to a nonphysicists mind in this respect.

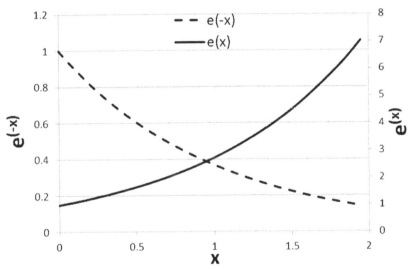

FIGURE 7.1 Plot of e and $1/e$.

Note that e^2 is just $e^1 \cdot e^1 = e^{1+1}$. The natural logarithm ln is defined with respect to base e such that

$$\ln(e^x) = x \tag{7.3}$$

The derivative is given by

$$\frac{d}{dx}\ln(x) = \frac{1}{x}; \frac{d}{dx}\ln(f(x)) = \frac{1}{f(x)}\frac{d(f(x))}{dx} \tag{7.4}$$

DOI: 10.1201/9781003123880-7

The exponential function e^x can be expanded in a Taylor expansion, from Equation 7.1 as

$$e^x = \sum_{k=0}^{\infty} \frac{x^k}{k!} = 1 + x + \frac{x^2}{2} + \frac{x^3}{6} + \frac{x^4}{24} + \cdots \tag{7.5}$$

7.2   DISTRIBUTIONS

There are many topics studied in the field of radiotherapy that are related to some sort of distribution. It could be a distribution that characterizes the survival of cells in tissue, or it could be a distribution of beam parameters in space. An understanding of the properties of these distributions is helpful to determine important parameters which will yield a successful radiotherapy process.

7.2.1   BINOMIAL DISTRIBUTION

When the success or failure of an event is considered and the number of times this outcome is tested is finite, the binomial distribution expresses the probability of the discrete, countable number of successful results. The event could be flipping a coin to get a result of three heads out of ten flips, rolling a pair of dice to get a snake eyes or it could be delivering dose to ten cells to get a result of two cell deactivations, it either happened or it didn't.

There are two parts to determining the probability of the number of successes (S) in a trial. If the result is not a success, it is a failure (F). One part is the probability of the success of one possible way the result can occur. The second is to determine the number of ways that it can occur. For example, if the chances of success are 50% for an event and it is of interest to know the chances of obtaining a result of three successes in six tries consider the ways this can occur. One way is if there are three successive successes: SSSFFF. The likelihood is the product of the probabilities (p) of each event or $p(S)*p(S)*p(S)*p(F)*p(F)*p(F)$ or 1.56%. It is convenient to define p to be the probability of a success and $q = 1 - p$ to be the probability of a failure, then the probability of this result is $p*p*p*q*q*q$. Thinking of it this way, then after the first three successes, each with probability p, the subsequent three events are characterized by the probability that it will fail or q.

However, this is not the only way three successes can be achieved. The number of different ways that n things can be arranged or the number of permutation is $n!$ For example, there are 6 (3!) permutations of ABC; they are ABC, ACB, BAC, BCA, CBA and CAB. Finally, the number of ways of selecting m distinct combinations of n objects is the binomial coefficient

$$\frac{n!}{m!(n-m)!} \equiv \binom{n}{m} \equiv {}_nC_m \tag{7.6}$$

(C stands for combinations in this context.) For example, the number of combinations of two of ABC is AB, AC, BC. Calculating this from the binomial coefficient gives $3!/(2! * 1!) = 3$.

Three parameters have been defined:

n = number of times the attempt is conducted (trials) or number of things examined
m = number of successes during those n trials or number of specific things found
p = the probability of a specific outcome for each trial

It is desired to know the probability of the occurrence of a number of events. There are four main conditions which must be satisfied if that probability is to be given by a binomial distribution. These are:

1. **Binary** = Each result is either a success or failure
2. **Independent** = Each trial is independent. No result of any trial depends upon the result of a previous or subsequent trial

3. **Number** = the number of trials is finite and is fixed and set in advance
4. **Success** = The probability of the event occurring is the same for every trial

The acronym **BINS** helps to remember this.

If BINS is satisfied and if b is the probability of a result (not just a single event = p) occurring with a binomial distribution, then b is the product of the number of ways the result can be achieved and the probability of a single result is given by

$$b(p,n) = \frac{n!}{m!(n-m)!} p^m q^{n-m} \qquad (7.7)$$

where in addition to the parameters defined earlier

$q = 1 - p =$ the probability of failure of an individual trial
$b =$ the probability of m successes to be realized in n tries

This distribution is sometimes confused with the Bernoulli distribution. The latter is a single trial that also has only two outcomes. The binomial distribution is a collection of Bernoulli trials. The expected value $E[x] = np$. For example if $p = 0.5$, it is expected that 10 out of 20 trials will be successful. This distribution includes the probabilities of multiple events happening or the product (since they are independent) of the probabilities of each individual event.

Binomial distributions are plotted in Figure 7.2 for four different numbers of trials. Since there are no values defined between integer numbers the distribution is plotted with bars, curves are included to guide the eye. The sum of all the probabilities within each distribution equals 1, so the maximum amplitude is reduced and the width increases as the number of trials increases.

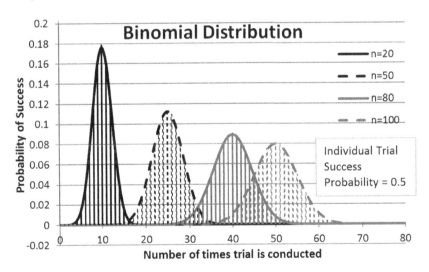

FIGURE 7.2 Binomial distribution examples.

The mean of the distribution is np as can be seen in Figure 7.2. The variance npq, related to the width of the curve, is very narrow when $p = 0$ or 1 and is widest when $p = 0.5$. The probability p is for a single event. The binomial distribution is used for a countable number of events.

7.2.2 POISSON DISTRIBUTION

It could happen that the number of trials is very high such as counting the number of trees in a forest or the number of grains of sand on a beach or the number of cells in an organ. Checking each of these items for an event or a characteristic is not practical. Also, the binomial expression does not lend itself easily to large numbers. Well, it wouldn't be necessary to use such a big sample if the probability of something were large enough that a smaller sample could be sufficient. If, however, the probability of the characteristic or event is very, very small a larger sample is needed and the binomial distribution can't be used. In this case, the outcome can asymptotically become a Poisson distribution. (How high is high and how low is low is not exactly defined.) The Poisson distribution is named after the mathematician Simeon Denis Poisson who used it in his work 'Research on the Probability of Judgments in Criminal and Civil Matters'.

In the case of a Poisson distribution:

n = number of times the experiment is conducted $\rightarrow \infty$
p = probability of a single event (in a trial) $\rightarrow 0$

Starting with the binomial distribution

$$b(p,n) = \binom{n}{m} p^m (1-p)^{n-m}$$ (7.7 repeated)

Defining $\lambda = np$ and taking the $\lim_{n\to\infty} b(p,n)$

$$= \lim_{n\to\infty} \frac{n!}{m!(n-m)!} \left(\frac{\lambda}{n}\right)^m \left(1-\frac{\lambda}{n}\right)^{n-m}$$ (7.8a)

$$= \lim_{n\to\infty} \frac{n!}{m!(n-m)!} \lambda^m \frac{1}{n^m} \left(1-\frac{\lambda}{n}\right)^n \left(1-\frac{\lambda}{n}\right)^{-m}$$ (7.8b)

$$= \left[\frac{\lambda^m}{m!}\right] \lim_{n\to\infty} \frac{n!}{(n-m)!\, n^m} \left(1-\frac{\lambda}{n}\right)^n \left(1-\frac{\lambda}{n}\right)^{-m}$$ (7.8c)

Given the limit, the rightmost term = 1. The first two terms after the limit together also reduce to 1. Going back to the definition of the exponential e, remember that $e = \lim_{k\to\infty}(1+1/k)^k$.

If k is defined to equal $= -n/\lambda$, the remaining term reduces to $e^{-\lambda}$. Therefore, the Poisson distribution can be expressed as:

$$P(\lambda,m) = \left(\frac{\lambda^m}{m!}\right) e^{-\lambda}$$ (7.9a)

where P is the probability that m is the number of occurrences of an event given that the expected average number of occurrences would be λ. The expected value $E(m)$ is given by

$$E(m) = \sum_{m=0}^{m=\infty} m\frac{e^{-\lambda}\lambda^m}{m!} = \lambda e^{-\lambda}\sum_{m=1}^{m=\infty}\frac{\lambda^{m-1}}{(m-1)!} = \lambda e^{-\lambda}e^{\lambda} = \lambda$$ (7.9b)

The examples suggested in the previous paragraph (beach, forest, organ) were all in some contained region. Perhaps even the universe could count as a contained region. In the binomial distribution, the success or failure of m single events in a trial is of interest. Yes, it's true that the binomial distribution is used to estimate the probability of multiple events happening in some sample, but still the concept of one at a time is valid. Given the formalism of the Poisson distribution, just derived, $\lambda = np$ is the important uniparametric quantity which is the average number of events in the contained region or group (blue trees, damaged cell etc.). The outcome could be six blue out of 1,000 trees inspected or 12 out of 2,000. It's just that in the Poisson formalism, the concept of an individual trial is no longer relevant, or it's no longer practical. The region can also be a region of time, or how many things happen in a span of time. The region of space or time can be called an interval. Therefore, the Poisson distribution is usually known for finding the probabilities of some number of things happening in an 'interval', but that interval can be defined in many ways. In some cases λ can be called a rate parameter and the product np is viewed as (events/time) * (time interval) or (events/area) * (area) or (events/number of things) * (number of things). The total number of expected events does depend on the sample size or interval. An average of six out of 1,000 is the same as 12 out of 2,000. A Poisson distribution will tell the probability of getting for example, four out of 1,000. The probability of getting m of those events in the interval is what the Poisson distribution calculates.

An event occurring a number of times equaling λ will have the highest probability. $P(\lambda,m)$ is the probability that something happens m times, which may be different than the average λ. If the total probability to get fewer than M successes is needed, it is necessary to sum all the discrete probabilities with $m < M$. In any given Poisson distribution, λ is constant, although it can also be a function of another variable – for example, it could be a function of dose which would determine the expected cell survival at that dose.

Figure 7.3 shows examples of 5 Poisson distributions with the probability of successful results from 0 to 100 given the average number of times the result happens (λ) is 1, 5, 10, 20 or 50. Note how the basic shape of the Poisson distribution changes for different curves. The plots are shown with bars as a reminder that the function has discrete values, but the curves over the bars are to help distinguish the five distributions. A low average probability results in a highly skewed distribution. With an average probability higher than 10, it begins to look more symmetric and can approximate a normal distribution, but it is not. The variance equals the mean in the Poisson distribution. That is not the case in a normal distribution.

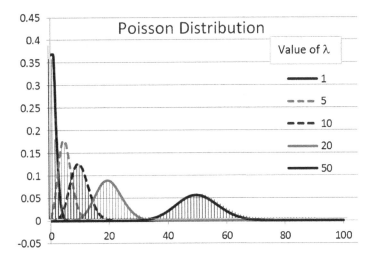

FIGURE 7.3 Poisson distribution examples.

The difference in Figure 7.3 between $\lambda = 10$ and $\lambda = 20$ may only be the sample size with the same probability. Therefore, the expected value changed from 10 to 20 and the variance (also λ) also scales. If there is a larger sample size, it's natural that the statistical spread will increase. Just because the probability may have been obtained with a larger sample size, doesn't mean that probability can't be used for a smaller sample size. For that matter, there is no requirement for a specific sample size with a Poisson distribution. Just knowing the average is sufficient.

Coming back to the question of how high is high, it is instructive to compare the binomial and Poisson distribution for equivalent expected values. Figure 7.4(left) contains five curves wherein the expected value is 5. This means that for the Poisson distribution $\lambda = 5$ and for the binomial distributions the product of $np = 5$ for values of n ranging from 10 to 500. The solid line is the Poisson distribution. Since λ is smallish, the Poisson distribution is skewed, and while the binomial distribution with $n = 10$ is centered about 5 ($p = 0.5$), the other binomial distributions are not. But the binomial distribution approaches the Poisson distribution when the value of $np \sim 200$ (some say 50 is the limit, some say $n > 100$ and $p < 0.01$). Figure 7.4(right) shows the differences with the Poisson distribution. So that sets a scale for how high an n has to be to be high. Of course, if λ is much higher, the Poisson distribution will become more symmetric as noted earlier.

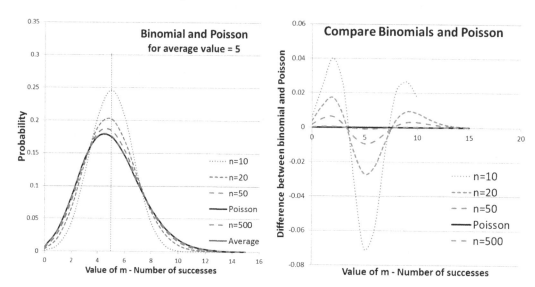

FIGURE 7.4   4 Binomials and 1 Poisson (left) and binomials and Poisson difference (right).

(The curves should be bars, not curves, since the distribution is discrete, but it's easier to compare the distributions this way.) The calculation is straightforward, mostly. But it can't always be done in Excel simply using the 'FACT' function since there is a limit to numbers up to about a power of 10^{170}. So it's helpful to use Stirling's approximation for factorials.

$$\log(n!) \approx \frac{\text{Log}(2\pi N)}{2} + N\text{Log}\left(\frac{N}{e^1}\right) \therefore n! \approx 10^{\log(n!)} \tag{7.10}$$

Sometimes λ may turn out to be large, for example, if $p = 0.001$ and $n = 1,000,000$, $np = 1,000$. But 1,000 out of 1,000,000 is close to 1 out of 1,000. Any events within any interval are independent of events during other intervals, so this can be used to help with calculations.

7.2.3   Gaussian Distribution

An important exponential function in this field is a normal distribution. It is a distribution function for independent, randomly generated variables. In this distribution, continuous variables can be used. There are no bounds and the normal distribution is symmetric and bell shaped. This distribution function has two parameters: the average (or mean) and the standard deviation (SD) with respect to the mean. A normal distribution encapsulates the tendency of random things to average out. While the Poisson distribution measures the rarity of something happening the normal distribution describes the overall distribution of all random events (so, effectively p does not have to be small). When independent random variables are measured and accumulated, according to the central limit theorem (CLT) of probability theory, their distribution approaches that of a normal distribution.

If that's the case, it is interesting to look again at a binomial distribution. When the success or failure of an event is considered and the number of times this outcome is tested is finite, the binomial distribution can be appropriate to determine the discrete, countable number of successful results.

If b is the probability of an event occurring with a binomial distribution, then

$$b(p,n) = \frac{n!}{m!(n-m)!} p^m q^{n-m} \qquad \text{(7.7 repeated)}$$

Assume that $p = 1/2$ (or the distribution could be reparametrized to be symmetric). To demonstrate the CLT, use of Stirling's approximation for the log of factorials and a Taylor series expansion results in the binomial distribution approaching the normal distribution for $n \to \infty$. The total integrated probability under the curve equals 1. How large does n have to be for the binomial distribution to look like a normal distribution?

The normal distribution is given by

$$p(x) = \frac{1}{\sigma\sqrt{2\pi}} e^{-\frac{(x-\mu)^2}{2\sigma^2}} \qquad (7.11)$$

This can be plotted with the binomial distribution by making the associations $\mu = np$ and $\sigma^2 = npq$. The result is shown in Figure 7.5 for two cases. The black solid curve on the left has $p = 0.1$ and $n = 50$, and the binomial is skewed since the skewness parameter equals $(q-p)/\sqrt{npq}$. Therefore, p must be half for a symmetric binomial distribution. The black solid curve to the right has $p = 0.5$ but also for $n = 50$. The difference with the equivalent normal distribution (gray dashed) is visible on the left and is near perfect on the right. Therefore, n doesn't have to be close to ∞ for this association.

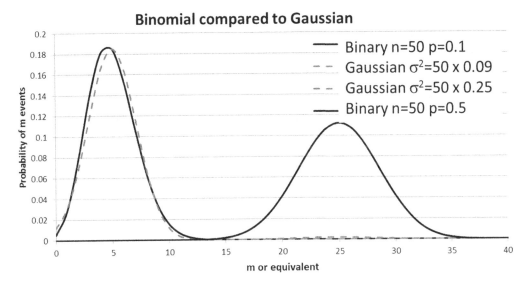

FIGURE 7.5 Binomial compared to Gaussian.

The total integrated probability under the normal curve has to equal 1. Therefore, it must be appropriately normalized as in Equation 7.11. This normalization results from an integration in which polar coordinates are used, which is where π comes from. Abraham de Moivre originally discovered this type of integral in 1733, and Carl Friedrich Gauss published the precise integral in 1809 when studying astronomical observation errors. The function is also now called a Gaussian distribution. The value, at any location, is the frequency of occurrence of something happening at that location. It is symmetric about its mean (average). That is the location at which most of the probability is centered. Its SD is σ and σ^2 is called the variance.

The value of σ determines the rate of change of the function. If σ is made smaller, the amplitude which is proportional to $1/\sigma$ is higher. Since the total integral under the curve is always 1, the curve must become narrower and therefore steeper. The mean and SD of a normal distribution control how tall and wide it is.

Some useful values of this function include

- The value of the exponent at which the Gaussian is half of its maximum value $(x-\mu)^2/2\sigma^2 = 0.693$ and if $\mu = 0$, $x = 1.177\sigma$
- This is the value of x when $p(x)$ is at its half width half max (HWHM). Its full width at half max (FWHM) = 2.355σ
- When $(x-\mu) = \sqrt{2}\,\sigma$, $p(x) = 1/e = 0.368$
- When $(x-\mu) = \sigma$, $p(x) = 1/e^{0.5} = 0.6065$

This is a probability distribution, and it is sometimes desired to know the probability that the result of a measurement will fall in a given range. The error function, erf(x), is defined as the probability that a sample value falls in the range between $[-x, +x]$ such as the hatched area in Figure 7.6(left). The probability density used for this function is defined for $\mu = 0$ and $\sigma = \sqrt{1/2}$ and has the form:

$$p(x) = \frac{1}{\sqrt{\pi}} e^{-(x)^2} \tag{7.12}$$

The integrated probability between $\pm x$ is given by

$$P(x) = \operatorname{erf}(x) = \frac{1}{\sqrt{\pi}} \int_{-x}^{+x} e^{-t^2}\,dt = \frac{2}{\sqrt{\pi}} \int_{0}^{+x} e^{-t^2}\,dt \tag{7.13}$$

where the fact that the Gaussian distribution is an even function has been used and capital P is the integral of the probability function $p(x)$. If $x = \pm\infty$, then this integral could be analytically evaluated (it's where the $\sqrt{\pi}$ comes from), but as it stands, it is not able to be evaluated in closed form. Numerical integration to arbitrary accuracy is done in software and tables. So, erf(x) is the integral within the shaded hatched area in Figure 7.6(left). For example, the integral from $x = \pm\sigma$ contains 68% of the observations of the thing happening and they are within ±1 SD from the mean. Within ±2 SDs 95.5% are contained.

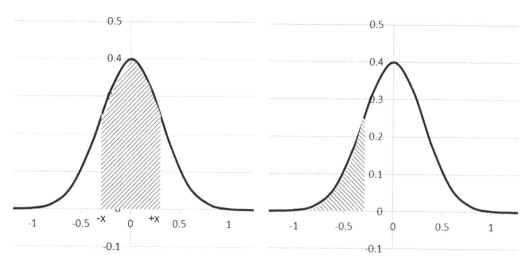

FIGURE 7.6 Gaussian hatched area = erf (left) and Gaussian hatched area = erfc (right).

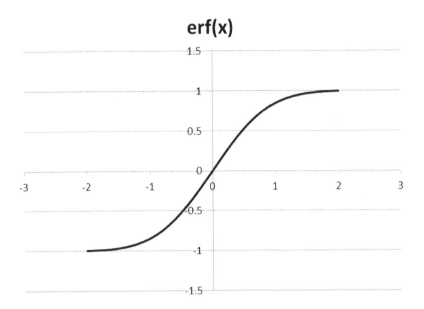

FIGURE 7.7 Graph of erf(x).

The area under the curve from ±∞ is independent of the parameters of the Gaussian since this is how it is normalized. The error function is plotted in Figure 7.7. Note that the error function is odd in that its values are negative for negative x. However, the error function is only interpreted for real statistics, for nonnegative values of x. So ignore the negative side of the function. Also, since the error function is defined for $\mu = 0$ and $\sigma = \sqrt{1/2}$, it is necessary to scale when using it for normal distributions with other parameters. For example, three Gaussian distributions are plotted in Figure 7.8 with $\sigma = 0.5$, $\sigma = \sqrt{1/2}$, and $\sigma = 1$. They are properly normalized so that the widest σ curve (longer

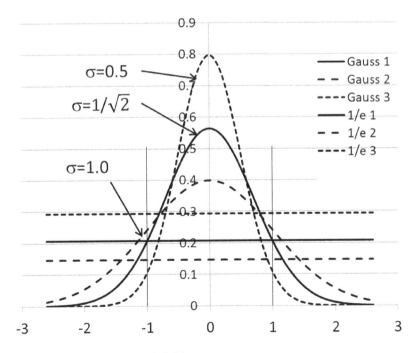

FIGURE 7.8 Three Gaussians for erf(x) definition.

dashed) has a lower amplitude than the smallest sigma curve (shorter dashed) which has a higher amplitude. Three horizontal lines are also plotted with the same characteristic (dashed, solid, dotted) as their respective Gaussian curves. They are at the $1/e$ height of these curves. The vertical lines are drawn at $x = 1$, which is the x value where the curve with $\sigma = \sqrt{1/2}$ is at $1/e$. Integrating the curves from $\pm x$ where x is the value at which the height of the curve is $1/e$ should all give the same result for each of them.

The integrated probabilities for three different values of σ $(1, \sqrt{2}, 1/\sqrt{2})$ follow:

$$P(x, \sigma = 1) = \frac{2}{1 \times \sqrt{2\pi}} \int_0^{+x} e^{-\frac{t^2}{2 \times 1}} dt \qquad (7.14a)$$

$$P\left(x, \sigma = \sqrt{2}\right) = \frac{2}{\sqrt{2} \times \sqrt{2\pi}} \int_0^{+x} e^{-\frac{t^2}{2 \times 2}} dt \qquad (7.14b)$$

$$P\left(x, \sigma = \sqrt{1/2}\right) = \frac{2}{\sqrt{1/2} \times \sqrt{2\pi}} \int_0^{+x} e^{-\frac{t^2}{2 \times (1/2)}} dt \equiv \mathrm{erf}(x) \qquad (7.14c)$$

These are general integrals (not all erf yet). For the proper scaling, define x' (the value to go into the erf(x')) to be $x/\sqrt{2}\sigma$ and $dx' = dx/\sqrt{2}\sigma$. Using this in Equations 7.14a–7.14c gives

$$P\left(x' = x/\sqrt{2}, \sigma = 1\right) = \frac{2}{1 \times \sqrt{2\pi}} \int_0^{+x} e^{-\frac{t^2}{2}} dt = \frac{2}{1 \times \sqrt{2\pi}} \int_0^{+x'} e^{-\frac{(\sqrt{2}t')^2}{2}} \sqrt{2}dt' = \frac{2}{\sqrt{\pi}} \int_0^{x'} e^{-t'^2} dt' \quad (7.15a)$$

$$P\left(x' = x/2, \sigma = \sqrt{2}\right) = \frac{2}{\sqrt{2} \times \sqrt{2\pi}} \int_0^{+x} e^{-\frac{t^2}{2 \times 2}} dt = \frac{2}{\sqrt{\pi}} \int_0^{x'} e^{-t'^2} dt' \qquad (7.15b)$$

$$P\left(x' = x, \sigma = \sqrt{1/2}\right) = \frac{2}{\sqrt{1/2} \times \sqrt{2\pi}} \int_0^{+x} e^{-\frac{t^2}{2 \times (1/2)}} dt = \frac{2}{\sqrt{\pi}} \int_0^{x'} e^{-t'^2} dt' \qquad (7.15c)$$

Therefore, all these distributions reduce to the form needed for the error function with the proper substitutions. The normalization also takes care of itself through the differential. If $\sigma = 1$, for example, it is necessary to enter $\mathrm{erf}\left(x/\sqrt{2}\right)$ instead of $\mathrm{erf}(x)$ or in general $\mathrm{erf}\left((x - \mu)/\sqrt{2}\sigma\right)$. For example, replace the x in erf(x) with $\mathrm{erf}\left(x/\sqrt{2}\sigma\right)$ (if $\mu = 0$).

The Gaussians scale, so all that is needed is the value of x relative to σ at the appropriate amplitude. If the value $\left((x - \mu)/\sqrt{2}\sigma\right)$ is used instead of x, it will scale the x position of the new normal distribution to the position with the same relative amplitude as the old distribution. That way the appropriate integration occurs. Looking at it from another direction, the one that is most relevant; with the proper scaling, the error function will return the integrated probability for any normal distribution.

The error function only evaluates in the range between $[-x, +x]$, it is not a cumulative probability distribution function which would integrate the curve from $-\infty$ to x as in the hatched region of Figure 7.6(right). In that case, notice that $1/2[1-\mathrm{erf}(x)]$ (the $1/2$ is because of symmetry) gives that result. The complementary error function $\mathrm{erfc}(x) = 1 - \mathrm{erf}(x)$ can also be used for such calculations.

7.2.3.1 Mean

The Gaussian distribution is a distribution of values that would be measured in a randomly generated system. For a set of measured data, the mean, also known as the average, is the sum of the values divided by the number of values. The arithmetic mean of a set of numbers $x_1, x_2, ..., x_n$ is typically indicated by $\bar{x}$. This may be distinguished from the expected value $-\mu$. The expected value of a random variable x is written as $\mathbf{E}[x]$. This is a sum of the product of every value measured times its probability. In the case where the probability of all measurements is the same, the $\mathbf{E}(x)$, expected value equals the mean. The mean is the usual 'expected value' of the measurements. The mean of a Gaussian is written as μ in the form $e^{-(x-\mu)^2/2\sigma^2}$.

7.2.3.2 Standard Deviation

The SD is a parameter which relates to the variation of a set of values from the mean. It is usually represented by the Greek letter σ. A smaller SD indicates a tighter distribution of measured values when compared to the mean. In practice, the SD is the square root of its variance. The variance is given by $\mathbf{V}[x] = \mathbf{E}[(x - \mu)^2]$. If all measurements are of equal probability, $\mathbf{V}(x) = \Sigma((x - \mu)^2) = \sigma^2$.

7.2.3.3 Skewness and Kurtosis

Real-life data rarely, if ever, follows a perfect normal distribution. The skewness and kurtosis coefficients measure how different a given distribution is from a normal distribution. The skewness measures the symmetry of a distribution. The normal distribution is symmetric and has a skewness of zero. If the distribution of a data set has skewness less than zero, or negative skewness, the left tail of the distribution is longer than the right tail; positive skewness implies that the right tail of the distribution is longer than the left.

The kurtosis statistic measures the lengths of the tail ends of a distribution in relation to the tails of the normal distribution. Distributions with large kurtosis exhibit tail data exceeding the tails of the normal distribution (e.g. five or more SDs from the mean). Distributions with low kurtosis exhibit tail data that is generally less extreme than the tails of the normal distribution. The normal distribution has a kurtosis of three, which indicates the distribution has neither fat nor thin tails. Therefore, if an observed distribution has a kurtosis greater than three, the distribution is said to have heavy tails when compared to the normal distribution. If the distribution has a kurtosis of less than three, it is said to have thin tails when compared to the normal distribution. The kurtosis is very sensitive to measurement noise or error.

7.2.3.4 Gaussian Algebra

There are several interesting properties of the Gaussian function. The exponents of the product of two Gaussians add: $e^A e^B = e^{(A+B)}$. The derivatives of a Gaussian are

$$\text{Gaussian} = \frac{e^{-\frac{x^2}{2\sigma^2}}}{\sqrt{2\pi}\sigma} \tag{7.16}$$

$$\text{First derivative} = -\frac{e^{-\frac{x^2}{2\sigma^2}} x}{\sqrt{2\pi}\sigma^3} \tag{7.17}$$

$$\text{Second derivative} = \frac{e^{-\frac{x^2}{2\sigma^2}} (x-\sigma)(x+\sigma)}{\sqrt{2\pi}\sigma^5} \tag{7.18}$$

Graphically these functions are displayed in Figure 7.9.

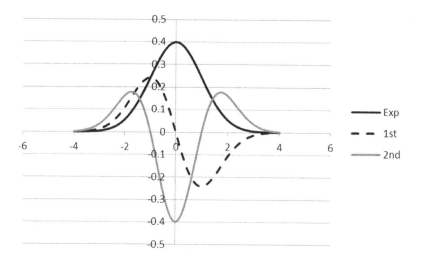

FIGURE 7.9 A Gaussian and its derivatives.

7.2.3.5 Summation of Gaussians

The behavior of Gaussians gives rise to practically magical properties. For example, adding two Gaussians, separated by some distance, the results are quite important. The resultant sum is practically flat for some amount of spacing between the two Gaussians. The graph of Figure 7.10(left) shows a spacing of 2 sigma, wherein the sum is flat between the Gaussians and the graph in Figure 7.10(right) shows how a larger spacing (2.5 σ) produces ripples (14%). Of course a spacing that is too small will result back to the original or slightly larger Gaussian (non-flat) shape. The resultant distribution in space for the dose applied to a radiotherapy target is quite important, and this will be used later.

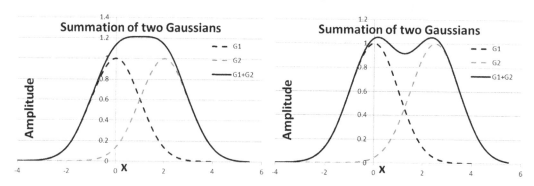

FIGURE 7.10 Two Gaussians well spaced (left) and two Gaussians spaced further apart (right).

It can be intuited, from the earlier graphs what the range of spacing should be to maintain a flat-ish sum. However, it's possible to do a little better than that. The derivative of the sum will tell the rate of change of the summed function at a particular location. The algebraic forms have been discussed in Section 7.2.3.4. The sum of the Gaussians (black solid curve), the two derivatives and the sum of the derivatives are plotted in Figure 7.11. Thus, the two derivatives (dashed lines) cross through zero at the Gaussian peaks. When spaced, the opposite signed derivatives cancel. When they are properly spaced, they are better canceled. Therefore, the spacing for a flat distribution, depending upon what degree of 'ripple' is desired (e.g. 2% total dose) is about two σ or where the HWHM (1.2σ) points on the two Gaussians cross. This will be even more important later. If they are too close the peak will have a smaller flat region. If more peaks are added, the spacing can be smaller with a flat result and typically the spacing, in that case would be about 1.2–1.5σ.

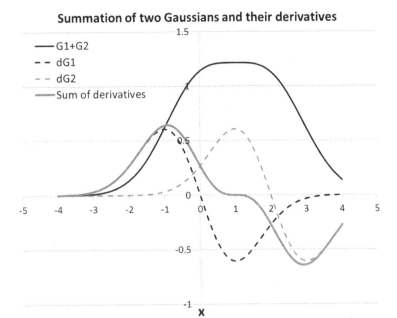

Summation of two Gaussians and their derivatives

- G1+G2
- dG1
- dG2
- Sum of derivatives

X

FIGURE 7.11 Gaussian sum and sum of derivatives.

7.3 INTERACTIONS

7.3.1 CAN HIT THE BROAD SIDE OF A BARN

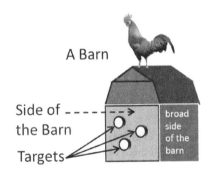

A Barn

Side of the Barn

Targets

broad side of the barn

FIGURE 7.12 An interaction barn.

It is useful to know the probability of an interaction between a projectile and a target such that some effect will be produced. Assume one is throwing a ball against the side of the barn in Figure 7.12. Three circular locations have been marked on the barn as targets. It is desired to identify the probability of hitting any of those targets with the ball. If it is assumed that the probability of hitting any place on the side of the barn is equal, the probability of hitting the circular targets would be the ratio of the cross sectional area of all of the circular targets divided by the full cross sectional area of the side of the barn. (One isn't assuming that the thrower is a professional pitcher.) Hence, the term cross section, measured in *barns,* is used for this probability. (1 barn = 10^{-24} cm^2)

Define
σ = area of a target (cm^2/target)
N = number of targets on the side of the barn
A = area of the side of the Barn (cm^2)
P = the probability that the projectile will hit a target = the fraction of the total area covered by the targets divided by the total area we have limited in this problem. In that case,

$$P = \frac{(\text{cm}^2/\text{target})(\text{number of targets})}{\text{cm}^2 \text{ of the barn side}} = \frac{\sigma N}{A} \tag{7.19}$$

This equation represents the probability that a projectile will hit (interact with) the outer wall of the barn (target).

7.3.2 INTERACTION TARGET

Now look beyond the surface. There may be target circles inside of the barn beyond the three of the surface wall such as those shown in Figure 7.13. If the ball, when it doesn't hit a target on the surface, can penetrate the surface, some of the balls will not interact right away but may go through a distance x and then hit targets inside the barn later. In this context, interaction and collision are interchangeable. The interaction cross section is related to the probability that the projectile will have 'an interaction' with a target inside an element of target thickness dx after having traveled a distance x in the barn side. In other terms, it is the effective area that a particular target offers to a particular projectile for a particular interaction of concern or for all interactions.

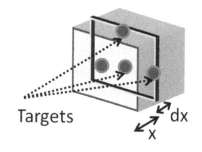

FIGURE 7.13 Internal interaction section.

In this case, the following parameters are defined:

n_0 = number of incident projectiles
σ = area of a target (cm²/target)
ρ_T = target density (number of targets/cm³)
x = linear distance to the interaction region
dx = differential thickness of interaction region
n = number of uncollided projectiles after a depth x
dn = number of projectiles that will interact in the region dx
P = Probability of an interaction in the interaction region dx

The target density thickness is defined basically as the number of targets per unit area within a given thickness which is the target density multiplied by the interaction thickness. The target density thickness is $(\rho_T dx) = \rho_T \left(\dfrac{\text{number of targets}}{\text{cm}^3} \right) dx \, (\text{cm})$. Therefore,

$$dn = -nP = -n\sigma\rho_T dx \rightarrow \sigma = -\frac{dn/n}{\rho_T dx} \left(\frac{\text{cm}^2}{\text{number of targets}} \right) \tag{7.20}$$

The negative sign indicates that this number of particles is lost (not available for further interactions) after having gone through a distance x since they have just interacted in the interval dx. Rearranging and integrating one obtains

$$\int \sigma\rho_T \, dx = -\int \frac{dn}{n} \Rightarrow n = n_0 e^{-(\sigma\rho_T)dx} \tag{7.21}$$

The number of projectiles that remain uncollided reduces exponentially as the distance travelled increases. The projectiles that interacted don't necessarily interact anymore.

One example of a projectile could be a proton. Some examples of targets could be electrons of an atom, neutrons or the nucleus. There can be various types of interactions, such as the Coulomb interaction, that may play a role in nature. The nature of the interaction may be different in different circumstances. The interaction described could involve the loss of that projectile, after the interaction. This does not happen for all interactions. The projectile, even after interacting, may continue along to interact yet again. For example, the reaction may cause an energy release and only be taken out of the game once all its energy has been released. In this case, the result of an interaction is an energy loss and the dn would count the number of energy losses in a region.

For example, think about an interaction which causes an energy loss. A beam of particles enters a thin layer of material, but instead of removing a particle, it is only slowed down. Let T_e denote the kinetic energy of the particle. Then write,

$$\Delta T_e = -n \, d\sigma_e(T_e) \rho_T dx \tag{7.22}$$

which indicates the change in kinetic energy is related to the number of targets per unit area times the differential stopping cross section ($d\sigma_e$). The particle will stop when all its energy is lost. The cross section can depend on the particle energy. Therefore,

$$\int d\sigma_e(T_e) = \sigma_e(T_e) = \int \Delta T_e \, dA \tag{7.23}$$

The cross section σ_e has units of cm^2 eV. The equation is for just one interacting target and one projectile, so n has also been removed. The $\rho_T dx$ which has units of e.g. number cm^{-2} has been moved to the other side of the equation and is now called dA which is units of area. The stopping cross section is defined as σ_e and it is within the length dx.

7.3.3 TARGET HITS

Consider an interaction that releases energy (which is approximately called dose) in the vicinity of a target site, which is sensitive to damage from this energy. Suppose the target represents a cell that has an undesirable function. The inactivation of this cell is the goal. Going back to the 'interaction target' formalism, recall that the number of inactivated targets dN, is proportional to the total number of targets N as well as the probability of the interaction being successful or causing the inactivation. In this case, the local energy deposition, related to the dose D, is related to that probability. Therefore,

$$dN \sim N * dD \tag{7.24}$$

which translates into

$$N \left(\text{number of surviving cells} \right) - N_0 \left(\text{initial number of cells} \right) e^{\, D/D_0} \tag{7.25}$$

This is only valid if one 'hit' (hit = collision = interaction) to a target is sufficient to cause the inactivation. The probability, or the cross section, is related to the probability of a 'hit' or the probability that the dose deposited resulted in a hit. The result of a hit is that one fewer cell survives. There are a large number of cells and a large number of particles that will interact with them in the body making effectively for a large number of events with a relatively low probability of any one particle hitting and causing inactivation of one cell. This can then be assumed follow a Poisson distribution representing the success of the interaction. The success of this goal increases as more cells are inactivated. However, in this case, instead of the probability of the result depending upon the number of attempts to 'hit' a single cell; it is also dependent on how hard or the amount of dose delivered locally to the cells.

The dependence on dose of cell survival is a little more complicated if you are not a biologist. For the purposes of this chapter, this can explored a little more. Define α as the linear component of cell deactivation. Let $\alpha \equiv 1/D_0$ from Equation 7.25. This represents the intrinsic sensitivity of the cell to radiation and is defined as the number of cells deactivated per unit dose (Gray). Therefore, α has a unit of Gray^{-1}. These cells are affected in a non-repairable way. The higher the dose, the higher the probability that more cells will be hit (deactivated) and the lower the probability that the cells in the field will survive.

However, it is the case that measurements show that the survival function does not look quite like this linear exponential curve. Most mammalian cells exhibit a 'shoulder' indicating that the inactivation fraction is lower with lower dose. There can be a variety of types of hits. For example, a hit with a definitive result of cell deactivation, such as described earlier, or one can have a hit with a result of damage but not full success, requiring additional dose to complete the job. In this latter case, if there is no additional dose in a biologically relevant time scale (e.g. hours), the cell could repair itself and survive as mentioned in Chapter 6. In the latter case, the survival curve shows a quadratic dependence on dose. Combining these two probabilities the mathematical cell survival (S) curve looks like

$$S(D) = \frac{N(D)}{N_0} = e^{-\alpha D} e^{-\beta D^2} \qquad (7.26)$$

There is a biologic rationale for using this model. DNA double-strand breaks are believed to be the primary mechanism leading to cell death. A single hit of radiation (one electron) can cause lethal injury by inducing breaks on two adjacent chromosomes (αD component). However, when two separate hits cause two single chromosome breaks, cumulative, but repairable injury can occur, and the probability of this occurrence is proportional to the square of the dose (βD^2).

Graphically, Figure 7.14 describes this behavior. These are survival curves representing the fraction of the targets that survive as a function of the dose received. In this semilog plot, showing the fraction of cells surviving, the solid black curve indicates the results for just the linear term, and the dashed curve shows the results for only the quadratic term. Of note is the curvature or shoulder of the quadratic form demonstrating that lower dose does not have the same degree of effect as higher dose. When combined, the solid gray curve shows a similar shoulder but has the benefit of the linear portion at the start. At dose $D = \alpha/\beta$ (in this case 3), the contributions from the linear and quadratic components of cell killing are equal as seen by where the black solid and black dashed curves cross.

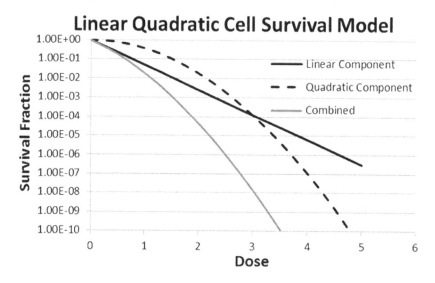

FIGURE 7.14 Survival fraction as a function of dose for different models.

The nonlethal damage of the quadratic dependence can be interesting for radiotherapy. This can become especially important if the radiation sensitivity of normal tissue is different than that for undesired cells. If the cells required to be deactivated are more dominated by the linear effect than the healthy cells, one can wait for repair of the healthy ones since the healthy cells may be more dominated by the quadratic effect. Thus, the dose is distributed over a sequence of fractions, and in between are periods of repair times.

If d is the single dose per fraction (now using small d) and n is the number of fractions given, then

$$\text{Total Survival} = S_{LQ}(D) = \prod_{k=1}^{n} S_{LQ}(d_k) = \prod_{k=1}^{n} e^{-\alpha d_k} e^{-\beta d_k^2} \tag{7.27}$$

where (now using capital D) D = the total dose over n fractions in which the dose at each fraction k was d_k. Using the wonderful properties of exponentials, the linear quadratic (LQ) model is formed:

$$S_{LQ}(D) = e^{\sum_{k=1}^{n}(-\alpha d_k - \beta d_k^2)} = e^{\sum_{k=1}^{n}\left(-\alpha d_k \left(1+\left(\frac{\beta}{\alpha}\right)d_k^2\right)\right)} \tag{7.28}$$

In the case where $d_k = d$ is constant for each fraction, this reduces to $= e^{-\alpha d(1+(\beta/\alpha)d)}$ per fraction. It's interesting how mathematically the β term retains the fraction dose d, representing a portion of nonlethal damage. This can allow full repair of sublethal damage between fractions. Figure 7.15 is an example of the survival fraction for such a fractionated dose delivery compared with a single fraction delivery. In this example, four fractions are shown with about 4 Gy of dose given to each fraction. If this were reduced to 3 Gy per fraction, the shoulder would begin again earlier on the survival curve resulting in a larger overall dose required to achieve the same overall cell survival fraction. So, the fractionated treatment requires an overall higher total dose for the same equivalent effect on the bad cells.

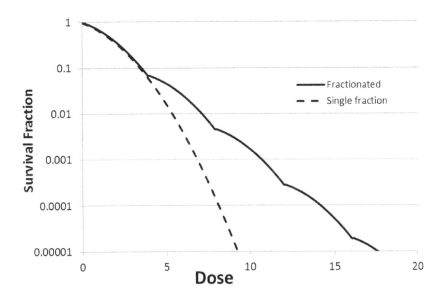

FIGURE 7.15   Single fraction vs. fractionated survival.

7.3.4   WHY RADIOTHERAPY WORKS MATHEMATICALLY

If it were possible to concentrate the entire required dose to the target cells, there would be no further consideration. The appropriate amount of dose would be delivered to deactivate all the bad cells and that would be that and the tumor control probability (TCP) would be equal to 1. However, since some of the dose might be delivered to healthy, normal tissue outside the volume of the target, that gives rise to normal tissue complication probability (NTCP).

Starting with the linear quadratic model as the basis of the cell deactivation as a function of dose, the fraction of cells $N(D)/N_0$, which survive for a dose D, is given by

$$N(D) = N_0 e^{-(\alpha D + \beta D^2)} \tag{7.29}$$

The probability of controlling a tumor $P(D)$ or the probability of no surviving tumor cells (TCP) can be approximated by a Poisson distribution (as noted earlier)

$$P(\lambda, m) = \left(\frac{\lambda^m}{m!}\right) e^{-\lambda} \tag{7.30}$$

with λ = average number of surviving cells for a given dose. This is given by pn (p, probability of cell survival [which depends upon dose] * number of cells [N_0]). So, each cell is a 'trial' and the probability of a cell surviving the dose is given by p. The probability of cell survival is small and the number of cells existing at the start of treatment is huge. For perfect control, there would be 0 cells left after a course of treatment. Therefore, it is necessary to calculate the probability that the result of the n trials (cells) gives 0 surviving cells, or $m = 0$ and λ is the number of surviving cells $= N$ from Equation 7.29. The number of surviving cells is related to dose D, and the TCP is also dependent on D. Thus,

$$TCP(D) = P(D) = e^{-N_0 e^{-(\alpha D + \beta D^2)}} \tag{7.31}$$

If the number of cells surviving is 0, the probability of the radiotherapy cure, or the TCP, is 1. Figure 7.16 includes three curves. The black solid and gray solid curves are the TCPs for two cases, one with ten times more initial cells (gray solid) than the other. The gray-dashed curve is the fraction of surviving cells as a function of dose (this is a semilog plot). It takes more dose to deactivate more cells. Dose is the energy deposited per unit mass. The same dose delivered to twice the mass results in the delivery of twice the energy, even though the dose unit number is the same. The point is that it is desired to get a certain value of the TCP for efficacy. That means that $e^{-\lambda}$ should be a certain number (e.g. greater than 0.5). If the number of cells increases by a factor of 10, the value of $e^{-(\alpha D + \beta D^2)}$ should be reduced by a factor of ten, which results in the need for a higher value of the

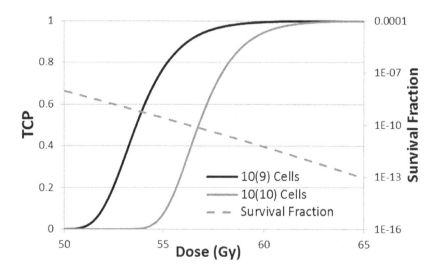

FIGURE 7.16   TCP and survival curves.

dose. This is not a matter of the volume or mass of the cells; it is a matter that a decreased survival fraction is required. Remember that the probability of 0 cells left is desired. In humans, the cell density is approximately 1×10^8 cells/ml. Figure 7.16 uses 10^9 and 10^{10} cells. The result indicates a total dose in the vicinity of 60 Gy is required for full control. This number comes from the α and β values chosen or, rather, measured from cell survival.

The shape of these TCP type curves is referred to as a sigmoid. However, a standard sigmoid function is given by $[1/(1 + \exp(-x))]$ which is not as meaningful in this case.

A plot of this TCP helps to visualize the dependence on dose. There is a semi-linear portion where the TCP increases more rapidly as a function of dose and where the largest fractional gain is realized if the dose to the tumor is increased. In radiotherapy, one would like to minimize the effect on the healthy tissue and maximize the effect on the target cells. The NTCP is the probability that the normal cells will be eliminated. So it is not the probability that the normal tissue will survive, it is the probability that the normal tissue is damaged (has a complication). The endpoint of the TCP and the NTCP are the same, but the former is good and the latter is bad.

There are a number of techniques that help to reach that goal. One such technique is perhaps biologically natural, in that the sensitivity to radiation of healthy cells is different than that of cancer cells. Therefore, the alpha beta parameters of those cells are different and have a different reaction to radiation therapy. For example, if the healthy cells had a larger β value or were half as sensitive as the cancer cells, the probability that they would succumb to the radiation is lower (depending on the dose given). Another technique is more technical, and this is a technology book. If one can reduce the dose delivered to the healthy cell when compared with the dose to the cancer cell, then even if the two cells had the same sensitivity to radiation, it stands to reason that the healthy cells would have a higher chance of survival compared with the cancer cell.

As an example of this, consider a reduction in the dose between the target cells and the healthy cells. Figure 7.17 contains four curves relevant to this. In this and subsequent figures, in this section, the dose scale has been reduced compared to Figure 7.16 by reducing the number of cells. This is to facilitate some aspects of the plots. The basic points are correct, even if the scale has changed. The solid gray curve is the target cell survival fraction as a function of dose. The solid black curve is the

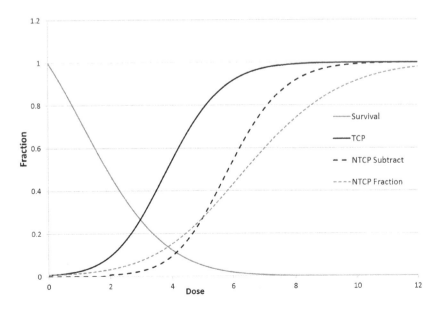

FIGURE 7.17 Comparison of different TCP types.

TCP dose dependence for this example. The dashed lines are the NTCP curves for two types of dose reduction. The black dashed curve is for the case where the dose to the healthy cell is a constant value less than the dose to the target cell (2 Gy less in this example), and the dotted line has the dose to the healthy cell 0.6 times that of the dose to the target cell. The two dashed curves have somewhat different shapes, but the main point is that the dose reduction (under the assumption of the same α/β ratio) separates the NTCP curves from the TCP curve. This means that there is some dose that can control the target cells while minimizing the damage to the healthy cells. Evaluating the product of the probability of controlling the tumor, times the probability of surviving healthy cells (1 – the probability of killing healthy cells) gives a sort of efficacy number. Maximizing that number can optimize the treatment plan. It can also be used to identify the tolerances of the treatment to errors in the delivery. The result is the dotted curve shown in Figure 7.18. Sometimes the ratio of the two curves at the ends of the arrow is called the therapeutic ratio. Medical expertise and experience is needed to weigh the relative risks and rewards and decide where along these curves to prescribe a dose for the treatment.

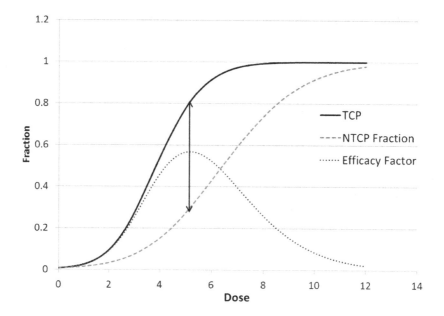

FIGURE 7.18   Therapeutic ratio example.

7.3.4.1   Tolerances

In anticipation of further commentary about this visual display of the effects of irradiation, it's useful to explore the formalism a little further. Figure 7.19 is a graph of the TCP curve as a function of dose; that is the solid line. The dashed line is the negative of the first derivative. The dotted lighter line is also the negative of the second derivative.

For those brave hearted souls, the derivatives can be found analytically. The first derivative looks like

$$\frac{d\left(\mathrm{TCP}(D)\right)}{dD} = \left[e^{-Ae^{-(\alpha D + \beta D^2)}} \right]\left[-Ae^{-(\alpha D + \beta D^2)} \right]\left[-(\alpha + 2\beta D) \right] \tag{7.32}$$

Looking at the location where the first derivative is at a maximum and the second derivative is zero the steepest part of the TCP curve is found. This would also be the worst case from the perspective of tolerances. At that place (depending on the α and β parameters), the TCP value changes by roughly 0.4 (or 40% control rate) per unit of D (for the α, β values chosen). In this particular

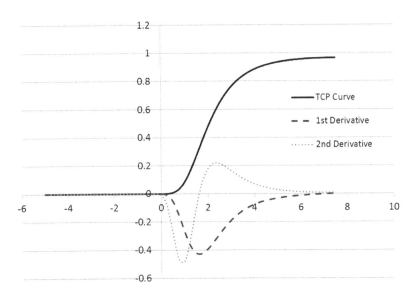

FIGURE 7.19   TCP curve and derivatives.

example, the D is about 1.75. If a 2% variation in control rate is tolerated, a 5% change in the D is tolerated. This is an interesting number. Other considerations, like more complicated statistics and parameters for clinical trials, would add to these considerations.

7.3.5   ATTENUATION

Attenuation is a situation in which a particle is lost after an interaction. There are many applications where attenuation is important. Consider a situation in which there are a number of things, say radioactive isotopes, from activated material, that disappear in an amount that is proportional to the number of these items existing at any given time. Define the time that it takes for half of them to be lost as the 'half-life' t_h, so that every t_h interval another factor of 2 is lost. (Or attenuated by a factor of 2.) In this case, the number of atoms N present at any time t is given by

$$N = \frac{N_0}{2^{t/t_h}} \tag{7.33}$$

While it might seem to be a non sequitur, it is noted that $e^{0.693} = 2.00$. Using this transforms Equation 7.33 into

$$N = N_0 \left(e^{0.693}\right)^{-t/t_h} = N_0 \left(e^{-0.693t/t_h}\right) \tag{7.34}$$

Now the dependence on the remaining number of things is cast in exponential form. This is related to the interaction cross section discussed earlier, or in this case, the capture cross section. Define the term λ_t (not the same as in the Poisson distribution) called the transformation constant to be the time it takes to get to $1/e$, or 36.78% of the original number of things leftover. This is the $1/e$ time (not the half-life). If at any time, there are N atoms present, and in a time Δt, ΔN decay, it is known that ΔN α N (if two times as many atoms are present, expect two times as many disintegrations in the same time interval). It is also proportional to the time interval for small time intervals, therefore

$$\Delta N = -\lambda_t \Delta t \left(\lambda_t \Delta t \ll 1\right) \rightarrow dN = -\lambda_t N dt \rightarrow \frac{dN}{N} = -\lambda_t dt \tag{7.35a}$$

$$N = N_0 e^{-\lambda_t t} \tag{7.35b}$$

where $\lambda_t = 0.693/t_h$.

One of the applications of attenuation is that of radiation shielding. Particles in a particle therapy beam are transported through most of the system in an evacuated beam pipe. Even so, there are restrictive apertures like collimators and slits with which the beam interacts and can cause beam scattering and portions of the beam may grow beyond the size of the beam pipe. The interaction of this primary beam with material can cause secondary particles to be generated such as prompt photons and neutrons as well as charged particles. The neutrons can have energies as high as the incident proton (assuming protons are the primary particle). The highest energy neutrons are forward peaked, relative to the direction of the primary beam at the loss point. Prompt radiation is that which only exists while the primary beam is on. Residual radiation is caused by the secondary radiation activating materials with a finite half-life. In the particle therapy application, perhaps the most important aspect is that this beam is directed toward the patient, and therefore, secondary particles are generated from that source. Some of these secondary particles contribute to small amounts of additional dose in the patient (much has been written about this in the literature), but much of it is released into the environment. Identification of the sources of these secondary particles is important. All personal in a radiotherapy facility must be protected from both the prompt and residual radiation, and therefore, the personal should be 'shielded' from the prompt radiation while the beam is on, and the residual radiation should be isolated away from personnel for a time given by the transformation constant to reduce the activity to an acceptable level. There are many modern and up-to-date methodologies that are used to calculate and design appropriate safeguards for this situation. They are the most accurate and useful. However, when a quick estimate, to give a general order of magnitude understanding of the requirements for shielding is possible, perhaps it can be useful.

The use of shielding is to attenuate the secondary radiation to a level that is safe, which is most of the time defined to be, within regulatory requirements. Note that these requirements fall over a wide range from country to country. A fast neutron is most readily attenuated by a combination of relatively high Z materials and materials with large hydrogen content. Hydrogen welcomes the company of neutrons. Concrete provides this combination and is relatively inexpensive. Alternatively, layers of iron or lead and concrete with polyethylene layers may be more attenuative than only concrete.

The dose at a point outside a shield from a source of particles inside a shield is given by H.

$$H = \sum_i \sum_\theta N_{i\theta} \times H_{\text{casc},i\theta} \times \frac{e^{-(\rho \times T \times \sec(\alpha)/\lambda_{i\theta})}}{d^2} \qquad (7.36)$$

where
$N_{i\theta}$ = the number of particles with energy E_i lost at orientation θ with respect to the shield
$H_{\text{casc},i\theta}$ = the cascade neutron source term (mrem m^2/proton) given particle energy E_i and
         orientation θ with respect to the shield. For short, call it H_0.
ρ = the shield density (g/cm^3)
T = the shield thickness (cm)
α = the angle between the wall and the ray connecting the source and the measuring point
$\lambda_{i\theta}$ = the shield attenuation length (g/cm^2)
d = the distance from the source to the measurement point (m). Note the $1/d^2$ dependence coming
         from the distance effect; however, not all cross sections are isotropic.

This equation essentially has two components. There is an exponential attenuation portion and a $1/d^2$ reduction. The latter comes from the fact that a small source can spread out radially as shown in Figure 7.20. In this case, a point source at the center of the sphere spreads out. The source, in this case, is secondary particles caused by interaction of the primary beam at the source location. By the time the secondary particles have reached the surface of the sphere, the surface area, A, for that section drawn in the figure has reached 1A. At a distance 2r, the surface area including the secondary particles has grown to 4A. So the same number of particles that passed through A at 1r are now

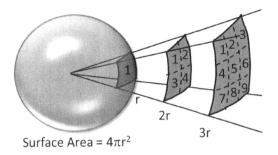

Surface Area = $4\pi r^2$

FIGURE 7.20 $1/r^2$ spreading.

passing through 4A at 2r and 9A at 3r. This results in the number of particles per unit area decreasing by $1/r^2$ even though the number of particles has not changed. The secondary particles could be peaked in a particular direction or be isotropic, but the $1/r^2$ dependence will still occur somewhat since there will be angular spread. If it is isotropic, then by the time it reaches the surface of the 1r sphere, it has already spread out by $4\pi r^2$. This is one of the two effects and is not attenuation but does contribute to a reduction in particle flux.

Now, temporarily call $\rho/\lambda = \mu$ (units of 1/cm) and ignore the d^2 and angular dependence in the attenuation equation (that can be added later). Then $H/NH_0 = \exp(-\mu T)$. The value of μ when this ratio is $1/e$ is the mean free path. The thickness of material required to reduce the secondary particle number by half is call a half value layer $= T_{HVL} = 0.693/\mu$. Finally the tenth value layer is the thickness of material required to reduce the number of secondary particles to 1/10 of their initial number of $T_{1/10} = 2.302/\mu$.

This quantity is summed over all energies and orientations. The material and distances are chosen so that H will be below the regulatory requirements, and more importantly below any concern for safety of the living creatures in the environment. H can include a constant called a quality factor which incorporates some measure of the biological damage caused by a particular particle (e.g. Q can add a factor of 20 for neutrons compared to dose calculated by local energy deposition). H also takes into account the area that the detector is integrating. Note that sometimes the ρ might be subsumed inside or outside the λ to change from thickness to mass thickness (see Section 9.3). Some rough values (again this is for back of the envelope calculations – real simulations require much more accurate values) are listed in Table 7.1.

TABLE 7.1
Concrete and Iron Attenuation Lengths

	Concrete	Iron	Concrete	Iron	Concrete	Iron
Energy (MeV)	100	100	150	150	200	200
Attenuation Length (g/cm²)						
0°	63.2	136	80.0	136	89.3	136
45°	59.6	126	75.5	126	84.7	126
90°	50.0	110	63.0	110	71.0	110

There is energy dependence to the attenuation length which must be factored in. Note that there are many other aspects of radiation shielding including soil, groundwater and air activation as well as other geometries, such as a maze which is structured to cause the secondary particles to literally 'bounce off walls' to calm them down, enabling an unshielded entrance. Given the above attenuation equation, it is possible to estimate (very roughly) some of these effects.

Consider the straightforward (literally) geometry of Figure 7.21. A source with NH_0 is on the left side of the wall and the secondary particles are tracked straight through the slab of concrete with thickness T and travel a total distance d ending up with H at the gray cylindrical shaped detector. Results will be geometry dependent since the secondary particle cross sections are energy and angle dependent as well as the attenuation parameters. For a given source caused by the interaction of N protons, together with the secondary particle cross section represented by H_0, the secondary particles travel through a slab of concrete of length T and are measured in a detector a distance d away. Taking $\lambda = 63.2$ g/cm^2, with the density of concrete $= 2.3$ g/cm^3, the attenuation ratio can be calculated as a function of T and d.

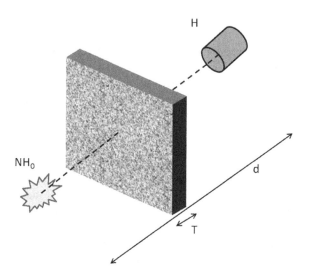

FIGURE 7.21   Attenuation by shielding.

Assume, for this, that the source is immediately before the slab and the detector is immediately after the slab or $T = d$ (including the $1/d^2$ effect here). The secondary particle attenuation as a function of thickness of material (slab) is shown in Figure 7.22. It might be necessary, in extreme cases,

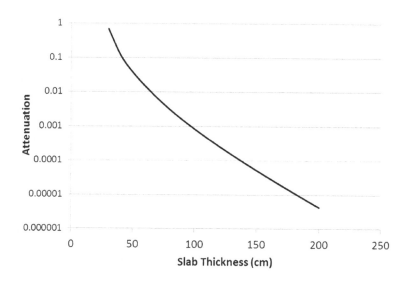

FIGURE 7.22   Secondary particle attenuation.

to reduce the radiation fluence by a factor of 10^8, and this can require on the order of 4 m of concrete. Other combinations of material can be better suited to different sources and different primary particle energies.

7.4 EXERCISES

1. The probability of 5-year survival with prostate cancer that includes distant spreading is 30% (SEER). What are the chances, out of 100 with this condition that 20, 30 or 40 will survive for 5 years?
2. According to SSA.Gov, the actuarial life tables include the following information

Age	0	30	60	90
Probability of dying 1 year	0.0063	0.00187	0.01152	0.1654
Samples number used	100,000	97,229	85,316	18,070
Future life expectancy	75.97	47.65	21.58	4.05

   a. Out of 1,000 newborns, what is the probability that 990 will survive 1 year?
   b. Out of 1,000 60-year olds, what is the probability that 985 will survive 1 year?
3. Golden doodles are very popular among millennials, 35% of whom own a pet. If 9 millennials are randomly chosen, what is the probability that 6 own a pet?
4. Patients enter a busy emergency room at the rate of 30 per hour.
   a. What is the probability that none (0) arrive in any given minute?
   b. How many are expected to come in, in 2 minutes?
   c. What is the probability that this expected number actually comes in, in a 2 minute interval?
5. What is the thickness of concrete needed to reduce a source of radiation by a factor of 10,000 (including the $1/d^2$ factor [d includes the concrete thickness]).
   a. What is the thickness if d = concrete thickness?
   b. What is the solution for d = an extra 3 m more than the concrete thickness?
6. What is the survival fraction for cells characterized by $\alpha/\beta = 3$ and $\beta = 0.075$, after they have been exposed to a
   a. Dose of 2 Gy?
   b. Dose of 20 Gy?
7. What is the dose (Gy) required to obtain a TCP of 0.5 for 5×10^{10} cells with $\alpha = 0.225$ and $\beta = 0.075$?

8 Relativistic Dynamics

Charged particles can be parameterized as follows:

- The transverse positions (x and y)
- The transverse angles (θ and φ)
- The transverse momenta (p_x and p_y)
- The momentum in the primary direction of motion (p)

What happens to that beam, or what happens to these parameters as the beam is produced, delivered and applied, is the rest of the story. But that will all depend upon the particle energy and direction. Particles in this context include any charged particle used for therapy.

The momentum of a particle in the beam that can be useful for therapeutic purposes is generally relativistic. Therefore, the basic laws of momentum and energy under the Lorentz transformations must be satisfied. The momentum of a particle is given by

$$\mathbf{p} = m\mathbf{v} \qquad (8.1)$$

where:

$\mathbf{p}$ = momentum
m = particle mass
$\mathbf{v}$ = particle velocity

The law of momentum conservation for particles traveling at high speed is refined by the basic postulates of special relativity. (It is assumed the reader is mostly familiar with this.)

- Physical laws are invariant in all inertial frames.
- The speed of light measured is the same in all inertial frames.

8.1 SPECIAL RELATIVITY (BRIEFLY IN THIS TIME FRAME)

The law of conservation of momentum is valid when considering an elastic collision between objects. The collision can be observed in reference frames that move or are at rest with respect to one or the other of the objects. While the mass of the object in each frame (with respect to the observer who believes they are at rest) is the same, special relativity shows that the timing between events is not.

This lack of simultaneity has interesting results. Feynman imagined a special clock shown in Figure 8.1. This clock has a light bulb under a shutter. There is a mirror placed a distance y above the clock. In the reference frame of someone who is not moving with respect to the clock, the shutter opens for a split second and a light pulse escapes. It travels up to the mirror, at the speed of light, (naturally) and is reflected back to the clock that detects the light pulse and increments the time. The time it takes for half the trip is ct_0. If the clock is now moved at a speed v relative to the same observer who is not moving with the clock, the light pulse appears to move in the path of the dashed black line. In a time experienced by the observer at rest, t_A, the clock has moved a distance x or vt_A to the right. However, the light moves along the diagonal, also at the speed of light, which is the same in all reference frames, a distance along the hypotenuse d. This distance is ct_A. So, Pythagoras,

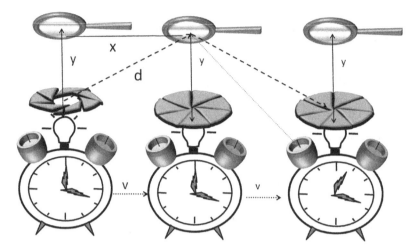

FIGURE 8.1 Feynman clock.

who evidently knew about special relativity, teaches us that $y^2 + (vt_A)^2 = c^2 t_A^2$. Since it was known that $y = ct_0$, then $c^2 t_0^2 = c^2 t_A^2 - v^2 t_A^2$.

Therefore,

$$t_A = \frac{t_0}{\sqrt{1-(v/c)^2}} \equiv \gamma t_0 \tag{8.2}$$

where v is the relative speed between the two frames and where γ is defined in the equation. It appears to the observer at rest, that the time for a clock increment is different if the clock is moving. This is known as the time dilation effect.

Now consider a simple example where a ball is bounced down, from a distance y, above a floor in one frame and thrown up, also a distance y, from below the floor in another frame. The observers are either moving or not moving along a direction parallel to the floor. One frame is denoted as A and the other as B, and there are a variety of ways of looking at the situation as indicated in Figure 8.2. In the leftmost scenario, observer A looks at his own ball and observer B looks at his own ball. In the next one, it is all as seen by observer B where reference frame A is moving with respect to reference frame B. In the next scenario all is as seen by observer A where reference

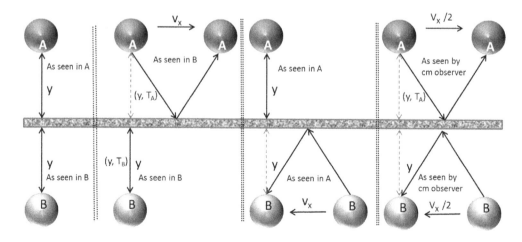

FIGURE 8.2 Reference frames for bouncing balls (separated by dotted vertical lines).

frame B is moving with respect to reference frame A. There is yet a fourth observer in a center-of-mass frame in which the two ball-bouncing events are seen as taking the same time. The two balls move an equal distance $2y$ in the rest frames (up and down or down and up).

Each of these cases can be analyzed separately and there are a number of paths to take. If time dilation (Equation 8.2) was not known, then the relative velocities using the Lorentz velocity transformation could be used and it would be learned that given, for example, the additional motion of the ball in reference frame A as seen by reference frame B, the only way to conserve momentum (a constraint) is to include a γ factor in the definition of momentum. But where is that factor associated? If time dilation is a known behavior then another simpler approach can be taken.

Looking at it from the perspective of the observer A at rest, the bouncing and catching in A are not synchronized with the bouncing and catching in B. In this frame, the change of momentum ΔP, in the y direction, of the ball in each frame is given by

$$\Delta P_A = m_A \left(\frac{2y}{T_A} \right) \tag{8.3a}$$

and

$$\Delta P_B = m_B \left(\frac{2y}{T_B} \right) \tag{8.3b}$$

Application of conservation of momentum results in these two quantities being equal. Therefore,

$$m_A \left(\frac{y}{T_A} \right) = m_B \left(\frac{y}{T_B} \right) \tag{8.4a}$$

and

$$m_A T_B = m_B T_A \tag{8.4b}$$

The ball in B appears to be moving parallel to the floor. If the time interval in frame A is $T_A = T_0$, then the time interval as seen from A in frame B is $T_B = T$. As in Equation 8.2, express T_A and T_B in terms of T_0 and use it in Equation 8.4b. Since the T_0s cancel, the time dilation factor, instead of redefining the momentum, appears to lead to an apparent change of mass of the moving ball so that

$$m_A = \frac{m_0}{\sqrt{1-(0/c)^2}} \text{ and } m_B = \frac{m_0}{\sqrt{1-(v_x/c)^2}} \tag{8.5}$$

In standard convention, m_0 is defined as the rest mass, or the mass of the particle in the stationary reference frame; m is defined as the mass in the arbitrarily moving reference frame. If you would like to spend more time on this section, go faster.

8.2 DYNAMICS

In general and in relativistic dynamics, the time rate of change of the relativistic momentum is due to a force acting on the particle. Newton's second law, in this case, looks like

$$\mathbf{F} = \frac{d\mathbf{p}}{dt} = \frac{d}{dt}(m\mathbf{v}) = m\frac{d\mathbf{v}}{dt} + \frac{dm}{dt}\mathbf{v} \tag{8.6}$$

where
   $t =$ time
   $\mathbf{v} =$ velocity (which has a direction and speed)
   $m =$ the mass in the arbitrarily moving reference frame

Note that if $dm/dt = 0$ (mass is not changing with time), then $\mathbf{F} = m\,d\mathbf{v}/dt = m\mathbf{a}$ as in classical, non-relativistic formalism. A special case of $\mathbf{F} = d\mathbf{p}/dt$ is when the speed is constant, but the vector velocity changes, as with circular motion. (Bold signifies a vector.)

(As a total aside, note that the assumption here is that the only two variables that may change with time are v, velocity, and m, mass. However, what if time changes with time (as when the velocity of the reference frame changes). Perhaps the concept of a changing mass is just a ramification of the evolving time dilation and no force can change momentum if time has stopped. (Not that this will change the results, but it's just a thought that may be wrong.)

The relativistic kinetic energy E_k of a particle is the total work done in bringing the particle from rest to a final speed v under a force F (F can vary). It can be expressed as

$$E_k = \int_0^s F(s)\,ds = \int_0^s \frac{dp}{dt}\,ds = \int_0^s \frac{d(mv)}{dt}\,ds = \int_0^t \frac{d(mv)}{dt}(v\,dt) \tag{8.7}$$

where ds is the distance traveled in a time dt, which is equal to $v(t)dt$. The integral can be reduced through differentiation by parts to

$$\int_0^t \left[(dm)v + m(dv)\right]v = \int_0^t v^2\,dm + \int_0^t mv\,dv \tag{8.8}$$

From the earlier equation, it is known that

$$m = \frac{m_0}{\sqrt{1-(v/c)^2}} \quad \text{or} \quad \left[1 - \left(\frac{v}{c}\right)^2\right] = \frac{m_0^2}{m^2} \tag{8.9}$$

Using this relation in Equation 8.8 results in $E_k = (m - m_0)c^2$ with the rest energy of the particle = m_0c^2. (Would you feel like Einstein if you solved this?)

The particles' total energy mc^2 is the kinetic energy $E_k = (m - m_0)c^2$ plus the rest mass energy m_0c^2.

$$E_T = \frac{m_0 c^2}{\sqrt{1-(v/c)^2}} \tag{8.10}$$

By convention, a subatomic particle's energy will be identified as its kinetic energy, since the rest mass of the particle should be known. Starting from the earlier equations, the relation between the momentum and the energy can be derived as

$$E_T^2 = (pc)^2 + E_0^2; \quad pc = \sqrt{E_k^2 + 2E_k E_0} \tag{8.11}$$

This enables the calculation of clinically relevant particle parameters. Well almost. A particle such as a carbon atom with atomic number $A = 12$ comprises 12 nucleons. Therefore, its rest mass is $12 \times m_0c^2$ where m_0 is the average nucleon rest mass. The convention of an atomic particles' kinetic energy is actually the kinetic energy per nucleon. So, for a carbon nucleus it would be the total kinetic energy of the nucleus divided by $A = 12$. Given E_k in terms of MeV/A, then the total kinetic energy of the nucleus is E_k multiplied by A.

The unit 'u' is a standard atomic mass unit. The value of 1 u is set such that a carbon-12 atom has a mass of exactly 12 u. In other words, 1 u is exactly 1/12 of the mass of a carbon-12 atom. The mass of a nucleon in u, 1 u = 931.48 MeV/c². This value comes from knowing a few conversions (given in Appendix B).

$$\text{Atomic mass unit}\left(amu, m_u\right) = 1.66 \times 10^{-27} \text{ kg}$$

$$1 \text{ amu} \times c^2 = m_u c^2 = 1.49 \times 10^{-10}\,\text{J} = 931.483 \text{ MeV}$$

$$\text{Speed of light } (c) = 2.997925 \times 10^8 \text{ m/s}$$

$$1\text{ MeV} = 1.6021 \times 10^{-19}\text{ J} \left(\text{kg m}^2/\text{s}^2 \right)$$

Since energy and mass are sometimes used interchangeably, this is a convenient unit. Also, be careful if given energy in terms of MeV/unit atomic mass as opposed to/nucleon.

Calculating a relativistic momentum for a particle with A nucleons results in

$$pc = \sqrt{\left(AE_k\right)^2 + 2\left(AE_k\right)\left(AE_0\right)} = A\sqrt{\left(E_k\right)^2 + 2E_k E_0} \qquad (8.12)$$

Being able to remove the A from the square root is pretty special. Using Equations 8.1 and 9.9, the following expression is obtained:

$$\frac{A^2 m_0^2 c^4}{\left(1 - (v/c)^2\right)} = A^2 m_0^2 c^4 + p^2 c^2 \qquad (8.13)$$

Solving for v/c

$$\frac{v^2}{c^2} = \frac{1}{1 + \left(\dfrac{A^2 m_0^2 c^4}{p^2 c^2}\right)} \qquad (8.14)$$

And pc is known from Equation 8.12.

$$\frac{v^2}{c^2} = \frac{1}{1 + \dfrac{A^2 m_0^2 c^4}{\left(A\sqrt{\left(E_k^2 + 2E_k E_0\right)}\right)^2}} = \frac{1}{1 + \dfrac{m_0^2 c^4}{\left(\sqrt{\left(E_k^2 + 2E_k E_0\right)}\right)^2}} \qquad (8.15)$$

What is very important here is that the A cancels out. Then, if E_k, the kinetic energy per nucleon, is the same for any particle, independent of the mass of the particle, the speed will be the same. For example, the graph in Figure 8.3 shows the speed of proton and carbon ions as a function of particle

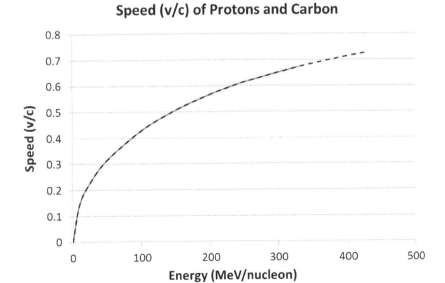

Speed (v/c) of Protons and Carbon

FIGURE 8.3   Speed vs. energy.

kinetic energy per nucleon. There is a solid and dashed curve and yes, there is no difference. The graph in Figure 8.4 shows the total momentum (not per nucleon) for these proton and carbon particles as a function of the kinetic energy per nucleon.

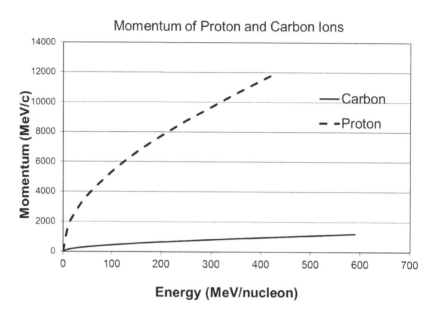

FIGURE 8.4   Momentum vs. energy.

A clinically relevant proton energy is about 200 MeV, which is achieved by a proton traveling at about 0.566 c with a momentum of about 644 MeV/c. A clinically relevant ^{12}C energy is about 430 MeV/nucleon with a momentum of 11,951 MeV/c and traveling at a speed of 0.73 c.

8.3   EXERCISES

1. Starting from Equation 8.8, derive $E_k = (m - m_0)c^2$.
2. Starting from Equation 8.10, derive Equation 8.11.
3. Show that $pv = E_{Total}\beta^2$
4. What is the momentum of a 230 MeV proton?
5. What is the momentum of a carbon Ion with the same speed as a 230 MeV proton?
6. Relative to the rest frame, what is the rate of time that passes for a 100 MeV proton?
7. What is the maximum speed of a particle above which the difference between the non-relativistic kinetic energy and the relativistic kinetic energy differs by more than 10%?
   a.   What is the kinetic energy of an electron moving at this speed?
   b.   What is the kinetic energy of a proton moving at this speed?
   c.   What is kinetic energy of lithium moving at this speed?

9 Charged Particle Interactions in Matter

An energetic (or lethargic) charged particle passing through matter interacts with the atoms of that material. The results of these interactions include energy loss and scattering, both of which are important for particle therapy. The term collision is sometimes used. This is not taken in the sense of billiard balls colliding physically, but the bodies coming close enough in proximity to experience a force between them such that one or both of the particles' properties change.

9.1 ENERGY LOSS

A moving charged particle with energy greater than thermal excitation loses some of that energy by a number of interactions with the atoms it passes, including

Ionization and excitation: Inelastic scattering with atomic electrons. – COMMON
The moving particle interacts with bound atomic electrons by the Coulomb interaction. The result is that one or more of the atomic electrons absorb energy and either get excited to a higher energy state or to an unbound state (becoming free). The latter case is called ionization. The atom is ionized and the unbound electron and the ion are called an ion pair. The energy required for this ionization depends on the strength with which the electron is bound. Such an energy exchange can cause the charged particle to change direction; thus, it is scattered too. A delta ray is an energetic electron that results from a collision between a heavy charged particle and an atomic electron. These delta rays have enough energy to travel far away and cause secondary ionizations and deposit energy away from the collision location.
Elastic Scattering: Elastic collisions with atomic electrons → Deflections only – RARE
In the case of elastic interactions in the Coulomb field of the atomic electrons, energy and momentum are conserved and the resulting energy transfer needed to satisfy the conservation of momentum is generally lower than what is needed for ionization to take place. Therefore, the atom may retain some of this energy. This interaction is only important for very slow-moving particle energy.
Nuclear short range: Elastic and inelastic collisions with the nucleus – RARER
If the path of the moving particle is close enough to the nucleus (an analogy might be a satellite passes inside the rings of Saturn and is closer to the planet), so that the forces from the outer electron charges no longer shield the moving particle from the forces of the nuclear components, the traveling particle experiences a deflection. It is possible that the interaction is strong enough to cause bremsstrahlung radiation to be emitted and therefore energy to be lost, but this is rare. If there is no radiation, the incident particle can still be deflected and it will effectively lose energy to be consistent with the laws of momentum conservation. Particles heavier than electrons don't experience this very often, but it is measurable.

While the interactions that will occur are of random probability, the nuclear strong force is much less probable than the Coulomb interaction. The dominant interactions are mediated by the Coulomb force between the moving particle and the atomic electrons and the most probable interaction is that of ionization. Take, for example, a particle of kinetic energy E_k, velocity v moving in the

DOI: 10.1201/9781003123880-9

x direction, as in Figure 9.1. Assume that this moving particle has a charge $q_1 = ze$ and the electron has a charge $q_2 = e$ ($e = 1.60E^{-19}$ C). The matter in which the particle is traveling is characterized by ρ_a atoms per unit volume with atomic number Z. Let the path the particle is following pass near an atomic electron with a minimum distance r_m. Thus, the distance between the charged particle and the electron is $r = r_m/(\sin\theta)$. The force between the two charged particles is given by the Coulomb force

$$F = k\frac{q_1 q_2}{r^2} = \frac{1}{4\pi\varepsilon_0}\frac{ze^2 \sin^2\theta}{r_m^2} \tag{9.1}$$

where

q_1 is the charge of the moving particle ze
q_2 is the charge of the electron e, and

the distance between them (at a given time) is $r = r_m/\sin\theta$.

The minimum distance, r_m, is sometimes called the impact parameter (b), which is the perpendicular distance between the initial path (diagonal dotted line in Figure 9.1) (this can change) of the moving particle and the center of the potential with which it will interact. This quantity is different for different interactions depending on the trajectory.

The change in momentum of the moving charged particle is in the direction perpendicular to its path, and thus the impulse given to the electron, is given by the following:

$$\Delta p_e = \int_{-\infty}^{+\infty} F \sin\theta \, dt \tag{9.2}$$

The change in the momentum results from the force between the projectile and the atom, dependent upon the distance between the particles and the *time* that the moving projectile spends near the atom. As shown in Figure 9.1, the projectile is coming from the left, getting closer to the atom, and then interacts, as shown by the deflection, by the force between them and moves on getting further away. This path has a hyperbolic shape and gives rise to a specific minimum distance between the incident particle and the atom. The curve in Figure 9.2 indicates the force that is between the charges as a function of the linear position x for five different positions 1, 2, 3, 4 and 5. The curve in Figure 9.2 is derived from the same force resulting in the trajectory of Figure 9.1 but is shown as a function of time t. There are two curves in Figure 9.2 showing the force seen at the same times. The lower curve is for a faster particle than the upper curve, which is a slower particle. Therefore, the

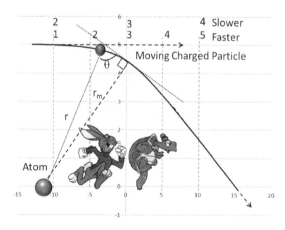

FIGURE 9.1 Charged particle trajectory – Coulomb force.

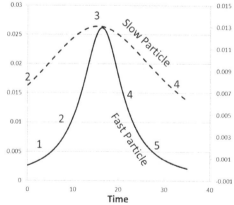

FIGURE 9.2 Force with time, two particles.

slower particle spends a longer time closer to the atom and experiences a higher force for a longer time and will therefore undergo a larger momentum change.

The outstanding question is of course which is 'better', or will the tortoise or the hare win?

This integral result leads to the change in the kinetic energy ΔT_e given to the atom's electron, at low speed as being described by

$$\Delta T_e = \frac{\Delta p_e{}^2}{2m} = \frac{z^2 e^4}{8\pi^2 \varepsilon_0{}^2 \left(r_m{}^2 \right) m_e v^2} \qquad (9.3)$$

or alternatively, it is the energy that has been transferred from the moving charged particle, the moving particle having lost that energy. This is the energy lost owing to an interaction with one atomic electron.

A stopping cross section can be defined as in Section 7.3.2

$$\sigma_e = \int \Delta T_e \, dA \qquad (9.4)$$

where ΔT_e is the energy loss of the moving particle while moving through material with a cross-sectional area dA. The traveling particles are slowed down. If the dimensions of the cross section are cm²-eV, it can be thought of as the probability of removing 1 eV from the particle per cm² of interacting material. During its travels, the moving particle passes many electrons. Consider a cylinder with radius r along a distance $dx \gg r$ as shown in Figure 9.3. The center of the cylinder is approximately where the particle travels along the dotted line. The number of electrons the moving particle passes is the number of electrons per unit volume (which is ρ_a (number of atoms/volume)·Z) multiplied by the volume considered. The area in which the targets sit

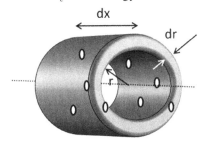

FIGURE 9.3 Interaction cylinder.

is $2\pi r_m dr_m$. They are at a range of minimum distances between r_{m1} and $r_{m2} = r_{m1} + dr_{m1}$ from the moving particle. Therefore,

$$\sigma_e = \int \Delta T_e \, 2\pi r_m \, dr_m = \int \frac{z^2 e^4}{8\pi^2 \varepsilon_0{}^2 m_e v^2} \frac{2\pi r_m dr_m}{r_m{}^2} \qquad (9.5)$$

The limits of integration are given by the range of impact distances in which the moving particle interacts or between r_{m1} and r_{m2}.

$$\sigma_e = \int_{r_{m1}}^{r_{m2}} \Delta T_e \, 2\pi r \, dr = \frac{z^2 e^4}{4\pi \varepsilon_0{}^2 m_e v^2} \ln \frac{r_{m2}}{r_{m1}} \qquad (9.6)$$

This is the probability of 1 eV loss per particle per electron interaction in that area per unit distance.

The number of atoms per mole of material is given by Avogadro's number $N_A = 6.022 \times 10^{23}$. The mass of 1 mole of material is A grams of that material, where A is the atomic number of the material. Finally, the mass density (ρ) of the material relates the mass to the volume. Therefore, the number of atoms per unit volume of the material is given by

$$\rho_a = \frac{N_a \left(\text{number of atoms/mole}\right)}{A\left(\text{g/mole}\right)} \cdot \rho\left(\text{g/cm}^3\right) = \frac{\text{number of atoms}}{\text{cm}^3} \qquad (9.7)$$

Sometimes N might be used for ρ_a. Since there are Z electrons per atom, the number of electrons per unit volume results from multiplying the previous equation by the Z of the material.

Define stopping power as the energy lost by a particle per unit length of distance through the stopping material. The energy loss per unit distance $\Delta T_e/dx$ per electron is

$$\frac{\Delta T_e}{dx} = \frac{\rho_a \sigma_e dx}{dx} \rightarrow \frac{dT_e}{dx} \tag{9.8}$$

The volume in which the interactions take place is included in Equation 9.8, but the area was already accounted for in the definition of σ_e, so all that remains is to multiply that by dx. However, the stopping power is defined per unit distance, so the dx is canceled. The stopping power, including the number of electrons (Z), is

$$\frac{dT_e}{dx} = -\sigma_e \rho_a Z = \left(cm^2 eV\right)\left(\frac{number\ of\ atoms}{cm^3}\right)\left(\frac{electrons}{atom}\right) \tag{9.9}$$

Changing from T_e to E, since that is the usual convention, the energy loss resulting from interaction with these electrons is then,

$$\frac{dE}{dx} = -\frac{z^2 e^4}{4\pi\varepsilon_0^2 m_e v^2} \ln\left(\frac{r_{m2}}{r_{m1}}\right)\rho_a Z \tag{9.10}$$

where $r_{m2} = r_{max}$ and $r_{m1} = r_{min}$ are the upper and lower limits of the integration. Now comes some physics hand-waving. A physically meaningful value for r_{min}, the closest point of approach, is the electron wavelength (size of the electron), $r_{min} = (h/m_e v)$ or related to Heisenberg's uncertainty of the electron position. The maximum distance is related to considering if it is too far away, there is insufficient energy to ionize and the effect is nonexistent. Alternatively, a meaningful time constant to consider is the speed of the orbiting electrons, or how long they are actually close enough to the traveling particle to interact. The time that the two are in 'proximity' when the moving particle is at a distance ~r from the atom is r_{max}/v where v is the speed of the moving particle. In the Bohr model, the electron orbits in a circle with at multiples of its half wavelength. The time that the orbiting electron oscillates in the atom is quantum mechanically preordained and is h/I. So $r_{max} = hv/I$. I is related to the energy of the electron orbit (nhv), or the ionization potential. Thus, $\ln(r_{m2}/r_{m1})$ becomes $\ln(m_e v^2/I)$. Depending on the moving particle energy, it can possibly excite any of the electrons surrounding the nucleus and so the I that is used is the mean ionization potential I_m (more on this in the next section).

From the earlier algebra and adding factor of 2 in the argument of the first logarithm, owing to quantum mechanical algebra and other terms from some more relativistic math, the energy loss per unit distance as it passes through matter is as follows:

$$\frac{dE}{dx} = -\frac{1}{4\pi\varepsilon_0^2}\frac{e^4}{m_e}\left[\frac{z^2}{v^2}\rho_a Z\right]\left[\ln\frac{2m_e v^2}{I_m} - \ln\left(1-\frac{v^2}{c^2}\right)-\frac{v^2}{c^2}\right] \tag{9.11}$$

Moving particle
  z = charge of moving particle
  v = speed of traveling particle

Material
  Z = atomic number of the atom
  ρ_a = number of atoms/unit volume
  I_m = mean ionization potential of the absorbing material (it 'absorbs' energy)

Constants
  e = unit charge
  m_e = mass of the electron
  ε_0 = vacuum permittivity (related to Coulomb constant $k = 1/(4\pi\varepsilon_0) = 8.9876 \times 10^9$ N-m^2/C^2)

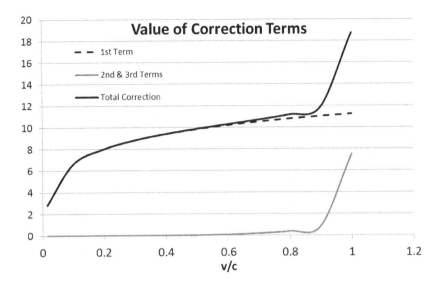

FIGURE 9.4 Stopping power correction term values.

Note that sometimes this equation is written without k or ε and/or without some of the corrections. So be careful of the units used in the equation chosen.

An energy loss per unit length has various names. Physicists call it the stopping power (S). Biologists call it the linear energy transfer (LET), and while these are not all exactly the same, their use can be confusing. The stopping power has two main terms, the first is the one to the left and the first square brackets and the second is in the rightmost square brackets. For a given particle and medium, the expression changes only with the particle speed. The second factor (in the square brackets) becomes important only at lower and higher energies, but not in the range of clinical utility. This is shown in Figure 9.4. The order of magnitude of the total is about 10. The value of the constants outside the square brackets (when v is replaced by v/c) is

$$\frac{1}{4\pi\varepsilon_0^2}\frac{e^4}{m_e c^2} = 8.17 \times 10^{-42} \text{ kg}\frac{m^4}{s^2} \tag{9.12}$$

Taking care with the electronic density and with the appropriate conversions from SI units to MeV, this equation will give MeV/cm. One of the factors in the square brackets is I, which is not known perfectly, although there are a lot of precise measurements. Differentiating the stopping power as a function of I results in the following fractional change in the stopping power energy loss due to errors in I:

$$\frac{dE}{E} = \frac{dI}{I}\frac{1}{\ln\left(2m_e c^2 \left(v^2/c^2\right)/I\right)} \tag{9.13}$$

For an error in I of 1 eV over a range from 5 to 18 eV, the average energy loss error is 0.2%.

In Equation 9.11, dE/dx gives the kinetic energy loss per unit length resulting from the main interaction, which is that of ionization. If the kinetic energy E is dissipated in ionizing collisions, then, $E = Nw$, and $S = wn$, where

N = total number of ion pairs created
n = specific ionization (ion pairs/unit length)
w = energy required to produce an ion pair (30–35 eV)

Oh, and was it noticed that this all started with the moving particle being deflected from its original trajectory? That's worth remembering. And finally, who won? What's the goal? The slower particle delivered more energy, the faster particle delivered less, in a given region!

9.2   IONIZATION POTENTIAL

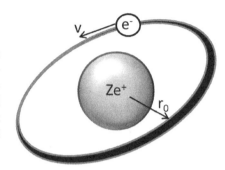

Imagine a single electron with charge e^- is orbiting a nucleus with charge Ze^+. Figure 9.5 can help the imagination. The force between the charges is a force derivable from a potential and the virial theorem can be used to identify a stable orbit. Alternatively, knowing that the orbit is stable means that the centripetal force and Coulomb force must balance so that

$$\frac{m_e v^2}{r_o} = \frac{kZe^2}{r_o^2} \tag{9.14}$$

FIGURE 9.5   An atom.

where m_e is the mass of the electron, r_o is the radius of the orbit, k is the Coulomb's constant and Z is the charge of the nucleus. This means that the kinetic energy of the electron, taking Equation 9.14 into consideration (using T in this section since that's normally used in this context), is

$$E_k = T = \frac{1}{2} m_e v^2 = \frac{kZe^2}{2r_o} \tag{9.15}$$

The total energy of the electron $E_T = T + U$ where U is the potential energy.

$$E_T = T + U = \left(\frac{1}{2} m_e v^2\right) + \left(-\frac{kZe^2}{r_0}\right) \tag{9.16}$$

The negative sign comes from the opposite signs of the electron and proton charges. Substituting the kinetic energy from Equation 9.15

$$E_T = T + U = \left(\frac{kZe^2}{2r_0}\right) + \left(-\frac{kZe^2}{r_0}\right) = \left(-\frac{kZe^2}{2r_0}\right) \tag{9.17}$$

And substituting the potential energy from Equation 9.14

$$E_T = T + U = \left(\frac{1}{2} m_e v^2\right) + \left(-m_e v^2\right) = \left(-\frac{1}{2} m_e v^2\right) \tag{9.18}$$

The angular velocity of the electron $= v/2\pi r_0$ where

$$\frac{v}{2\pi r_0} = \frac{\left(-2E_T/m_e\right)^{\frac{1}{2}}}{2\pi\left(-kZe^2/2E_T\right)} \tag{9.19}$$

Niels Bohr, who was a student of Ernest Rutherford, disproved his mentors plum pudding model of the atom replacing it with one in which the electrons orbited the nucleus with stable orbital periods. In this model, the angular momentum of the electrons were quantized in units of $h/2\pi$. An approximation of the energy (not including the mass) of an electron orbiting a nucleus, given by the Bohr model, is written as $E_n = n\,h\nu$, where n is the orbital number, $h\nu$ is the energy of the lowest orbital and v is the frequency of the electron orbit that was also expressed in Equation 9.19. Therefore, with $E_T = E_n$

$$E_n = nh\frac{\left(-2E_n\right)^{3/2}}{k\sqrt{m_e}Ze^2} \tag{9.20}$$

The difference in energy between two electron orbits is given by

$$E_n - E_{n-1} = \Delta E_n = \frac{\hbar\left(-2E_n\right)^{3/2}}{k\sqrt{m_e}Ze^2} \text{ or } \frac{dE}{dn} = \frac{\hbar\left(-2E_n\right)^{3/2}}{k\sqrt{m_e}Ze^2} \tag{9.21}$$

Integrate this to find the energy required to free the electron

$$\int\frac{dE}{\left(-E\right)^{3/2}} = \frac{\hbar}{k\sqrt{m_e}Ze^2}\int dn \tag{9.22}$$

Note that as $n \to \infty$ the energy of the electron $E_n \to 0$ (electron is free). The energy difference between a bound (starting from level n) and an unbound electron is given by

$$E_n = \frac{k^2 m_e Z^2 e^4}{2\hbar^2\left(n+C\right)^2} = 13.6\frac{Z^2}{\left(n\right)^2}\,\text{eV} \tag{9.23}$$

where C is a constant of integration. For completeness sake, since this derivation is based on the Bohr model, it shouldn't be surprising that the orbit radius r_o is the Bohr radius a_o.

$$a_o = \frac{\hbar^2}{km_e e^2} \tag{9.24}$$

The ionization potential derived here is the energy required to release one electron from orbit n in a nucleus with a charge Z. That is the definition of this parameter. The energy for the single electron bound to a hydrogen atom is 13.6 eV. In real-life measurements, the atoms are filled with electrons and the energy required to release an electron depends upon several things, some of which include the following. As the atomic number increases, the Coulomb force increases, but as the atom gets bigger, some of the electrons are further away from the moving particle and the force decreases. Also as the number of electrons increases, some electrons shield the other electrons from a portion of the Coulomb force. The measured ionization energy to remove the most loosely bound electron as a function of Z is plotted in Figure 9.6.

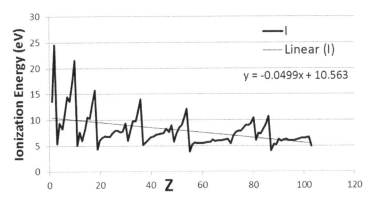

FIGURE 9.6   Z dependence of ionization potential.

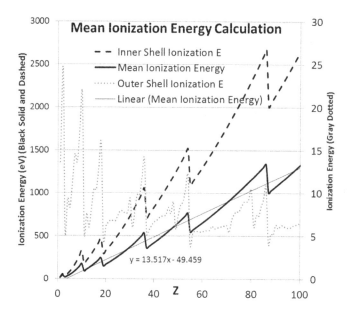

FIGURE 9.7   Mean ionization energy.

The mean ionization energy I_m used in Equation 9.11 accounts for all possible ionization energies in the outer shell of the atom. The curve in Figure 9.6 is the lowest possible energy needed to free those electrons, but the electron configuration may be different at the time the moving particle is close and it can be in a range of impact parameter distances and is also capable of transferring higher energies. So, many ionizations are possible. The ionization energy required to free an electron, not including all the shielding and distance considerations, is given by Equation 9.23, which is different than the curve in Figure 9.6. It remains to determine the mean ionization energy. For help with that consider Figure 9.7. The black dashed line is Equation 9.23. The sharp drops are indications of when the electron orbital number increases with the periods in the periodic table. (The constant is taken as 0.) The gray dotted line is the measured energies from Figure 9.6. Take these two as the highest and lowest ionization energies possible for removing the outer shell electron. Their average results in the black solid curve. The scale for the black curves is on the left. The mean ionization energy, I_m, is fitted to a linear curve (gray line) with the equation ($x = Z$) shown under the black solid line. This is an approximation since the initial condition can be adjusted. Typically I_m is approximated by

$$I_m \cong 13(\text{eV})Z \tag{9.25}$$

9.3   LINEAR THICKNESS OR MASS THICKNESS

The energy loss relation is normally seen as the energy loss per unit length. However, the amount of material that a particle will pass through a given length, of course, depends on the material. A total of 1 cm of lead has a different number of atoms than 1 cm of beryllium.

Consider the thin slab of material in Figure 9.8. It has a volume of abt. Call t the linear thickness. Its mass is ρabt where ρ is the density of the material. Given this material, if the mass of the slab is known, and the area of the slab is known, the thickness can be determined. A mass-equivalent thickness t_m can be

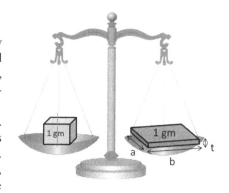

FIGURE 9.8   Mass thickness examples.

defined to be $t_m = \rho t$ with units, for example g/cm². By choosing a mass thickness of, say, 1 g/cm², then the linear thickness required for a unit area (1 cm²) of the slab to have 1 g of mass is chosen. Any material that is selected with a mass thickness of 1 g/cm² will have 1 g of material in a 1 cm² area with the appropriate thickness. Thus, there will be N_A/A atoms in that slab and the number of atoms within each slab will depend only on the atomic number. Since energy loss depends on Z, there will be $N_A Z/A$ electrons to interact with. If the energy loss through a linear thickness of 1 cm of Pb is to be calculated, it will be different than the energy loss through 1 cm of Be, but if the energy loss is calculated through 1 g/cm² mass length of these two elements, the results would be more similar (but not the same) (particularly since the relevant ratio, in that case, would be Z/A which is fairly constant). So, this is a convenient calculation tool.

The energy loss equation can be rewritten. dE/dx could be replaced by $dE/(\rho dx)$ and the energy loss per unit mass length (thickness) (e.g. MeV/g/cm²) would be calculated. It would perhaps be easier to compare the energy loss for a particle penetrating an orange against the same particle penetrating an apple per unit mass length (but not as intuitive). *Be careful when reading tables or performing calculations to know if the values used are per unit linear length or mass length.*

9.4 RANGE

If a number of charged particles pass through a substance, the particles lose energy gradually until only a little energy remains. At a low enough energy, they are captured by some atoms and do not escape. The Bragg number curve is presented in Figure 9.9. Due to the randomness of the interactions, the particles lose different amounts of energy as a function of range, so some have more and some have less at any given distance, even if they all started out with the same energy. The ranges of the individual particles are statistically spread around a mean range R. This spread of final ranges in a medium is called range straggling. (This is not the only effect that spreads the range but is the only one related to what has been discussed up to this point.)

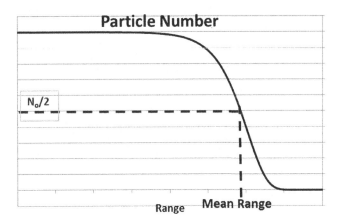

FIGURE 9.9 Particle number with range.

The range is defined as the mean value of the range of the particles in the medium when they stop their trip. Given the expression for energy loss of a particle in the medium, it is possible to compute the range. Until they start stopping, the number of particles remains unchanged. The range is given by

$$\bar{R} = \int_{E_0}^{0} -\frac{dx}{dE} dE = \int_{0}^{E_0} \frac{dE}{-dE/dx} = \int_{0}^{E_0} \frac{dE}{S} \tag{9.26}$$

where S, as noted earlier, is defined as the stopping power of the medium for that particle.

Recall the energy loss equation and consider particles with lower speed (relativistic corrections not necessary)

$$-\frac{\mathrm{d}E}{\mathrm{d}x} = \frac{1}{4\pi\varepsilon_0^2}\frac{z^2e^4}{m_e v^2}\rho_a Z\left[\ln\frac{2m_e v^2}{I}\right] \tag{9.27}$$

Recognizing that in the nonrelativistic case, the kinetic energy of the moving particle with mass M is $E = 0.5\,Mv^2$ and inverting the equation begets

$$-\frac{\mathrm{d}x}{Mv\mathrm{d}v} = 4\pi\varepsilon_0^2\frac{m_e v^2}{z^2e^4}\frac{1}{\rho_A Z}\left[\ln\frac{2m_e v^2}{I}\right]^{-1} \tag{9.28}$$

$$-\mathrm{d}x = 4\pi\varepsilon_0^2\frac{m_e v^2}{z^2e^4}\frac{1}{\rho_A Z}\left[\ln\frac{2m_e v^2}{I}\right]^{-1}Mv\mathrm{d}v \tag{9.29}$$

$$\int \mathrm{d}x = C\int \frac{v^3\mathrm{d}v}{\ln(2m_e v^2/I)} \tag{9.30}$$

$$\text{If }\; y = \left(\frac{2m_e v^2}{I}\right)^2; \int \mathrm{d}x = C\int \frac{\mathrm{d}y}{\ln(y)} \tag{9.31}$$

C is a constant accumulating the other constants and the last integral in Equation 9.31 is called an exponential integral (Ei), which is not analytically solvable but well tabulated.

The graph in Figure 9.10 compares the nonrelativistic kinetic energy (gray line) with the relativistic kinetic energy (black line) as a function of the speed for protons. They are within 10% of each other below about 75 MeV proton kinetic energy. So the range of validity of these approximations is better clarified.

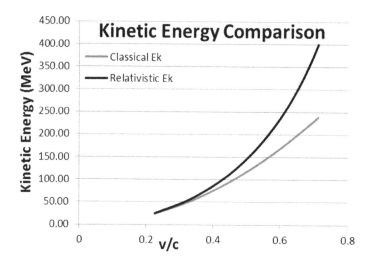

FIGURE 9.10 Compare relativistic and classical kinetic energy.

While the integral wasn't evaluated, the earlier relations are rich in some details. From Equation 9.28, $\mathrm{d}x/\mathrm{d}v$, the change in distance for a given speed loss is related to M/z^2 and is a function of v. If two different particles have the same initial speed, the incremental difference in distance traveled between particles 1 and 2 is given by

$$\frac{dx_1}{dx_2} = \frac{\left(M/z^2\right)_1}{\left(M/z^2\right)_2} = \frac{R_1}{R_2} \tag{9.32}$$

This can approximately scale to their relative full ranges R_1 and R_2 (even as they slow down). The relation holds better as the energy decreases, but it's not bad for back of the envelope estimates.

In the relativistic case, the z^2/v^2 term is the key for the moving particle. Recall from Chapter 8 that any particle with the same kinetic energy/nucleon has the same speed. Therefore, a carbon ion with, for example, 430 MeV/nucleon has the same speed as a proton with kinetic energy 430 MeV. This carbon ion with a charge $z = 6$ will be losing 36 times more energy per unit distance as a proton with the same speed. However, the total kinetic energy of the carbon ion starts out at 12 times that of the proton (A_p = 12 for carbon). (A_p for the moving particle.) Therefore, the carbon has to lose 12 times the kinetic energy before it stops. Since the energy is lost 36 times more per increment, all the energy will be lost in about one-third of the total distance when compared to a proton. A 430 MeV/nucleon carbon ion has a range of about 30 cm in water, compared to a proton with a kinetic energy of 430 MeV that has a range of about 90 cm in water. This ratio is related to M/z^2, or the same ratio derived in the nonrelativistic case. Using the fact that particles with the same kinetic energy/nucleon have the same speed, and this range ratio relationship, it is magically easy to estimate the range energy relationships for different particles as shown in Figure 9.11. Note that He is a dashed line and proton is a solid line, but they overlap. Therefore, if two particles have the same speed and the range of particle 1 is known, the range of particle 2 can be determined. In the special case comparing helium with hydrogen, the (M/z^2) ratios are the same and the ranges will be the same if the initial speeds are the same. For example, the range of a He ion with a total kinetic energy of 600 MeV or 150 MeV/nucleon will be the same as a proton with a kinetic energy of 150 MeV. In practice, there are some differences in the ranges of these two, given that helium has two protons to lose while penetrating the medium, compared to one. The horizontal dashed line in Figure 9.11 is at a range of 30 cm.

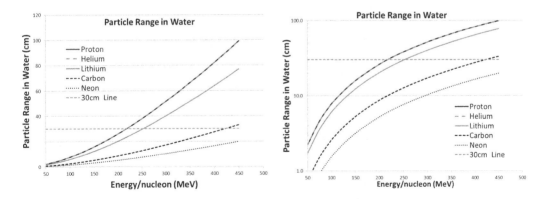

FIGURE 9.11 Particle range in water (linear) (left) and particle range in water (semi-log) (right).

In the clinical energy range, the proton range *in air* is given by the approximate relation (for proton, energies less than a few hundred MeV).

$$R(\text{cm}) = 100 \left(\frac{E(\text{MeV})}{9.3} \right)^{1.75} \quad \text{in air} \tag{9.33}$$

(Although the author has sometimes found that the proton range is better fit with an exponent of 1.7235, even though the fit in air with 1.75 is still good.)

Thus far, the range of different particles in the same material has been considered. Taking into account, different materials involves consideration of the material-related terms in the formulae. The relevant terms that are proportional to dx are ρ_a and Z of the material. There does not appear to be a simple exact expression, but the Bragg-Kleeman rule makes use of Bragg's early measurements showing the dependence of the atomic stopping power to be proportional to $\sqrt{A}$. Remember atomic density from Equation 9.7 and define an atomic stopping power

$$S_a \equiv \frac{dE}{dx\rho_a} \propto \left[\frac{z^2}{v^2}\right] Z \left[\ln \frac{2m_e v^2}{I_m}\right] \tag{9.34}$$

Consider two materials 1 and 2 for the same moving particle. The ratio of the atomic stopping powers gives a ratio of ranges, if $R = \int dx$

$$\frac{dE/\left(dx_1 N_a \rho_1/A_1\right)}{dE/\left(dx_2 N_a \rho_2/A_2\right)} = \frac{\sqrt{A_1}}{\sqrt{A_2}}, \tag{9.35}$$

which simplifies to

$$\frac{dx_2}{dx_1} = \frac{\rho_1\sqrt{A_2}}{\rho_2\sqrt{A_1}} = \frac{R_2}{R_1} \tag{9.36}$$

If a reference medium is taken to be air, knowing the range in air (e.g. Equation 9.33) can be used for range scaling to determine the range of the same particle in other media characterized by A and ρ (in units of g and cm).

$$R = 3.2 \times 10^{-4} \frac{\sqrt{A}}{\rho} \times R_{\text{air}} \tag{9.37}$$

If the material is a mixture of materials, it is possible to define an equivalent material stopping power. Define an effective stopping power $<S> = \Sigma n_i S_{ai}$, e.g. for water $<S> = 2/3\ S_H + 1/3\ S_O$.

9.4.1 RANGE SENSITIVITY

It may be necessary to know how important the incident beam energy is to the setting of the beam's range. From Equation 9.33, the range sensitivity to beam energy in air is

$$\frac{\Delta R}{R} \sim 1.75 \frac{\Delta E_k}{E_k} \tag{9.38}$$

The required energy adjustment of the beam production equipment can be determined from this relation. However, the key parameter in the beam handling systems is momentum (P). This can be the resolution of the beam transport system, or the resolution of the accelerator. Therefore, this should be taken into account as follows:

$$\frac{\Delta P}{P} = \frac{\Delta E_k}{E_k}\left[\frac{E_k + E_0}{E_k + 2E_0}\right]; \text{ For low } \frac{E_k}{E_0} \text{ then } \frac{\Delta P}{P} = \frac{\Delta E_k}{2E_k} \tag{9.39}$$

9.4.2 ENERGY LOSS, DISTAL DOSE FALLOFF OR RANGE SPREAD

A graph of the energy lost by a single proton starting with 230 MeV and traveling through water, using only Equation 9.11, is shown in Figure 9.12. The particle energy is shown decreasing with the depth of penetration (gray dashed line and scale to left), while the energy lost per unit distance (related to stopping power) is shown as a plateau (flattish) initially and then increasing as the depth increases and the particle's kinetic energy (and velocity) is reduced. At the right-hand side of the graph, the particle's energy has been depleted. The energy loss curve is known as the Bragg curve (BC) or the Bragg peak (BP). Both can be used almost interchangeably, but BP is sometimes used more when referring to the distal peak of the BC.

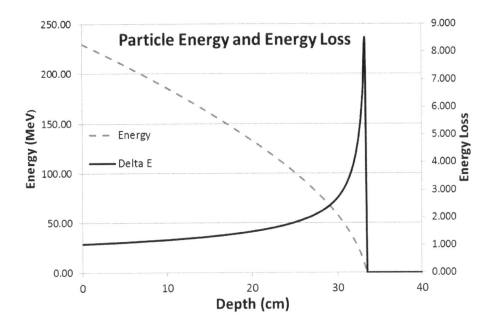

FIGURE 9.12 Energy loss and particle energy.

There are many particles in a beam incident upon the matter. There is also a spread in the energies of the beam, in that different particles have different energies. This can be from the initial energy spread of the beam from the beam production system, which may be close to a Gaussian distribution. Added to this can be the statistical nature of range straggling, as discussed previously, which can also be close to a Gaussian distribution. Then multiple particles will give rise to BPs of different ranges according to the statistical spread in ranges and this can round out the BP curve. If the beam has an energy spread, for example 1%, the sharp BP is rounded out and looks like the dashed curve as shown in Figure 9.13.

For protons, the statistical process of range straggling results in a final spread in ranges, which is Gaussian (remember the central limit theorem) with a sigma (in range) of

$$\sigma_R \sim 1.2\% \, \text{Range}. \tag{9.40}$$

This distribution (only due to straggling), (with R = range), looks like

$$\frac{1}{n_0} \frac{dn}{dx} = C e^{-[(x-R)/\sigma_R]^2} \tag{9.41}$$

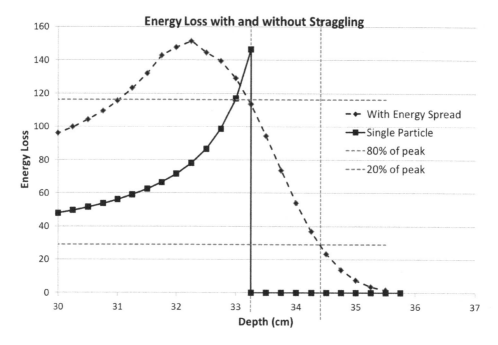

FIGURE 9.13   Energy loss with and without range straggling.

The distal dose falloff is usually defined as the falloff from 80% of the dose after the BP to 20% of the dose. (These are shown in Figure 9.13 by the dashed horizontal and vertical lines.) This is related to the Gaussian sigma (range sigma) of the beam by the following:

$$\Delta R_{80/20} \sim 1.3 * \sigma_R \tag{9.42}$$

An interesting fact is that the depth at which the end of range of the single-particle BP intersects with the range straggled BP is at 80% of the height of the range straggled curve. (Note the upper dashed horizontal line.)

The range straggling of the beam will increase due to material placed in the beam to degrade its energy or scatter the beam. The range straggling will add in quadrature to the energy spread of the beam. The straggling in the patient cannot be avoided; anything else is excess.

$$\sigma_{Total} = \sqrt{\sigma^2_{Total\ Range\ Straggling} + \sigma^2_{Beam}} \tag{9.43}$$

$$\sigma_{Excess} = \sigma_{Total} - \sigma_{Range\ Straggling\ in\ patient} \tag{9.44}$$

It is a goal of the system designer to minimize this excess according to the clinical parameters desired (usually < 1 mm), but depending upon the depth of the target, a considerable portion of this could result from the straggling in the patient. Be warned however, BPs can sometimes be too sharp and this will be explored further.

9.5   SCATTERING

Charged particles interact with the Coulomb field of the nucleus, as discussed in Section 9.1. One of these interactions is deflection as shown in Figure 9.1, partially reproduced here in Figure 9.14. The larger ball at the bottom is the atom and the smaller ball at the top is the moving particle (which could be a proton or another particle). The dashed arrow through the moving particle at the top is its initial trajectory and the lower arrow is the trajectory as a result of the Coulomb force interactions

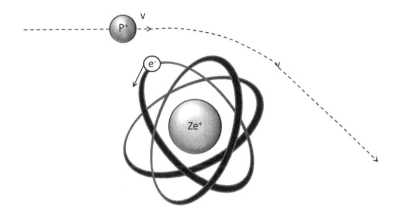

FIGURE 9.14 Coulomb deflection.

with the electrons. The root mean square (rms) scattering angle in a thick target is derived from the rms scattering angle from a succession of thin targets using a lot of atomic physics techniques which are beyond the scope of this book. This succession of scattering from multiple layers and multiple targets is called multiple scattering. There are multiple derivations and multiple contributions to this.

Equation 9.45 shows the relation of the growth of the rms angle θ of a beam when it is scattered through a (thin) target.

$$\theta_{rms}(\textbf{Radians}) \approx \frac{\textbf{Constant}}{pv} Z \sqrt{\frac{L}{L_R}} \left[1 + \frac{1}{9}\textbf{log}_{10}\frac{L}{L_R} \right] \tag{9.45}$$

The quantities are as follows:

L = length of penetration in the medium
L_R = radiation length of the material (a constant of the material)
p = momentum of the incident particle
v = velocity of the incident particle
Z = charge of the atoms in the medium
Constant = 14.1 MeV
Remember (if you did the exercises), $pv = E_{Total}\beta^2$ (MeV)

This mechanism is basically the result of the Coulomb interaction between the charged particle and the charged atom. Each of the individual interactions with atoms results in a change of direction of the incident ion, and the statistical average of all these particle angles results in a spread, which is summarized by the rms divergence angle of the beam. An example of the statistical nature of this is shown in Figure 9.15 with hundreds of individual particles tracked through a material. The result of this approximates a Gaussian distribution in angles sometimes with extended tails.

Multiple Scattering through a medium

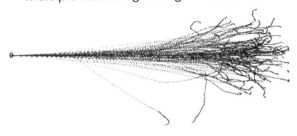

FIGURE 9.15 Multiple scattering example.

The dependences of the rms Gaussian divergence, in Equation 9.45, are approximately

$$\theta_{rms} \approx Z \sqrt{\frac{L}{L_R}} \text{ and } L_R \ \alpha \frac{A}{Z^2} \xrightarrow{\text{yields}} \theta_{rms} \approx Z^{3\!/\!2} \sqrt{L} \tag{9.46}$$

The previous equation is only for one thin target; it is not appropriate to use it as the basis for multiple thin targets stacked. The particle loses energy in the target as it penetrates and this must be taken into account if the target is thick enough. Bernie Gottschalk et al. have generalized this equation to a form that can be stacked. There is much literature exploring the origin and derivation of these equations in depth. The Highland formula is appropriate for thin targets. For thick targets, it is necessary to integrate the formula. However, a reasonable approximation for medium thickness targets is to replace pv in Equation 9.45 with $\sqrt{p_1 v_1 \times p_2 v_2}$ where 1 and 2 refer to the initial and final quantities. More accurately, it's appropriate to integrate the expression

$$\theta_{rms} = 14.1 \ Z \left(1 + \frac{1}{9} \log_{10} \frac{L}{L_R} \right) \left[\int_0^L \left(\frac{1}{pv} \right)^2 \frac{dl}{L_R} \right]^{\frac{1}{2}} \tag{9.47}$$

L_R, or as it is sometimes written X_0, the radiation length, is the distance that a high-energy electron must travel so that its energy is reduced by a factor of $1/e$. For $Z > 4$, the radiation length is approximately given by

$$X_0 \approx \frac{1}{4(\hbar/m_e c)^2 \alpha^3 N_a} A \left[Z(Z+1) \log \left(\frac{183}{Z^{1\!/\!3}} \right) \right]^{-1} \text{ cm} \tag{9.48}$$

where
N_a = Avogadro's number
h with a bar = Planck constant/2π
m_e = electron rest mass
c = speed of light
α = fine-structure constant
A = the atomic number of the medium

This expression can be written as

$$\approx 716.4 A \left[Z(Z+1) \ln \left(\frac{183}{Z^{1\!/\!3}} \right) \right]^{-1} \left(\text{g/cm}^2 \right) \tag{9.49}$$

Or also sometimes written as

$$\approx 716.4 A \left[Z(Z+1) \ln \left(\frac{287}{\sqrt{Z}} \right) \right]^{-1} \left(\text{g/cm}^2 \right) \tag{9.50}$$

Equations 9.49 and 9.50 are plotted in Figure 9.16 (they are on top of each other, except at the beginning) and greater than $Z = 4$ are within a few percent of each other. The radiation length has an A/Z^2 dependence since the ln function varies slowly over most of the range. The measured radiation length of carbon 12 is $L_R = 42.7$ g/cm^2. (It is closer to the value calculated with the formula of Equation 9.50.)

For combinations of materials, the combined radiation length can be approximated by

$$\frac{M_{\text{Total}}}{X_0} = \sum_i \frac{M_i}{X_i}$$ (9.51)

where M_{Total} is the total mass of the material and M_i is the mass of an individual component.

Figure 9.17 plots the transverse extent of a beam (starting with a 0 width) as it is being scattered in some material. The black solid lines (indicative of the outer extents of the beam) show the growth of the beam size. After scattering through a target, the beam of particles can be thought of as having a new effective origin (virtual source) as shown in Figure 9.17 with the black dashed arrows pointing to the new origin location. This is a helpful tool for calculations using the scattered beam.

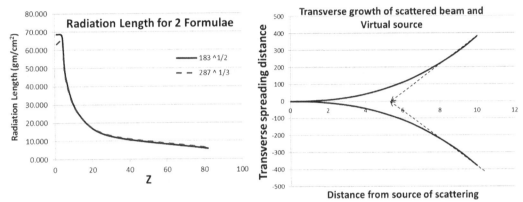

FIGURE 9.16 Radiation length formulae.

FIGURE 9.17 Transverse growth and virtual source.

The curves in Figure 9.18 are also examples of the growth of the transverse dimension of a beam passing through multiply stacked or a thick target(s) if, for every thin section, the scattering angle is the same. The black curve is the cumulative angular growth and the dashed is the effect on one half of the beam size. It doesn't grow linearly because the rms angle is added as the square of the angle. However, as the beam passes through matter, it loses energy. This results in a scattering angle, for each successive layer that is larger and larger. This is shown approximately in Figure 9.19, starting

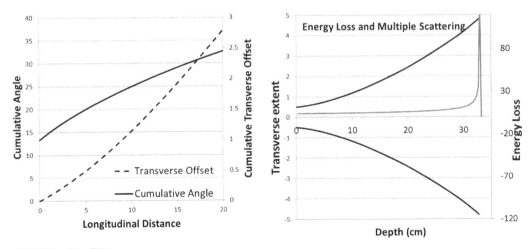

FIGURE 9.18 Effect of scattering.

FIGURE 9.19 Scattering with energy loss.

with a nonzero size. That figure includes a BC as an indication of the degree of penetration of the beam and shows how the angular divergence increases with energy loss.

9.6 DEPENDENCE OF THE BRAGG PEAK ON THE BEAM SIZE

There are a number of other features that are interesting to explore. As the beam is scattered away from the center, the trajectory of extreme rays is curved, the total length traveled by all particles with similar energy is similar. The curve then means that the extreme particles will stop at a shorter longitudinal position compared to one on the central axis (see Figure 9.15). Therefore, as shown in Figure 9.20, the depth-dose distribution will exhibit a bulge in the middle (note the dashed bulge line on the right). The three rays shown in the plot have traveled to different depths in the target, despite having traveled, in their minds, an equal distance. In addition, consider the rectangular blocks in Figure 9.20. There are four shorter unfilled blocks and four longer, hatched blocks. Those blocks are detectors, with a transverse extent given by the height of the rectangle. They measure the energy loss from the beam that passes through them. The scattering affects that result. First, consider the taller detectors. In real life, this is called a Bragg peak detector. These integrate the entire beam of particles. Moving them along the depth direction, the BC measured will look normal through the plateau until closer to the peak. Because of the scattering, the effective range of the beam has been spread so the peak will be a little less sharp. Now consider the smaller detectors. The fact that the smaller detectors cover a smaller and smaller fraction of the beam gives rise to a much different response curve. Nothing has changed in the beam, but the detection results are different. The detector could be replaced by a clinical target, with the same result.

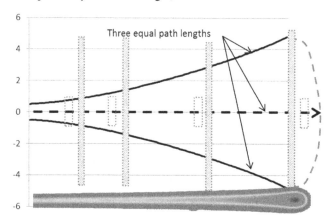

FIGURE 9.20 Measurement with different detectors.

The effectiveness of the BP is reduced as the initial beam size is smaller. This is one reason why electrons, which are charged particles, do not exhibit the same properties as heavier charged particles. While electrons have an end of range, they do not have a sharp BP. Figure 9.21 includes two simulated measurements of BCs (the gray curves). The incident transverse sizes are different (top is greater than bottom). The top beam spreads less and the top depth-dose curve as measured by the small detectors is fairly normal. The bottom beam spreads more and the proportion of beam that passes through the small detectors reduces with depth. Therefore, the peak is much smaller, and the plateau is not as flat as a normal plateau with depth. The smaller proportion of beam measured weighs against the increasing energy released, enabling a vestige of the BP to remain.

In the case when multiple beams are delivered next to each other, they scatter into each other; thus, the fluence measured along the central axis reflects the sum of all the beams. If the beams are at an appropriate distance apart, the particles that scattered out from one beamlet are compensated by particles scattered in from other beamlets, and the BP shape is maintained.

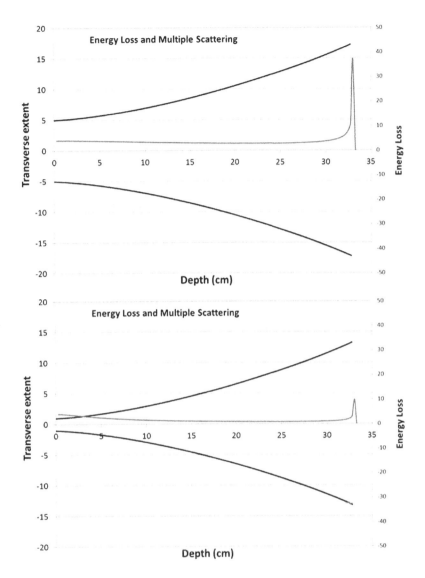

FIGURE 9.21   Effect of small beam on Bragg peak.

9.7   ENERGY LOSS AND SCATTERING DEPENDENCIES

Both the energy loss and scattering equations include quantities related to the material through which the moving particle is traveling. The parameters included have different dependencies. Consider the equations below, reproduced here:

$$-\frac{dE}{dx} = \frac{1}{4\pi\varepsilon_0^2}\frac{z^2 e^4}{m_e v^2}\rho_a Z\left[\ln\frac{2m_e v^2}{I}\right] \qquad \text{(9.27 repeated)}$$

$$\theta_{rms} \approx \frac{14.1}{pv}Z\sqrt{\frac{L}{L_R}}\left[1+\frac{1}{9}\log_{10}\frac{L}{L_R}\right] \qquad \text{(9.45 repeated)}$$

$$L_R \approx 716.4A\left[Z(Z+1)\ln\left(\frac{287}{\sqrt{Z}}\right)\right]^{-1} \text{g/cm}^2 \qquad \text{(9.50 repeated)}$$

It could be useful to know the scattering angle and energy loss in 1 g/cm² of material. These equations can be approximated as follows:

$$dE \propto \rho_a Z dx \propto N_A \frac{Z}{A}(\rho dx) \tag{9.52}$$

$$\theta \propto Z\sqrt{\frac{L}{L_R}} \propto Z\sqrt{\frac{Z(Z+1)}{A}}\sqrt{\rho dx} \approx Z^{\frac{3}{2}} \tag{9.53}$$

$$L_R \propto A\left[(Z)(Z+1)\right]^{-1} \tag{9.54}$$

Using these formulae, Figure 9.22 is a plot of the energy loss (160 MeV protons in 1 g/cm²) (solid black line) and multiple scattering angles (gray dashed) for different materials. Roughly speaking, the dependence of θ is close to quadratic and increases with Z (but the radiation length gets smaller – so it is plotted against radiation length). The energy loss, given that the ratio of Z to A does not change very dramatically, is fairly constant. This provides useful design guidance. It is seen that material of higher Z, relative to lower Z, provides proportionally more multiple scattering. Thus, if material is desired to minimize multiple scattering with roughly the same energy loss, it is helpful to choose material of lower Z and vice versa. Also, combinations of materials can be found that will maintain a constant scattering angle for different energy losses.

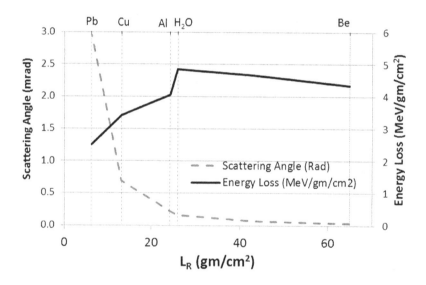

FIGURE 9.22 Comparison of scattering and energy loss for materials.

9.8 WHAT COULD GO WRONG – 1?

This is the first of a series of sections with this title. This is not meant to engender paranoia, but to enhance critical thinking that can contribute to a culture of safety and a positive contribution to the design and measurement requirements of a facility.

This chapter has described aspects of charged particle interactions in matter. These are elements of the fundamental physics, which contribute to the dose distribution in the patient, and also contribute to the parameters of the beam that arrive at the patient. The quantities identified and derived have some explicit or implied assumptions. Proper evaluation of potential errors motivates making these all explicit so that they can be assessed. These can include:

- The range in material depends upon knowing what that material is, what its density is and how much material the particle is passing through. If any of these are not what is expected, the range of the particle will not be what is expected.
- The range in the patient depends upon knowing the particle velocity, or going one step back, knowing the type of particle and the kinetic energy of the particle. It's hard to have the wrong particle (although heavier ion species that pass through material could generate multiple ions). However, if something (actually it takes a lot of things in some systems and only one thing in other systems) in the system is not set correctly, the kinetic energy of the particle transported could be incorrect.
- The energy spread in the beam could be different than expected and this would result in a distal falloff different than desired.
- The same uncertainty in materials could result in multiple scattering different than expected, changing the beam size and the transverse penumbra.

This is only Chapter 9, and for the most part, only material in the beam has been discussed, but incident beam energy is an input to all the outputs that result from the interactions in the matter.

9.9 EXERCISES

1. Starting from Equation 9.2, derive Equation 9.3 (see Figure 9.23).
2. Who wins, the tortoise or the hare? Why?
3. Calculate the energy loss of a proton going through 1 g/cm² Pb compared to 1 g/cm² Be? What about going through 1 cm of each of the materials?
4. Define (a) an atomic thickness (atoms/cm²) in terms of the appropriate quantities and (b) electronic thickness (electrons/cm²).
5. Add the energy loss and scattering angle for carbon to Figure 9.22. (1 g/cm² of material, 160 MeV protons).

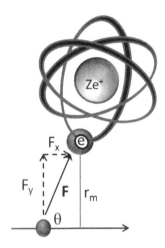

FIGURE 9.23   Exercise 1.

10 Review of Charged Particle Motion

A beam of charged particles is produced by an accelerator system that uses electric and magnetic fields to impart energy to the particle. The beam is then transported toward the patient and prepared for appropriate treatment use. This transportation and preparation can involve particle beam interactions in matter as well as manipulation of the charged particle beam by magnetic and electric fields. The precise combination of these interactions is determined by the modality of the beam delivery that will be used. The previous chapter discussed elements of what happens when charged particles interact in matter. This chapter will discuss how charged particles interact with magnetic fields.

There is a clear distinction between the properties of a beam and the effects that elements of a beam transport system have on that beam. The differences are often overlooked or confused. In fact, the standard terminology itself sometimes makes it hard to distinguish. An understanding of the primary factors that influence charged particle motion is helpful to understand the design constraints and tolerances. Much of this understanding comes from the derivations of the equations governing charged particle motion. Some of this material will be reviewed herein, but not in depth. Further information can be found in the literature.

A particle beam is a collection of particles that have been generated by some source and satisfy Liouville's theorem (most of the time). Liouville's theorem says that the volume and number of particles inside a phase space of canonical coordinates remain unchanged. The trajectory of an individual particle in that beam is called a ray. The ensemble of rays is a beam whose properties evolve in a beam transport system. The motion of the beam centroid behaves like a ray whose trajectory is modified through interactions with the beam transport equipment. Particles that deviate from this centroid are also affected. These concepts are illustrated in Figure 10.1.

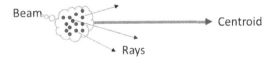

FIGURE 10.1 Concepts of rays and beam.

A ray is characterized by five spatial coordinates. The energy of the particle is primarily in the direction of motion, which will be called s. Other particles in the beam may be in different positions along s relative to some centroid. This relative position, along s, will be called l. Transverse to this main central axis are the directions x and y. Particles that deviate from the central trajectory in these directions can result from a transverse momentum of the particles p_x and p_y in those directions. This terminology will evolve as the tools of charged particle motion are developed.

10.1 MANIPULATION OF LIGHT RAYS

It is sometimes useful to start with intuitive concepts. In this case, most familiar are the effects that lenses and prisms have on light as in the Figures 10.2 and 10.3. Figure 10.2 demonstrates how white light is refracted through a prism. White light includes all colors of the spectrum. The degree to which the light is refracted or bent depends upon its frequency and the angle between the light ray

DOI: 10.1201/9781003123880-10

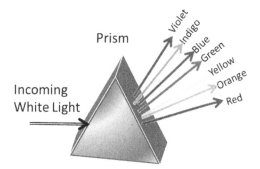

FIGURE 10.2   Refraction of light.

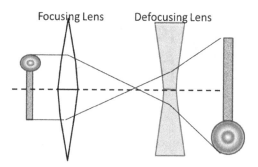

FIGURE 10.3   Focusing light with lenses.

and the normal to the glass interface. Different colors of light travel at different speeds inside the prism. ($v = f\lambda$ and the wavelength λ changes, not the frequency f). Violet is the slowest so its index of refraction is highest and is the most bent. This is an example of electromagnetic interactions in a medium. Glass shaped in this form results in the rainbow of colors that emerge from the prism. Light of a single frequency – e.g. a laser light – would be bent through a specific angle. Glass can be shaped for the appropriate use. Figure 10.3 shows glass formed in the shape of a focusing lens and a defocusing lens. There is a candle on the left-hand side. The lenses focus and defocus the light rays coming from the candle. What does focus mean? A light ray coming from the top of the candle enters the focusing lens and is bent down, since the glass is at an angle relative to the light ray. The refracted light exits the focusing lens, travels straight to the defocusing lens where it is bent further away from the axis. The combination of the two lenses can create an image which is inverted and magnified when viewed at a location downstream of the second lens. Thus, through the use of refraction and the properties of glass, the effects on light can be that of focusing and bending. These properties help to identify features to look for in charged particle optics.

10.2   ELECTROMAGNETIC FORCES

In physics, the **Lorentz force** is the force exerted on a charged particle in an electromagnetic field. The particle will experience a force due to an electric field of $q\mathbf{E}$, (vector $\mathbf{E}$) and due to the magnetic field $q\mathbf{v} \times \mathbf{B}$ (vectors $\mathbf{v}$ and $\mathbf{B}$). Combined, they give the Lorentz force equation/law

$$\mathbf{F} = q\mathbf{E} + q\mathbf{v} \times \mathbf{B} \qquad (10.1)$$

In this equation, the vector force $\mathbf{F}$ on a charged particle with charge q is determined by the electric field $\mathbf{E}$ and the magnetic field $\mathbf{B}$ in which the charge q is introduced. In the case of an electric field, the force is in the direction of the electric field, in which case, the charged particle can be accelerated in the direction of the electric field and the particle's energy may be increased. In the case of a magnetic field, the force is perpendicular to both the direction of the magnetic field and the velocity, $\mathbf{v}$. Since the force is perpendicular to the direction of initial motion, this force does not accelerate the particle in the direction of its initial motion but bends the particle trajectory into a circular arc. Yes, it's also true that an electric field can point perpendicular to the motion of a charged particle and also cause a bending force, but the electric field force is much weaker than the magnetic field force and is not used for this purpose, for beams of clinical energy. Magnetic fields cannot be used to accelerate particles in this energy range, although they were used for lower energy electrons in a betatron. The technology of delivering a charged particle of the appropriate energy to a patient involves accelerating that particle, then possibly bending and focusing the particles toward the target.

The behavior of a moving charge in a magnetic field is described by the 'right-hand rule'. As shown in the Figure 10.4, moving charged particles flowing from the lower left to the upper right (see the direction of *v*), represented by a current *I* in a cylinder, are in a magnetic field *B*. The direction of the field is to the right, resulting in a force *F* pointing down. Pointing the fingers of the right hand in the direction of motion and then curling them from the direction of the charged particles toward the magnetic field, the thumb points in the direction of the force. (Don't strain your wrist.) The beam is then bent down as indicated by the black solid line.

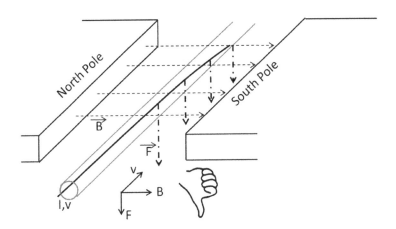

FIGURE 10.4 Bending of a moving charged particle.

In much the same way as light is manipulated by the properties of the glass, charged particles can be manipulated by shaped magnetic and electric fields. Applying these forces can result in adjustments to the particle trajectory. Allowing the particles to drift without any forces is also permissible. Thus, in analogy with light optics, through the use of electromagnetic forces, the effects on a charged particle can be to

1. Focus particles
2. Bend particles
3. Change the energy of the particles

And without any fields, the beam will drift.

One can generalize the idea of the force on a particle to be the external force on a system of particles. Assuming the internal forces between or among the particles can be excluded (which is not always the case), then the system of particles can be treated as one particle at the center of mass of that system. The motion of this center of mass will be represented as the motion of an individual ray. A ray has momentum **P**, which is a vector moving in a particular direction. As indicated in Figure 10.5, the direction of the larger component of momentum is taken as the direction of travel of the particle (*s* direction). The orthogonal direction represents the momentum transverse to the direction of motion (*x* or *y*) and appears as the angle θ the particle is traveling with respect to the primary momentum.

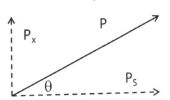

FIGURE 10.5 Transverse momentum.

In the absence of any external forces, the transverse position will continue to increase linearly according to its transverse momentum. If a force in the direction parallel to the main motion is applied (accelerating the particle), then the ratio of the transverse momentum to the main momentum is reduced. This reduces the effective angle of the ray and the transverse position increase will slow. Figure 10.6 shows the transverse position of three rays with different accelerations. The upper gray curve has no acceleration but a nonzero initial angle. The black middle curve has acceleration. Therefore, in the same distance travelled (in the s direction), the effective angle is reduced and the rate of beam size growth is also reduced since the momentum in the s direction has increased relative to the transverse momenta. This effect is even greater with increased acceleration as shown in the dashed curve. It can be described as a change in the effective drift length when compared to a ray without acceleration. Alternatively, getting ahead a little, the relative reduction in transverse momenta can be viewed as a reduction of the beam emittance (there is an external force, so it doesn't violate Louiville's theorem), which is why a normalized emittance ($\beta\gamma\varepsilon$), which does not change as a function of energy, is sometimes defined when discussing acceleration. Alternatively, if a magnetic force is applied, which acts in the transverse direction to the direction of motion, it will directly affect the transverse momentum and can either increase or decrease it according to the force direction.

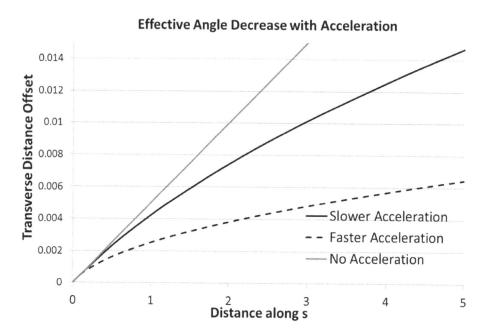

FIGURE 10.6 Decrease angle with acceleration.

10.3 EQUATIONS OF MOTION

10.3.1 BENDING IN A MAGNETIC FIELD

As discussed earlier, while moving charged particles are in a magnetic field, they exhibit an arc of circular motion as shown in Figure 10.7(left). In this case, the magnetic field B (black dots) is pointing out of the paper (or screen) and the particle is moving from the left side to the right side. When the particle enters the magnetic field, according to the right-hand rule, it will experience a force F at right angles to its motion, in this case, pointing toward the bottom of the figure (be careful, don't strain your wrist again verifying this) and thus bends the particle in the clockwise direction with a radius ρ. Physically, this magnetic field could be generated between two flat pieces of iron as shown in Figure 10.7(right). The field would be constrained to be uniform and parallel. (Well this is not as

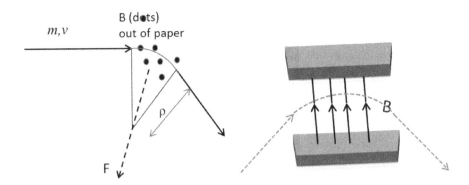

FIGURE 10.7 Bending in magnetic field (left) and magnetic field (right)

simple as it sounds, but this is not a course on detailed magnet design – although I will admit, one of the author's more satisfying moments was measuring an octupole I designed using conformal mapping techniques taught to me by Klaus Halbach – a magnet great.)

Recalling the centripetal force, this motion is described by the following equations:

$$F = qvB = \frac{mv^2}{\rho}; \; B\rho = \frac{1}{q}P \tag{10.2}$$

Some useful parameters include;

1 Tesla = 1 kg/(Coul sec)
1 GeV/c = 5.34 × 10⁻¹⁹ kg m/s
1 elementary charge = 1.6 × 10⁻¹⁹ Coulombs

This allows the creation of a very useful favorite formula, which the author has used all his professional life, as follows:

$$B(\text{Tesla})\rho(m) = \frac{1}{q(\text{coul})}P(\text{kg}(\text{m/s})) \tag{10.3}$$

$$B(\text{kG})\left[\frac{1}{10}\right]\rho(m) = \frac{1}{1.6 \times 10^{-19}}P(\text{GeV/c})\left[5.34 \times 10^{-19}\frac{\text{kg}(\text{m/s})}{\text{GeV/c}}\right] \tag{10.4}$$

$$\mathbf{B(\text{kG})\rho(m) = 33.356P(\text{GeV}/c)} \tag{10.5}$$

Equation 10.5 assumed a single charge. A more general form of the expression would include a particle with a charge of zq. This results in

$$B\rho = \frac{m}{zq}v = \frac{P}{zq} = \frac{\left(A\sqrt{(E_k)^2 + 2E_kE_0}\right)/c}{zq} \tag{10.6}$$

Therefore, for a given magnetic field B, the bending radius ρ depends on the speed of the particle and the charge-to-mass ratio. This is an extremely important constraint in the design of beam production systems.

10.3.2 THE FORM OF THE FORCE

Generally speaking, equations of motion are reduced to Newton's second law; 'Force is equal to the change in momentum per change in time. For a constant mass, force equals mass times acceleration'. Otherwise written as

$$F = \frac{d(mv)}{dt} = ma \tag{10.7}$$

Work is done by a force **F** moving something from a point A to point B, as follows:

$$W_{AB} = \int_A^B F \cdot ds \tag{10.8}$$

where s is the coordinate representing the direction of motion. If the total work done starting from point A and returning to point A is 0, then the force is called a conservative force, or:

$$\oint F \cdot ds = 0 \tag{10.9}$$

If there was friction, this would not be conservative, since friction is always going against the motion no matter the direction of the motion. Given the closed loop integral equals 0, Stokes theorem indicates that $\nabla \times F = 0$ and, therefore, **F** can be represented by the gradient of a potential $F = -\nabla V$. Since the force is conservative, this will lead to an expression indicating the conservation of energy, or T (kinetic energy) + U (potential energy) = Constant. What is missing here is the nature of the force to determine the appropriate equation of motion.

All the knowledge of electrodynamics that is necessary for this book and more is contained in Maxwell's equations and the Lorentz force law. Maxwell's equations can be written as

$$\nabla \cdot \vec{E} = \frac{\rho}{\epsilon_0} \tag{10.10a}$$

$$\nabla \cdot \vec{B} = 0 \tag{10.10b}$$

$$\nabla \times \vec{E} = -\frac{\partial B}{\partial t} \tag{10.10c}$$

$$\nabla \times \vec{B} = \mu_0 \vec{J} + \frac{1}{c^2}\frac{\partial E}{\partial t} \tag{10.10d}$$

And the Lorentz force law is

$$\vec{F} = q\left\{\vec{E} + \frac{1}{c}\left(\vec{v} \times \vec{B}\right)\right\} \tag{10.11}$$

where E is the vector electric field, B is the vector magnetic field, J is the vector current density, ρ is the scalar charge density, μ_0 is the vacuum permeability, c is the speed of light = $1/\sqrt{\epsilon_0\mu_0}$ and v is the vector velocity of particle.

Since $\nabla \cdot \vec{B} = 0$, the magnetic field can be written as $\vec{B} = \nabla \times \vec{A}$ due to the identity $\nabla \cdot \nabla \times A = 0$. A is called the magnetic vector potential. In the general case, since $\nabla \times \vec{E} \neq 0$, then E is not the gradient of a scalar function owing to the nonconservative force being applied. On the other hand, if the magnetic field does not vary with time, then $\nabla \times \vec{E} = 0$ and $E = -\nabla\phi$ can be written, where ϕ

is the electric potential. If the constraint on the magnetic field is not applied (it has a nonzero time-varying component), then it's possible to define another potential that makes use of the $\nabla \cdot \vec{B} = 0$ equation. In this more general case,

$$\nabla \times \vec{E} + \frac{\partial \vec{B}}{\partial t} = 0 = \nabla \times \vec{E} + \frac{\partial}{\partial t}\left(\nabla \times \vec{A}\right) = \nabla \times \left(E + \frac{\partial A}{\partial t} \right) = 0 \qquad (10.12)$$

And so one can define something equivalent to a potential (ϕ) (even though $\nabla \times \mathbf{E} \neq 0$).

$$E + \frac{\partial A}{\partial t} = -\nabla \varphi \qquad (10.13)$$

The Lorentz force law can then be written as

$$\vec{F} = q\left\{ \left(-\nabla \varphi - \frac{\partial A}{\partial t} \right) + (v \times \nabla \times A) \right\} \qquad (10.14)$$

where

$$E(\text{Electric field}) = \left(-\nabla \varphi - \frac{\partial A}{\partial t} \right) \qquad (10.15)$$

Now look at just the x component of the force:

$$F_x = q\left\{ -\frac{\partial}{\partial x}(\varphi - v \cdot A) - \frac{d}{dt}\left(\frac{\partial}{\partial v_x}(A \cdot v) \right) \right\} \qquad (10.16)$$

Define $U = q\varphi - qv \cdot A$ and recall the total energy of a particle is $E = m_0 c^2 \sqrt{1 - \beta^2}$.
After some algebra on Equation 10.16, and still considering only the x component of the force, this can be written in the form

$$F_x = -\frac{\partial U}{\partial x} + \frac{d}{dt}\frac{\partial U}{\partial v_x} \qquad (10.17)$$

where U is a generalized potential (not the potential energy) that enables a Lagrangian to be written as $L = T - U$. The system is then amenable to Hamiltonian dynamics and the equations of motion are derivable. Of particular note is the Hamiltonian concept of canonical variables, and particularly the relationships between particles positions and their momenta. From here, using Hamiltonians and Hamiltonian dynamics as well as a sprinkling of the magic dust known as symplectic matrices that embody some of the physics of the system, the equations of motion can be generated. However, this is beyond the scope of the book and the systems under consideration are somewhat less complicated. Perhaps a more intuitive derivation could be helpful. (It is, however, still amazing how the physics contained in the formalism follows through the algebra into predictable motions.) What was intended earlier was to give some feel for the math behind the physics behind the equations and how the math propagates that physics.

10.3.3 THE EQUATIONS OF MOTION IN A MAGNETIC FIELD

The Lorentz force on a charged particle is

$$\vec{F} = \frac{dp}{dt} = q\left(\vec{E} + \vec{v} \times \vec{B} \right) \qquad (10.18a)$$

For the purposes of what will be discussed in this book, it is helpful to consider some special cases of the general equation. It will be assumed that:

- For beam optics purposes in this section, there is no electric field.
- For beam optics purposes herein, the magnetic field is not time varying.
- The magnetic field potential is symmetric about the midplane.
- There are no magnetic fields along the central trajectory motion direction, on the midplane.
- A ray travels along a central path, which may include drift sections and sections where the particle is bent by a magnetic field. Of interest is deviations of particles from that central path.
- The magnetic fields in the y and x direction do not vary in the y and x directions, respectively.

It is convenient to explore the position of a charged particle along the central path of its motion. This is the dark black curve in Figure 10.8, which is in the midplane of the 3d axes indicated by the gray-shaded plane region. The path along the central trajectory is represented by the coordinate s that follows along with the path of the particle (even when it bends). The x-axis is perpendicular to the s-axis at any given point. A random path above the plane (y-axis) is also shown as the dashed gray line. The dotted line in the midplane refers to the path of a small deviation from the central trajectory.

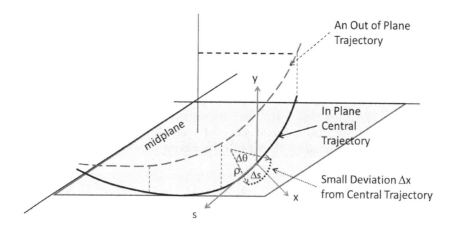

FIGURE 10.8   Beam path coordinate system.

The central path trajectory is influenced by the force of magnetic field B as given by

$$\vec{F} = \frac{d\vec{p}}{dt} = q\left(\vec{v} \times \vec{B}\right)$$

(10.18b)

where B is in the y direction only (at least for starting). The centroid particle is moving along a path that may or may not have a magnetic field – it is either moving straight or bending through an arc and either or both are in fact the central trajectory. To repeat, the nominal, non-deviated path of the particle is the central trajectory whether or not there is a B field in the y direction or zero field. What is of interest is to examine the path of a particle that deviates slightly from that central trajectory. It is desired that this particle is not lost, or that there be some form of restoring force to keep it near the central trajectory. If, for example, a particle were kicked in the y direction, it could keep going and going if something didn't correct it. In particular, a gradient magnetic field will be included, because it will be seen that this provides a restoring force. For the purposes of this chapter, there are no non-restorative forces considered.

Already the line between physics and math, if there is one, is blurred and examination of Maxwell's equations shows that $\nabla \times \boldsymbol{B} = 0$. (No current and no electric field.) The magnetic field can look like (okay, a component in the x direction has been included for completeness)

$$\boldsymbol{B} = B_x \boldsymbol{x} + B_y \boldsymbol{y} \qquad (10.19)$$

Writing the curl in matrix form

$$Det \begin{vmatrix} \mathbf{x} & \mathbf{y} & \mathbf{s} \\ \dfrac{\partial}{\partial x} & \dfrac{\partial}{\partial y} & \dfrac{\partial}{\partial s} \\ B_x & B_y & B_s \end{vmatrix} = \mathbf{x} \left(\dfrac{\partial B_s}{\partial y} - \dfrac{\partial B_y}{\partial s} \right) - \mathbf{y} \left(\dfrac{\partial B_s}{\partial x} - \dfrac{\partial B_x}{\partial s} \right) + \mathbf{s} \left(\dfrac{\partial B_y}{\partial x} - \dfrac{\partial B_x}{\partial y} \right) \qquad (10.20)$$

There are no fields along the s direction so terms with B_s are 0 and there is no variation of the fields in the s direction, so those terms also go away. That just leaves terms in the s direction, which also must be zero to satisfy Maxwell's equations. This latter point requires that

$$\frac{\partial B_y}{\partial x} = \frac{\partial B_x}{\partial y} \qquad (10.21)$$

Since only small deviations from the central trajectory are being considered, the magnetic field can be expanded about the central trajectory near the particle

$$\mathbf{B}(x,y) = \left(B_x(0,0) + \frac{\partial B_x}{\partial y} y + \frac{\partial B_x}{\partial x} x \right) \mathbf{x} + \left(B_y(0,0) + \frac{\partial B_y}{\partial x} x + \frac{\partial B_y}{\partial y} y \right) \mathbf{y} \qquad (10.22)$$

where (0,0) is the coordinate on the central trajectory (not the origin of the coordinate system in Figure 10.8); x and y (bolded) are unit vectors and x and y are small deviation from the central trajectory. The $B(0,0)$ terms are effectively zero, since by definition they are not causing deviations and are part of the constraints of the particle on the central trajectory. The third term in the x direction is zero as is the third term in the y direction since that force is in the y direction but has no y dependence. (It also does not create a restorative force.) The second coefficient terms have already been determined to be equal.

Given that, in principle, a gradient restorative force is the only one remaining (Maxwell's equations and some constraints helped with that), if the particle deviates, for example, in the x direction from the central trajectory, the angle of that particle with respect to the s path can change. As a result of $\partial B_y / \partial x$ this particle (ray) is experiencing a force that the centroid doesn't. This change in direction is caused by a magnetic field in the y plane. Given by what's left in Equation 10.22, define $B_y = (\partial B_y / \partial x) x = B'x$. As seen in Figure 10.8, the change in the offset particle's (single charge) angle due to the force it experiences, $\Delta \theta = \Delta x' = dx/ds$, is given by

$$\Delta x' = -\frac{\Delta s}{\rho} = -\Delta s \left(\frac{eB_y}{P} \right) = -\left(\frac{eB'\Delta s}{P} \right) x \qquad (10.23)$$

As $\Delta s \to 0$ and recognizing how x deviates from the central trajectory as the angle changes, then

$$\frac{dx'}{ds} = \frac{d^2 x}{ds^2} = -\left(\frac{B'}{B\rho} \right) x \qquad (10.24)$$

$$x'' + \frac{B'(s)}{B\rho} x = 0 = x'' + K(s)x = 0 \qquad (10.25)$$

This equation $x'' + K(s)x = 0$ is the form of a simple harmonic oscillator equation. Therefore, the goal of including a restoring force results in the derivation of equations of motion that will enable a particle to stay close to the central trajectory is successful. This will, in practice, depend on the actual force used. A more general form can be written including momentum deviation (ΔP) from the central beam momentum and higher order magnetic gradient terms (in this case S = sextupole and O = octupole) magnetic fields. These higher order terms may not be so restorative.

$$x'' + K(s)x + S(s)x^2 + O(s)x^4 = \frac{1}{\rho}\frac{\Delta P}{P} \tag{10.26}$$

Therefore, the equation of motion looks like the classical oscillator equation with driving terms. Taking the simpler result ($S = O = 0$ and $\Delta P = 0$) of a simple harmonic oscillator, a classic solution is that of a sinusoidal trajectory so that

$$x(s) = \cos\left(\sqrt{K(s)}L(s)\right)x + \frac{1}{\sqrt{K(s)}}\sin\left(\sqrt{K(s)}L(s)\right)x'(s) \tag{10.27}$$

where $L(s)$ is a length term in the s direction of the transport system.

The solution can also be written in matrix form

$$\begin{pmatrix} x_f \\ x'_f \end{pmatrix} = \begin{bmatrix} \cos\left(\sqrt{K}L\right) & \frac{1}{\sqrt{K}}\sin\left(\sqrt{K}L\right) \\ -\sqrt{K}\sin\left(\sqrt{K}L\right) & \cos\left(\sqrt{K}L\right) \end{bmatrix} \begin{pmatrix} x_0 \\ x'_0 \end{pmatrix} \tag{10.28}$$

where f refers to the final coordinates of the particle, 0 refers to the initial coordinates, and the s dependence of K and L are assumed and removed for brevity.

10.4 EFFECTS OF BEAM TRANSPORT ELEMENTS

A good way to understand the terms in Equation 10.28 is to consider types of measurements. There are two types of measurements. Passive measurements are those in which some beam properties are measured with the beam in the beamline system as it is. An example is measuring the beam size or the beam centroid position with an ionization strip chamber. This is different than active measurements, or gedanken active measurements that measure the response function of some beam property as a function of the variation of some parameter of the beamline system. The gedanken measurement can measure things that aren't measurable. An example of the active measurement is applying a bending force to a particle at a particular location and measuring the resulting change to the particle position at a downstream location. The starting and ending points are dependent upon the particular beam transport system implementation and the specific quantity of interest.

To determine the transformation property (reaction to a change), it is necessary to determine the appropriate quantity, which must be varied (and how to vary that quantity), and the resulting effect that must be measured (and how to measure that quantity). For example, it may be that it is necessary to vary the input ray transverse offset and measure the final ray angle. How can the force 'knobs' identified earlier accomplish this?

Examine the trajectory of a ray under different beam transport conditions. The trajectory of a ray can be represented beginning with coordinates (x_o, θ_o) and after traveling through a beam transport element, finishing with coordinates (x_f, θ_f) (where now θ is being used instead of x'). The x coordinate represents the transverse offset of the ray from a reference axis (the central trajectory). The θ variable represents the angle the trajectory makes with the reference axis, or it is a measure of the momentum transverse to the main reference direction (Figure 10.5).

10.4.1   DRIFT SPACE

Figure 10.9 shows a representation of the trajectory of a ray that begins with coordinates (x_o, θ_o) (offset from the reference central axis) and, after traveling along a drift length L, winds up with coordinates (x_f, θ_f). The drift length is a region in which no external forces are applied to the beam. The equations that relate the final coordinates to the initial coordinates can be written (Equations 10.29 and 10.30). In this case, the transverse displacement of the ray will change according to the angle its trajectory makes with the reference axis. Since there are no outside forces acting on the particle, the transverse momentum of the ray does not change and its angle remains constant. A matrix representation can also be written as in Equation 10.31.

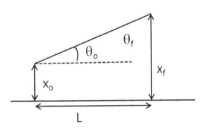

FIGURE 10.9   A ray in a drift.

$$x_f = x_0 + L\theta_0 \qquad (10.29)$$

$$\theta_f = \theta_0 \qquad (10.30)$$

$$\begin{pmatrix} x_f \\ \theta_f \end{pmatrix} = \begin{pmatrix} 1 & L \\ 0 & 1 \end{pmatrix} \begin{pmatrix} x_0 \\ \theta_0 \end{pmatrix} \qquad (10.31)$$

10.4.2   THIN LENS FOCUSING ELEMENTS

Figure 10.10 is a representation of the trajectory of a ray that begins with coordinates (x_o, θ_o) (where $\theta_o = 0$) and, after traveling a through a focusing element with focal length f, winds up with coordinates (x_f, θ_f). A focal length is the length of travel, from the focusing element, that it takes for a ray with an initial finite offset from and is parallel to the central trajectory, to cross the central trajectory after the lens – note the length f in Figure 10.10. No drift lengths are considered here and the coordinate transformation represented is that immediately before and immediately after the element. Therefore, the effect of the focusing force is only to change the angle of the beam. Since no drift length was traversed, there has been no opportunity for the transverse offset to change. The equation that relates the final coordinates to the initial coordinate, as well as the matrix for that, is given in Equations 10.32–10.34. By the way, the word focusing can be replaced with defocusing if needed, but in that case, the focal length would be negative because it would be necessary to drift backward to see where the virtual ray crosses zero.

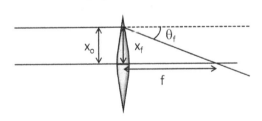

FIGURE 10.10   Ray in a focusing element.

$$x_f = x_0 \qquad (10.32)$$

$$\theta_f = -\left(\frac{1}{f}\right)\theta_0 \qquad (10.33)$$

$$\begin{pmatrix} x_f \\ \theta_f \end{pmatrix} = \begin{pmatrix} 1 & 0 \\ -\frac{1}{f} & 1 \end{pmatrix} \begin{pmatrix} x_0 \\ \theta_0 \end{pmatrix} \qquad (10.34)$$

The thin focusing element is an example of the charged particle ray being bent. How is that done? This is not quite like a light-focusing lens. Consider the specific case of a magnetic quadrupole. There are four poles, with magnetic fields (North [N] and South [S]) as shown in Figure 10.11. The magnetic field arrows (solid black lines) are pointing from N to S. Five particles are shown traveling into the paper (or the screen). The dashed lines indicate the gap size between the poles of the quadrupole. The shape of the poles is special, so that the magnetic fields exit and enter the iron as close to normal to the boundary as possible.

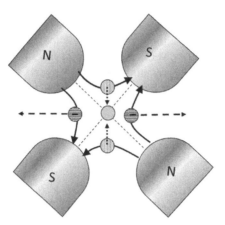

FIGURE 10.11   Quadrupole with 5 particles.

The five particles represented by circles with various types of fills are shown in the center, to the right and left of center and above and below the center. The motion of a charge particle between each of the poles can be analyzed using the right-hand rule. The dashed arrows in the figure indicate the resulting horizontal forces and similarly for the dotted vertical arrows. Thus, the left and right particles, filled with horizontal lines, move left and right, respectively, and the upper and lower particles, filled with vertical lines, move down and up, respectively. The left and right particles are diverging and the upper and lower particles are converging toward the axis. The upper and lower particles are experiencing a restorative force since they are away from the central trajectory and being directed back toward that path. The center particle feels no force. The left and right particles experience the opposite of a restorative force. A magnetic quadrupole, unlike a glass lens for light, diverges in one plane (defocuses) and converges (focuses) in the other plane.

Comparing the location of the particle in the center (filled with solid gray) with the particles to the left, right, up and down, it is seen that those outer particles are between the poles in a location where the distance between those poles is smaller. The N-S field line through the particle on the right is stronger than the field from the S-N poles on the left that is further away from the rightmost particle (the field from the left poles can extend that far). In that position, the magnetic field experienced is stronger than if it were in the center. The particle in the center is equidistant from the N and S poles; therefore, the fields from the right pair cancel with those from the left pair and it feels no force. In fact, a magnetic quadrupole is designed in a specific way so as to ensure that the field strength felt by a particle increases linearly as a function of the distance away from the center of the poles. Therefore, dB_y/dx = constant and is called the gradient. This is shown in a plot of the magnetic field strength as a function of the distance x from the center in Figure 10.12. The slope of the line,

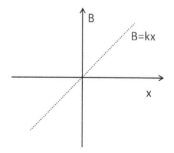

FIGURE 10.12   Quadrupole field.

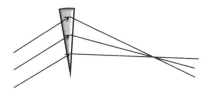

FIGURE 10.13   Wedge dipole.

representing the strength of the quadrupole gradient field, or the gradient, is given by k here. If the particle is offset from the center, it will experience a bending force.

Particles can also be focused with a dipole. Shown in Figure 10.13 is a nonphysical element; a thin wedge. This is a representation of the fact that the ray bends through a larger angle if it enters the upper (thicker – well thin, but effectively thicker) section than if it enters the lower part of the wedge. The result, if this were a physical component with length, is simply because of the path length considerations, not like the quadrupole whose focusing effect comes from the increasing strength of the magnetic field as a function of distance. The bend angle is proportional to $\int Bdl$. So both B and L work. The way it is drawn in Figure 10.13, the three rays intersect, as if there were a focal length (in the bending plane). This would mean that $(\sin\theta/\rho) = 1/f$, where θ (in this case) is the angle of the bend (not the particle offset) and ρ is the bending radius.

But this is an artificial construct for now, which is helpful for quick calculations.

10.5   GENERAL RAY COORDINATE TRANSFORMATION

The earlier two cases showed that it is possible to represent the coordinate propagation of a ray through a set of linear equations and represent the transformation of coordinates from one longitudinal position to another through a matrix relationship. It has also been shown that the general solution of the equations of motion can be represented in matrix form (Equation 10.28). Assuming small transverse motion (as in Equation 10.25), this can be done for any linear optical system. To obtain further insight into the meaning of the terms, expand the final x offset in a Taylor series in terms of the initial coordinates

$$x_f = \frac{\partial x_f}{\partial x_0}(x_o) + \frac{\partial x_f}{\partial \theta_0}(\theta_o) + \cdots \tag{10.35}$$

Writing the differentials as a matrix element symbol, for example, $\partial x_f / \partial x_0 = (x/x)$ gives

$$x_f = (x/x)x_0 + (x/\theta)\theta_0 + \cdots(other\ coordinates) \tag{10.36}$$

Using this terminology, it is possible to rewrite Equation 10.35 in matrix form as

$$\begin{pmatrix} x_f \\ \theta_f \end{pmatrix} = \begin{pmatrix} (x/x) & (x/\theta) \\ (\theta/x) & (\theta/\theta) \end{pmatrix} \begin{pmatrix} x_0 \\ \theta_0 \end{pmatrix} \tag{10.37}$$

Or

$$\begin{pmatrix} x_f \\ \theta_f \end{pmatrix} = R \begin{pmatrix} x_0 \\ \theta_0 \end{pmatrix} \tag{10.38}$$

R is defined as the transfer matrix. (**Bold** can mean a matrix or a vector, check the context.) The matrix elements can be indicated by their row and column, so that, yet another way to write (x/x), for example, is R_{11}. An important and fundamental quantity that must be known in almost any beam transport system is the transfer matrix. This is a measure of the effect that the beam transport system has on the beam particles. It will answer all questions about the gedanken type measurements that may be posed. Here, again, it is important to distinguish the *effects of beam-modifying devices on the particles in a beam* from the *properties of particles in a beam and the overall beam*. The two component vectors (vertical matrices), indicating the final (*f*) and initial (0) coordinates of a particle ray, are related to the properties of a particle or a ray. The transfer matrix R is related to the properties of the beam transport system and it sometimes changes the properties of a particle.

Through examination of the matrix elements, and using the insight gained in the preceding examples (Equations 10.31 and 10.34), the following associations can be made:

$(x/x) = R_{11}$ = transverse magnification
$(x/\theta) = R_{12}$ = effective drift distance
$(\theta/x) = R_{21}$ = −1/focal length
$(\theta/\theta) = R_{22}$ = angular magnification

These are terms most familiar in the terminology of light optics. Note that the magnification here is NOT the magnification of the beam size but the magnification of the offset of the trajectory of the ray. A beam has not yet been fully defined. Similar statements can be made of the other parameters.

The matrix representation of beam transport had several parents. Karl Brown, with whom I had the pleasure of interacting, is one of the most recognized, with many accomplishments. In addition, it's perhaps not so well known that in 1961 Sam Penner published a paper about the Calculations of Properties of Magnetic Deflection Systems – one of, if not the first of its kind. I will only add that when Sam arrived in the morning after a night of intensive nuclear physics data taking (while I was a graduate student), one had better have had everything in order.

10.6 OPTICAL MATRIX EXAMPLES

10.6.1 MOMENTUM

Thus far a 1 × 2 matrix, which includes the transverse offset and angle of a particle, has been considered. It is also important to consider the longitudinal parameters including the longitudinal momentum and the relative position along the central trajectory of a particle and understand how magnetic elements affect those coordinates. Consider first, the momentum. Define $\delta = \Delta p/p$, as the fractional deviation of a particle's momentum (p) from a reference momentum. Such a reference momentum can be defined by the magnetic field set in a dipole that is required to bend a particle of momentum p a certain desired angle. Rays of different momentum will bend different amounts in the thin wedge dipole shown in Figure 10.14. (Note how this is drawn to look like a prism.) The transfer matrix for a thin wedge (lengthless) dipole can be written as

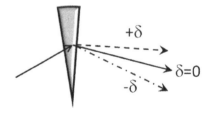

FIGURE 10.14 Wedge dipole.

$$
\begin{bmatrix} (x/x) & (x/\theta) & (x/\delta) \\ (\theta/x) & (\theta/\theta) & (\theta/\delta) \\ (\delta/x) & (\delta/\theta) & (\delta/\delta) \end{bmatrix} = \begin{bmatrix} R_{11} & R_{12} & R_{16} \\ R_{21} & R_{22} & R_{26} \\ R_{61} & R_{62} & R_{66} \end{bmatrix}
$$

$$
\begin{bmatrix} (x/x) & (x/\theta) & (x/\delta) \\ (\theta/x) & (\theta/\theta) & (\theta/\delta) \\ (\delta/x) & (\delta/\theta) & (\delta/\delta) \end{bmatrix} = \begin{bmatrix} 1 & 0 & 0 \\ -\sin\theta/\rho & 1 & \sin\theta \\ 0 & 0 & 1 \end{bmatrix} \tag{10.39}
$$

where θ (in this case) is the angle of the bend (not the particle offset angle) and ρ is the bending radius. The matrix term representing the effect of the initial δ on x_f is (x/δ).

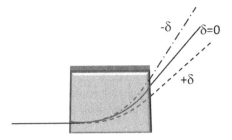

FIGURE 10.15   Physical dipole.

Consider a more realistic dipole that looks like the sketch in Figure 10.15. This is not a wedge shape, so be warned. In this figure, the paths of three particles are shown from left to right. They all enter together. The dashed line is that for a higher momentum particle while the dash-dot line is that for a lower momentum particle. Ignoring edge effects for now, the result is that the three different particles come out of the dipole at different x positions with different angles. So (x/δ) is not zero. In fact, the matrix element for this actual, finite length (L) element looks like

$$
\begin{pmatrix}
\cos\dfrac{L}{\rho} & \rho\sin\dfrac{L}{\rho} & \rho\left(1-\cos\dfrac{L}{\rho}\right) \\[2mm]
-\dfrac{1}{\rho}\sin\dfrac{L}{\rho} & \cos\dfrac{L}{\rho} & \sin\dfrac{L}{\rho} \\[2mm]
0 & 0 & 1
\end{pmatrix}
\tag{10.40}
$$

The length of the dipole is L. The $R_{16} = (x/\delta)$ matrix element is the degree to which the x offset of the ray changes with respect to its fractional change from the central momentum of the collection of particles. This is called the dispersion, and the $R_{26} = (\theta/\delta)$ term is called the angular dispersion. It is seen that the dispersion depends upon the bending radius and angle. In the extreme case where $L \to 0$, (or $\theta \to 0$), the matrix reduces to the matrix of Equation 10.39. Note in Figure 10.15 how (x/δ) keeps growing after the dipole since (θ/δ) is not zero at the output of the dipole.

10.6.2   LONGITUDINAL POSITION

Consider the prism-like effect of the dipoles shown in the Figures 10.14 and 10.15. This can be exploited in the system shown in Figure 10.16. In this case, again there are three particles traveling together. They are being transported in what is called a chicane (owing to its crookedness and

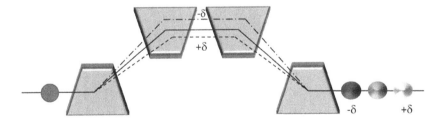

FIGURE 10.16   Four dipole chicane.

maybe the trickery to follow). They start (from the left) all together in a location indicated by the flat gray circle. They each have different momenta. The particle with the higher momentum, the dashed line, travels a shorter distance than the others (because it is bent less – or follows an arc in the magnetic field with a longer radius of curvature) while the particle with the lower momentum, the dash-dot line, travels a longer distance. The particle trajectories in the dipoles are drawn as straight lines for convenience but are really curved as in Figure 10.15. Clever use of reverse bends (bending in one direction, then the other direction) is used to combine these particles in the transverse direction, at the end. However, although they are together in the transverse direction, the three particles, as shown by the three gray balls, are no longer at the same longitudinal place along the reference trajectory. The higher energy charged particle gets to the bar first (lightest gray), but unlike the neutron (who went straight, because it has no charge), it had to pay. This can become very important when the timing of arrival of a particle is considered, such as when recirculating a beam in a radiofrequency accelerator, in which case, it's position timing is called phase. The longitudinal deviation of the particle from the reference is given by (l/δ), which is R_{56} in the transfer matrix.

10.6.3 TRANSVERSE FOCUSING

If there are multiple transport elements in a system, their combined effect on a particle's coordinates is given by the matrix multiplication of each of the elements. Assume there are three elements labeled 1, 2 and 3. Then

$$\begin{pmatrix} x_f \\ \theta_f \end{pmatrix} = \begin{pmatrix} (x/x)_3 & (x/\theta)_3 \\ (\theta/x)_3 & (\theta/\theta)_3 \end{pmatrix} \begin{pmatrix} (x/x)_2 & (x/\theta)_2 \\ (\theta/x)_2 & (\theta/\theta)_2 \end{pmatrix} \begin{pmatrix} (x/x)_1 & (x/\theta)_1 \\ (\theta/x)_1 & (\theta/\theta)_1 \end{pmatrix} \begin{pmatrix} x_0 \\ \theta_0 \end{pmatrix} \qquad (10.41)$$

The order of multiplication is not commutative; the system must be evaluated in this order. (Associative and distributive properties hold.) Multiple elements are combined to create a desired effect. Some effects can be accomplished by assembling optical building blocks.

10.6.3.1 Point-to-Point Focusing

One important optical condition is point-to-point focusing. In this case, any particle that starts at the same transverse offset x_o ends with the same transverse offset x_f, although x_o may be different than x_f. This is independent of any initial angle θ_o. Since particles that start at the same transverse 'point' and end at the same 'point', this is point-to-point focusing. Figure 10.17 shows such a system. The solid lines start at $x_o = 0$ and end at $x_f = 0$. This means that the position of the particle at the end will not move, even if the angle of the particle is changed at the start. An active measurement would

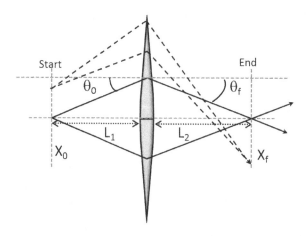

FIGURE 10.17 Point-to-point focusing.

show this. Since $x_f = (x/x)x_o + (x/\theta)\theta_o$, in order for this condition to be fulfilled $(x/\theta) = 0$. In the case for this system, with the ray traveling from the left, the following elements are identified

$$\begin{pmatrix} (x/x)_1 & (x/\theta)_1 \\ (\theta/x)_1 & (\theta/\theta)_1 \end{pmatrix} = \begin{pmatrix} 1 & L_1 \\ 0 & 1 \end{pmatrix} \tag{10.42}$$

$$\begin{pmatrix} (x/x)_2 & (x/\theta)_2 \\ (\theta/x)_2 & (\theta/\theta)_2 \end{pmatrix} = \begin{pmatrix} 1 & 0 \\ -\dfrac{1}{f} & 1 \end{pmatrix} \tag{10.43}$$

$$\begin{pmatrix} (x/x)_3 & (x/\theta)_3 \\ (\theta/x)_3 & (\theta/\theta)_3 \end{pmatrix} = \begin{pmatrix} 1 & L_2 \\ 0 & 1 \end{pmatrix} \tag{10.44}$$

and the matrix representing the transport through these three elements is

$$\begin{pmatrix} (x/x) & (x/\theta) \\ (\theta/x) & (\theta/\theta) \end{pmatrix} = \begin{pmatrix} 1 - L_2/f & L_1 + L_2 - L_1L_2/f \\ -1/f & 1 - L_1/f \end{pmatrix} \tag{10.45}$$

For this particular case, there is one condition $(x/\theta) = 0$ and three variables (L_1, L_2 and f). This is an important aspect of charged particle beam optics – to be sure that there are sufficient variables to achieve the desired goal. The goal is not always to just stuff a beam inside a pipe, there are optical conditions that are worthwhile to achieve. In this case, $L_1 + L_2 - (L_1L_2/f) = 0$, or $f = L_1L_2/(L_1 + L_2)$. Because there were three variables, this condition only established a relationship of one to the other two, thus there is some range of L_1 and L_2 that can satisfy this condition. It's possible to constrain the system so that $L = L_1 = L_2$, which then creates a relationship between f and L ($f = L/2$), which also creates the condition that $x_f = -x_i$. Thus two constraints have been used. A final constraint would be choosing a desired value of L_1 or f, which would then fix f and the magnetic field strength in the quadrupole and all the variables have been used up. Please note that this is for one and only one plane. The x plane has been chosen for this example. As explained earlier, the orthogonal plane (y) has a focal length with the opposite sign and the result for this will not be $(y/\phi) = 0$.

This condition works for any initial offset x_o as shown by the dashed lines in Figure 10.17. While x_f may not equal x_o, all particles starting at a nonzero x_o point end at a nonzero x_f point independent of their angle. Considering only the x plane, this condition states the following: No matter the angle of a beam at the start of this specific system, the final position of the beam will only depend upon the initial position of the beam. Thus, in the practical terms defined in Section 10.5,

(x/x) = transverse magnification = depends upon the system
(x/θ) = effective drift distance = 0
(θ/x) = −1/focal length; focal length = $L_1L_2/(L_1 + L_2)$.
(θ/θ) = angular magnification = depends on the system

This can be thought of as an 'imaging' system, representing a magnified (magnification of 1, less than 1 or larger than 1, upside down or right side up) image of whatever it is at the start (e.g. a beam of particles or a collimator). Specifying the magnification would be another condition.

10.6.3.2 Point-to-Parallel Focusing

Another important condition is point-to-parallel focusing. In this case, for any particle starting with the same x_o, the final angle θ_f for those particles is independent of the angle θ_o at the start but is dependent on the initial x_o. To fulfill the condition, $(\theta/\theta) = 0$. For this particular case, working out the transport matrix (Equation 10.45) shows that $1 - (L_1/f) = 0$, or $f = L_1$. The final angle is 0 if the particles start at $x_0 = 0$, so the drift length, L_2, after the lens is not relevant in that case. The condition established a relationship between two variables and L_2 is free. The solid lines of Figure 10.18 depict the case when $x_0 = 0$, showing that the final angles are all 0. The dashed lines indicate what happens when x_0 is not zero. While the final angle may not be 0, the angles for all particles emanating from x_0 are the same, but x_f does depend on θ_0.

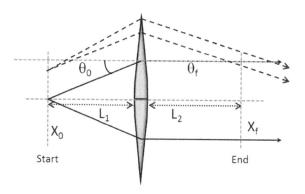

FIGURE 10.18   Point-to-parallel focusing.

Physically this has quite the opposite result when compared with the point-to-point focusing condition. This will not produce an image of the initial collection of rays. However, it will create extreme stability of the ray angles at the end of the system, should there be instability of a magnetic element at the start whose magnetic field may be varying. In this case, the practical matrix element words are

(x/x) = transverse magnification = depends upon the system
(x/θ) = effective drift distance = depends on system
(θ/x) = −1/focal length; focal length = L_1
(θ/θ) = angular magnification = 0

10.6.3.3   Parallel-to-Point Focusing

The opposite of point-to-parallel would be parallel-to-point. The transverse offset x_f is independent of the initial transverse offset x_0 for a given incident θ_0. The conditions and relationships are left as an exercise. Simply stated, as shown in Figure 10.19, in this situation, the position of a ray at the end

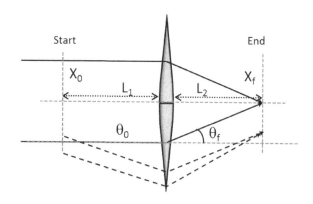

FIGURE 10.19   Parallel-to-point focusing.

does not depend upon the position of the ray at the start and is, therefore, independent of any beam position instability at the start but does depend on the angle.

10.6.3.4 Achromatic Combined System

Now it's time to bring in the momentum. The term chromatic relates to how a system treats the momentum of the particles. Chroma indicates the purity of a color in light (lack of white or gray) and is therefore used referring to the momentum of charged particles. The term geometric relates to the properties of the transverse coordinates (independent of momentum). When particles pass through a magnetic element, the force of the magnetic field on a particle will depend upon the particle's momentum. If the result of this action is a spatial or angular separation of different particles as a function of their momenta, then there is a chromatic effect on the particle behavior. If a system is designed in such a way that all particles, independent of their momenta, are following the same trajectory coinciding with the reference trajectory, then the system is said to achromatic (not having a chromatic dependence). Such a system is useful.

Examine the system in Figure 10.20. There are seven elements. Yes, there are. With the particles coming from the left, they include

1. Drift L1
2. Dipole
3. Drift L2
4. Lens
5. Drift L3
6. Dipole
7. Drift L4

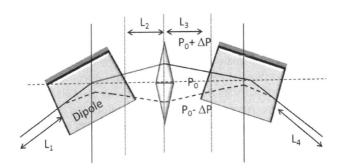

FIGURE 10.20 Achromat.

This system makes use of one of the most powerful techniques in beam optics. As shown in Equation 10.28, the beam trajectories take on some aspects of sine and cosine behavior. Therefore, symmetry can be a very important tool in the design of a system (e.g. search 'Symmetry corrected second-order achromat') If it is chosen that L1 = L4, L2 = L3 and the two dipoles are identical, there will be a symmetry point at the center (halfway into the quadrupole). In this case, the four matrix elements are

$$\begin{pmatrix} (x/x)_1 & (x/\theta)_1 \\ (\theta/x)_1 & (\theta/\theta)_1 \end{pmatrix} = \begin{pmatrix} 1 & L_1 \\ 0 & 1 \end{pmatrix} \tag{10.46}$$

$$\begin{pmatrix} (x/x)_2 & (x/\theta)_2 & (x/\delta)_2 \\ (\theta/x)_2 & (\theta/\theta)_2 & (\theta/\delta)_2 \end{pmatrix} = \begin{pmatrix} \cos\dfrac{L}{\rho} & \rho\sin\dfrac{L}{\rho} & \rho\left(1-\cos\dfrac{L}{\rho}\right) \\ -\dfrac{1}{\rho}\sin\dfrac{L}{\rho} & \cos\dfrac{L}{\rho} & \sin\dfrac{L}{\rho} \\ 0 & 0 & 1 \end{pmatrix} \tag{10.47}$$

$$\begin{pmatrix} (x/x)_3 & (x/\theta)_3 \\ (\theta/x)_3 & (\theta/\theta)_3 \end{pmatrix} = \begin{pmatrix} 1 & L_2 \\ 0 & 1 \end{pmatrix} \tag{10.48}$$

$$\begin{pmatrix} (x/x)_4 & (x/\theta)_4 \\ (\theta/x)_4 & (\theta/\theta)_4 \end{pmatrix} = \begin{pmatrix} 1 & 0 \\ -\dfrac{1}{f} & 1 \end{pmatrix} \tag{10.49}$$

The goal of this system is to be achromatic. Actually achromaticity is a stronger condition than may have been interpreted earlier. Both the spatial and angular dispersion must be zero. If the system is simply dispersionless, $(x/\delta) = 0$ at the end of the system but the angular dispersion is not, then the dispersion will grow further downstream. Thus, there are two main conditions. If another constraint is that the dipoles must bend the particles through a certain angle (e.g. to direct the beam to a room), then the only variables that are still free are the L2 and the lens strength (ignoring L1 for now since that will not help achieve the goal). That is two variables with two conditions. Is the symmetry idea an added constraint or a help? The answer is left as an exercise for the reader. As an aside, an achromat has another benefit. If a single energy particle at P_0 is transported and the magnetic fields of the achromat are scaled, then that particle will exit the achromat with its trajectory unchanged, so the achromat is less sensitive to power supply fluctuations if properly wired.

10.6.4 VARIABLES AND CONDITIONS

Sprinkled in among the various examples earlier are words related to the goals of these systems. It is important to define the appropriate goals and to understand what constraints are imposed and how many degrees of freedom exist. Sometimes a novice may use a software package and try to twiddle virtual knobs to stuff a bunch of particles through a beamline, only to find that squeezing them here will make them bigger elsewhere. Applying the optical constraints described earlier (and others) will help to ensure that a collection of particles will have the desired, stable and measurable behavior.

10.7 THE ORTHOGONAL DIRECTION

Most of the earlier discussion has dealt only with one transverse plane (x, θ, δ), generally called the bend plane since it is in this plane that the particles are bent. Until now the matrix elements have not included a 3rd or 4th row and column. However, unlike lenses used for light optics, both transverse planes are not the same in charged particle optics. As shown earlier, magnetic quadrupoles focus in one plane and defocus in the other plane. The transformation matrix can be expanded to include the other plane and for the transverse dimensions it looks as follows for a thin magnetic quadrupole:

$$\begin{bmatrix} x_f \\ \theta_f \\ y_f \\ \varphi_f \end{bmatrix} = \begin{bmatrix} R_{11} & R_{12} & R_{13} & R_{14} \\ R_{21} & R_{22} & R_{23} & R_{24} \\ R_{31} & R_{32} & R_{33} & R_{34} \\ R_{41} & R_{42} & R_{43} & R_{44} \end{bmatrix} \begin{bmatrix} x_0 \\ \theta_0 \\ y_0 \\ \varphi_0 \end{bmatrix} = \begin{bmatrix} 1 & 0 & 0 & 0 \\ -1/f & 1 & 0 & 0 \\ 0 & 0 & 1 & 0 \\ 0 & 0 & 1/f & 1 \end{bmatrix} \begin{bmatrix} x_0 \\ \theta_0 \\ y_0 \\ \varphi_0 \end{bmatrix} \tag{10.50}$$

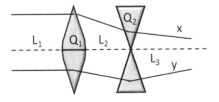

FIGURE 10.21   Quad doublet.

Most systems will require a focus in both transverse planes simultaneously. This is accomplished by the judicious use of multiple quadrupoles of different polarities. One example is a quadrupole doublet as shown in Figure 10.21. This comprises three drift lengths and two quadrupoles. The lenses are drawn as per convention based on their function in the horizontal (x) plane. The ray in the upper half shows the trajectory of a ray in the x plane. But the ray in the lower half shows the trajectory of a ray in the y plane. That's why it seems to bend it in the wrong direction. It's helpful to get used to this typical display of x and y rays. The relative lengths and magnetic field strengths are adjusted to achieve a double focus. The transfer matrices describing these elements are summarized in Equation 10.51. These can be combined to create the desired focusing effect, which would be a net positive focus in both planes. These are two conditions, but with five variables. So, judicious compromises are possible.

$$\begin{pmatrix} x_f \\ \theta_f \end{pmatrix} = \begin{pmatrix} 1 & L_3 \\ 0 & 1 \end{pmatrix} \begin{pmatrix} 1 & 0 \\ 1/f_2 & 1 \end{pmatrix} \begin{pmatrix} 1 & L_2 \\ 0 & 1 \end{pmatrix} \begin{pmatrix} 1 & 0 \\ -1/f_1 & 1 \end{pmatrix} \begin{pmatrix} 1 & L_1 \\ 0 & 1 \end{pmatrix} \begin{pmatrix} x_0 \\ \theta_0 \end{pmatrix} \tag{10.51a}$$

$$\begin{pmatrix} y_f \\ \varphi_f \end{pmatrix} = \begin{pmatrix} 1 & L_3 \\ 0 & 1 \end{pmatrix} \begin{pmatrix} 1 & 0 \\ -1/f_2 & 1 \end{pmatrix} \begin{pmatrix} 1 & L_2 \\ 0 & 1 \end{pmatrix} \begin{pmatrix} 1 & 0 \\ 1/f_1 & 1 \end{pmatrix} \begin{pmatrix} 1 & L_1 \\ 0 & 1 \end{pmatrix} \begin{pmatrix} y_0 \\ \varphi_0 \end{pmatrix} \tag{10.51b}$$

The resulting combined matrix is left as an exercise to the reader.

10.8   DIPOLE FOCUSING

As mentioned in Section 10.4.2 in addition to quadrupoles, dipoles can provide focusing. The way they do that may not seem quite as obvious.

10.8.1   Sector Focusing

One type of focusing is sector focusing as shown in Figure 10.22(left). In addition to the strength of the magnetic field, it is the length of travel in the dipole that determines the amount of the bend. The bend angle is proportional to $\int B \cdot dl$. A sector dipole is wedge shaped and the bend angle depends on the entrance position. This can result in a net focusing effect shown by the three rays crossing in Figure 10.22(left). This can be compared with a rectangular dipole in which both the B and the length are the same and no focusing effect is seen in the bend plane as in Figure 10.22(right).

On the other hand, if the sector dipole were misaligned up or down (in the plane of the paper), then the bend angles of any given incident ray would change. This is not the case with the rectangular dipole which is less sensitive to this misalignment.

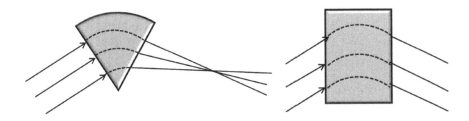

FIGURE 10.22   Sector dipole (left) and rectangular dipole (right).

10.8.2 POLE EDGE FOCUSING

One aspect of the sector (wedge) dipole shown in Figure 10.22(left) is that the charged particle rays enter perpendicular to the dipole edge, while in the rectangular dipole of Figures 10.22(right) and 10.23(left), there can be an angle between the dashed line (perpendicular to the charged particle's direction) and the dipole edge. Figure 10.23(left), a view from above, shows the angles α and β made between the normal to the entrance and exiting particles with respect to the pole edge. (Note α and β can be independent.) A view from the side is seen in Figure 10.23(right). It shows both the uniform field, indicated by B_y in the center of the dipole and the fringe fields sticking out the edge of the dipole. These fringe fields become important.

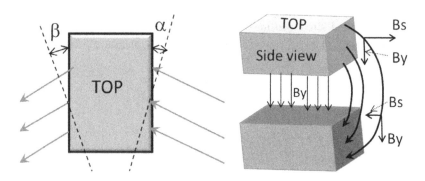

FIGURE 10.23 Dipole pole edge rotation (left) and dipole fringe field (right).

Decomposing one of the fringing field lines in Figure 10.23(right) into components along the axes of interest results in a component in the y direction B_y (just a continuation of the uniform field that decreases in magnitude as the distance from the edge of the dipole increases). This continues to bend the beam along the central trajectory. There is also a component in the direction normal to the magnet yoke. Of course, this would normally not have an effect if the beam trajectory was normal to the magnet edge since $\mathbf{v_s} \times \mathbf{B_s} = 0$. However, remember that there is an angle (e.g. α) between the direction of the beam and the pole edge. Look again at the direction of this magnetic field from the top view shown in Figure 10.24. The B_s field is not in the same direction as the particle and can be decomposed into components in the beam direction and perpendicular to the beam direction (dashed line with arrow). It has already been shown that the component in the beam direction does not affect the beam but the perpendicular component does. This results in a force either to the bottom (or top) of the dipole, which is a focusing (or defocusing) force. This is pole edge focusing. Note also, that owing to the symmetry of the midplane, the direction of this magnetic field in the top half

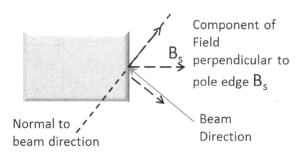

FIGURE 10.24 Dipole top view fringe field.

is opposite to that in the bottom half (follow the curved fringe field line) and zero in the middle. Also, the proportional strength of this force increases as the distance from the midplane increases (as the field lines curve more). This analysis will be continued in Section 13.3.2.

10.9   MISALIGNMENTS

Up until this point, a ray on the central trajectory has followed that central trajectory. Other rays, which departed a small amount (in position and angle) from the central trajectory, have followed their own nearby, semi-controllable, path. It is possible that the position of a magnetic element is not

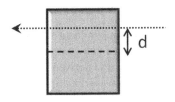

FIGURE 10.25   Element misalignment.

at the intended position. While installers will work hard to try to accomplish good alignment, some magnets are just not in the exact right spot. (The author once aligned a 200 m linac – it wasn't easy.) Sometimes that's okay and sometimes that makes a difference. Figure 10.25 shows an arbitrary element dislocated a distance d from the nominal central trajectory indicated by the dotted line. The center of the element is the dashed line, which should be coincident with the dotted line.

How could this affect the trajectory? It's possible to do a little trajectory gymnastics. Imagine a ray coming from the right (not left this time). Normally it would be collinear with the dashed center of the arbitrary element but it isn't. Now shift the coordinates of that ray by d.

$$\begin{pmatrix} x_1 \\ \theta_1 \end{pmatrix} = \begin{pmatrix} d \\ 0 \end{pmatrix} + \begin{pmatrix} x_0 \\ \theta_0 \end{pmatrix} \tag{10.52}$$

So, with respect to the arbitrary element, it looks like the incoming ray has been shifted and any effect of this misalignment will be propagated. However, if the rest of the beamline is perfectly aligned, this gymnastic offset should not be continued, while the effect of the misalignment should be continued. Therefore, a matrix transfer element can be used to propagate the effect of the misalignment and then the offset can be removed (−d) as in Equations 10.53 and 10.54.

$$\begin{pmatrix} x_f \\ \theta_f \end{pmatrix} = \begin{pmatrix} -d \\ 0 \end{pmatrix} + R\left[\begin{pmatrix} d \\ 0 \end{pmatrix} + \begin{pmatrix} x_0 \\ \theta_0 \end{pmatrix} \right] \tag{10.53}$$

$$\begin{pmatrix} x_f \\ \theta_f \end{pmatrix} = R\left[\begin{pmatrix} x_0 \\ \theta_0 \end{pmatrix} \right] + [R-1]\begin{pmatrix} d \\ 0 \end{pmatrix} \tag{10.54}$$

As an example, take a thin quadrupole. For that quadrupole

$$R - 1 = \begin{pmatrix} 1 & 0 \\ -\dfrac{1}{f} & 1 \end{pmatrix} - \begin{pmatrix} 1 & 0 \\ 0 & 1 \end{pmatrix} = \begin{pmatrix} 0 & 0 \\ -\dfrac{1}{f} & 0 \end{pmatrix} \tag{10.55}$$

Therefore,

$$\begin{pmatrix} x_f \\ \theta_f \end{pmatrix} = R\begin{pmatrix} x_0 \\ \theta_0 \end{pmatrix} + \begin{pmatrix} 0 \\ -\dfrac{d}{f} \end{pmatrix} \tag{10.56}$$

The ray, which was coming in on the central trajectory, has now experienced an angular kick with a magnitude proportional to the offset, d, from the center of the quadrupole. This adds the equivalent of a dipole component to the quadrupole, for the central trajectory. But otherwise, don't worry, the focusing effect is unchanged (it's a linear element). There is no need to correct the focusing; there is only a need to correct the offset ray angle or subsequent offset ray position.

If an element is rotated, the angular term can be included as well, even if a rotation matrix is needed for more complicated misplacements. Thus, the magnitude of the effect of a misalignment can be estimated and the effect of corrector elements can also be handled this way.

10.10   WHAT COULD GO WRONG – 2?

So far, this chapter has described aspects of how electromagnetic components influence the behavior of charged particles. These are also elements of the fundamental physics that contribute to the parameters of the beam that arrive at the patient. The quantities identified and derived have some explicit or implied assumptions. Proper evaluation of potential errors motivates making these all explicit so that they can be assessed. These can include:

- There is a central trajectory ray and trajectories of particles that deviate from that central trajectory. These were discussed in this chapter. The implied assumption is that the central trajectory ray goes where it is intended. However, it is possible that some magnetic elements are not positioned where they are expected to be, or the magnetic fields in those elements are not what are expected. In principle, they are all measured (well, not really) and accurately aligned (well, not really); this process isn't perfect – that's why there are corrector elements in beam lines.
  - A quadrupole field is linear crossing through zero at the center. So the central ray is unaffected if it passes through the center. If the quadrupole magnet is misaligned, then that ray will experience a nonzero magnetic field that will cause the trajectory to deviate from the desired path. While such a misalignment does not usually change with time (periods of months or years, unless there's an earthquake), the fact that there can be a misalignment can cause the beam to deviate from the desired location in the target and thus must be measured and corrected. This means there should be sufficient instrumentation to enable this process.
  - All of the behaviors discussed in this chapter depend upon the appropriate magnetic field being produced in the magnet. These are electromagnets powered by current in conductors energized by a power supply. That power supply can be in error. The conductors can be shorted. If conductors are shorted, the correct voltage will cause the wrong current to be generated. There are several things that can happen if the desired magnetic field is not applied. In the case of a quadrupole, an incorrect field can modify the beam size. In the case of a dipole, an incorrect field can change the beam central trajectory. (The author can't count the months spent measuring magnets as his family came to feed him.)
  - A variety of types of dipole magnets were discussed in this chapter. Some were relatively insensitive to displacements in the plane of the bend (e.g. rectangular outline), but some were sensitive (e.g. wedge outline). Thus, alignment is important for these elements as well.
  - Knowing the transfer matrix from any one point to any other point in a beam line will enable the results of an error at the first point to be calculated at the second point. Without appropriate instrumentation in the system, this cannot be verified. Instrumentation is not a cost to be cut.
  - The alignment concern is amplified with a rotating gantry. The support structure carrying the magnetic elements cannot be as rigid as a stand on a floor and the system

must be designed to allow for misalignments that will happen from one gantry angle to another. Fast corrections would be needed to position the beam accurately to the patient target.

- That being said, the author had the experience of being in an office right next to a building construction site. Hundreds of piles were driven into the ground. Things were falling off my desk. My office actually got demolished. In all that time, the beam downstairs was on and never deviated from accurate patient treatment. Considerable effort at the start of the construction was devoted to properly engineer the concrete floor slab and to design and measure the resonant frequencies of the magnet stands, along with some creative testing. (Dropping dumpsters in the backyard.)
- It was shown how dipoles can focus in the vertical plane using their fringe fields. This means that there will be sensitivity to magnet alignment also in the vertical plane. And don't forget placement along the longitudinal beam path.

- An input to all the magnetics is the incident particle momentum. If this is incorrect, then the bending and focusing of the magnetic elements will not behave as expected.

In this part of Chapter 10, the magnetics have been added to the materials discussed in Chapter 9 and this is a new class of components that can contribute to errors in the beam parameters.

10.11 A BEAM; A GAUSSIAN BEAM

Previously the effects that transport system elements have on a particle or ray or the centroid of a collection of particles have been discussed. A beam is an ensemble of many particles that satisfy the condition that the longitudinal and transverse momenta are relatively close enough to each other to be transported through a beam transport system. (Depending upon the size and number of particles, there can even be an interaction among the particles that affect the particle trajectories.) The transverse properties of a beam can be characterized with a phase space diagram in which the transverse particle position and transverse angle (transverse momentum) of each particle in the beam is plotted. These are the canonical Hamiltonian phase space coordinates that satisfy Liouville's theorem.

Generally, a beam coming from an accelerator has a spatial position number (N) distribution of Gaussian form such as given by Equation 10.57 (or review Section 7.2.3).

$$N(x) = \frac{1}{\sigma\sqrt{2\pi}} e^{-\frac{x^2}{2\sigma^2}} \tag{10.57}$$

The beam size is characterized by the value of sigma (σ), which is related to the full width at half maximum amplitude (FWHM) by FWHM $= 2.35\ \sigma$. Each of the coordinates (e.g. x and θ, or y and ϕ), have their own Gaussian distribution. A two-dimensional particle number density can be represented as a product of two Gaussians, as in Equation 10.58.

$$N(x,\theta) = e^{-\frac{x^2}{2\sigma_x^2}} e^{-\frac{\theta^2}{2\sigma_\theta^2}} = e^{-\left(\frac{x^2}{2\sigma_x^2} + \frac{\theta^2}{2\sigma_\theta^2}\right)} \tag{10.58}$$

Figure 10.26 shows a plot of the 2d phase space diagram. The ensemble of these particles is called a beam. Each gray particle is plotted with its own coordinates. The projection of the two-dimensional number densities in each coordinate is plotted below and to the side of the 2d graph in histograms. These number distributions were generated randomly and within statistical error, they are Gaussian. It can be seen from Equation 10.58 that the locus of constant particle number distribution is

$$\frac{x^2}{2\sigma_x^2} + \frac{\theta^2}{2\sigma_\theta^2} - \text{Constant} \tag{10.59a}$$

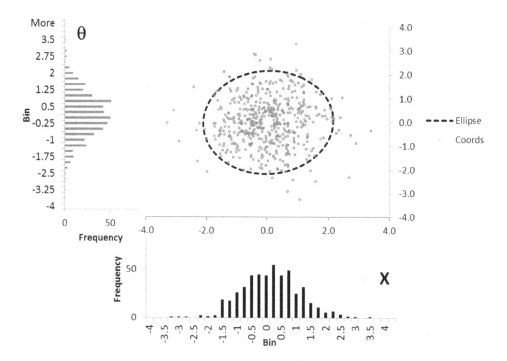

FIGURE 10.26   2d ellipse and profiles.

Note that this is the equation of an ellipse, and this is the form normally used for the outline of the phase space. Actually, in Figure 10.26, the black dashed ellipse is the outline containing 90% of the particles. In this case the constant is about four, if the constant were one, then the ellipse would contain about 40%. The area of this ellipse (with constant = 1) is $\pi\left(2\sigma_x\sigma_\theta\right)$ otherwise written as $\pi\varepsilon$, where ε is defined as the 'emittance' of the beam. When an emittance is quoted, it is important to ask what fraction of the particles is included within the ellipse, it is not always $1/e$, so values quoted could be different. Also area and emittance are different, although sometimes that distinction is confused so ask if the π is included or not.

The beam is a collection of particles and can be described not only by the phase space but also by its 2d transverse extent. So a similar locus for the (x, y) phase space is

$$\frac{x^2}{2\sigma_x{}^2} + \frac{y^2}{2\sigma_y{}^2} = \text{Constant} \tag{10.59b}$$

A typical phase space emittance for the canonical variables from an accelerator that uses a degrader is about 20 mm-mrad, and a typical emittance for a beam directly from the accelerator for particle therapy is about 5 mm-mrad or less. This results in a beam size from mm to cm in a beam-line. A smaller beam transported to the patient results in a smaller and sometimes less expensive beam transport system.

10.11.1   THE ELLIPSE

The equation of a more general (possibly rotated) ellipse can be written as:

$$ax^2 + bx\theta + c\theta^2 = 1 \tag{10.60}$$

The equation can always equal one since any other value can be absorbed into the a,b and c constants. This equation has to be positive definite or $4ac - b^2 > 0$, otherwise, the equation would not represent an ellipse (it could, for example represent a hyperbola). With some algebra and integration, the area of this ellipse A is found to be:

$$\text{Area} = A = \frac{2\pi}{\sqrt{4ac - b^2}} \quad (10.61)$$

This reinforces the condition on $4ac - b^2$. The same algebra results in the location of the maximum x and y of this ellipse as

$$x_{max} = \frac{A}{\pi}\sqrt{c} \text{ and } \theta_{max} = \frac{A}{\pi}\sqrt{a} \quad (10.62a,b)$$

The equation of this ellipse can also be written in matrix form (well, why not) as follows:

$$\begin{pmatrix} x & \theta \end{pmatrix} \begin{pmatrix} a_{11} & a_{12} \\ a_{21} & a_{22} \end{pmatrix} \begin{pmatrix} x \\ \theta \end{pmatrix} = 1 = a_{11}x^2 + (a_{12} + a_{21})x\theta + a_{22}\theta^2 \quad (10.63)$$

Without any condition to the contrary, it's possible to choose a symmetric matrix so that $a_{12} = a_{21} = b/2$. Then, the area of the ellipse can be written in terms of the matrix quantities as

$$A = \frac{\pi}{\sqrt{\det a}} \quad (10.64)$$

$$x_{max} = \frac{A}{\pi}\sqrt{a_{22}} \text{ and } \theta_{max} = \frac{A}{\pi}\sqrt{a_{11}} \quad (10.65a,b)$$

This is all well and good, but it's never this easy. Knowing what's to come in subsequent steps, it's easier here to replace the **a** matrix with an inverse matrix $\mathbf{a} = \boldsymbol{\sigma}^{-1}$, which is made symmetric so that a later result will not have an inverse matrix. If

$$\boldsymbol{\sigma} = \begin{pmatrix} \sigma_{11} & \sigma_{12} \\ \sigma_{12} & \sigma_{22} \end{pmatrix}; \text{ then } \boldsymbol{\sigma}^{-1} = \frac{1}{\det\sigma}\begin{pmatrix} \sigma_{22} & -\sigma_{12} \\ -\sigma_{12} & \sigma_{11} \end{pmatrix} \quad (10.66a,b)$$

and the ellipse equation now looks like

$$\sigma_{22}x^2 - 2\sigma_{12}x\theta + \sigma_{11}\theta^2 = \det\sigma \quad (10.67)$$

with the ellipse area A

$$A = \pi\sqrt{\det\sigma} \equiv \pi\varepsilon. \quad (10.68)$$

10.11.2 PHASE SPACE REPRESENTATION AND EQUATION OF THE BEAM

As shown in Equation 10.67, the equation of the more general ellipse can use canonical variables where x is the transverse position and θ is the transverse angle. Equation 10.68 effectively defines the term ε^2 which is $\det\sigma$. Owing to the properties of the Hamiltonian, it can be shown that σ is a symplectic matrix, but that will not be done here. The properties of symplectic matrices are beyond the scope of the book, but simply stated, they encapsulate the appropriate properties of the charged particle beam propagation in a conservative system.

Following Equation 10.63, Equation 10.67 can be rewritten as a matrix as follows:

$$x^T\boldsymbol{\sigma}^{-1}x = 1 \quad (10.69)$$

Given this formalism, some physical interpretations of the parameters of the ellipse can be identified. Figure 10.27 indicates these important properties of the beam ellipse (Equation 10.65).

x_{max} is related to beam size $= \sqrt{\sigma_{11}}$
θ_{max} is related to the beam divergence $= \sqrt{\sigma_{22}}$

The orientation of the ellipse is related to the beam position and angle correlation, $\sigma_{21} = \sigma_{12}$.

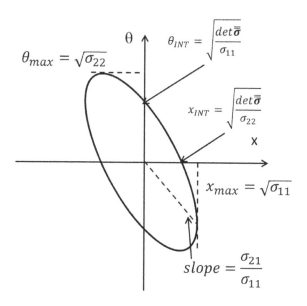

FIGURE 10.27 Beam ellipse parameters.

The plot of Figure 10.27 is not a physical representation of an image that an observer would view of the beam on a monitor (for example, on a fluorescent screen or a film). Rather it is a result of a gedanken measurement or a mathematical representation of the position and angle of the particles of a beam in a particular plane. It is related to the position and transverse momentum of the ensemble of particles. These are conjugate coordinates (x, p_x) and have a special relationship. This plot is called a phase space diagram. A projection of this ellipse, on the x-axis would give an indication of the x extent of the beam (as shown in Figure 10.26) and the particle distribution could be measured in this direction, but that is all that can be directly measured.

A plot of the x and y coordinates of a beam at a particular length along a beam line could look quite different, but that is what would be seen on a film. The phase space representation (x,θ) is useful since the angle of a particle will help determine the offset of that particle downstream (and upstream for that matter) in a beamline. It is necessary to properly design a beamline. Simply exploring the x (e.g. $\sqrt{\sigma_{11}}$) and y extents of the beam is not sufficient for a good beamline design. In any case, all calculations will depend on these beam phase space parameters.

Up to this point, nothing has been said about the behavior of the beam. The ellipse, which encloses a desired percentage of the beam, has been parameterized and parameters that could characterize some aspects of the beam at this one location have been identified. All the particles that are collectively called the beam are inside this ellipse (and outside depending upon the percentage of the beam specified). The parameter x_{max}, for example, is a parameter of the overall beam, not an individual particle. There may also be a particle at exactly that point, or there may not be one there. Determining how this beam ellipse will propagate through a transport system is one of the goals. After all, the beam does actually have to fit in the vacuum pipes and be the appropriate size for clinical use. Do not confuse this beam propagation with the transformation of the ray coordinates that were discussed in the previous sections. While a specific particle inside the beam, e.g. one at x_{max} and $\theta = 0$ can be transported with the techniques previously discussed, those same techniques cannot be used for the beam.

10.12   PROPAGATION OF THE BEAM

The behavior of a beam depends both on the properties of the beamline transport elements and the initial properties of the beam that is injected into that beamline. A mathematical concept has been defined to represent the parameters of a beam. It is useful to determine how the beam propagates through a beam transport system. Propagation of the coordinates of a particle has been discussed and is summarized as follows:

$$x(1) = Rx(0); \; x(0) = R^{-1}x(1) \tag{10.70}$$

Note that $R\,R^{-1} = 1$
where
   $x(0)$ are the vector coordinates at position 0 – the start of the system.
   $x(1)$ are the vector coordinates at position 1 – the end of the system.
   R is the transport matrix from position 0 to position 1.

Starting from Equation 10.69, the equation of the ellipse, it is recognized that the vectors that are part of that equation are actually the coordinates of arbitrary particles (which may or may not be physical, but will be *on* the ellipse). This can be helpful. Some algebraic matrix manipulation (good practice for those rusty) produces the following:

$$x^T(0)\,\sigma(0)^{-1}\,x(0) = 1 \tag{10.71a}$$

$$\left[R^{-1}x(1)\right]^T \sigma(0)^{-1}\left[R^{-1}x(1)\right] = 1 \tag{10.71b}$$

$$x(1)^T\left\{\left[R^T\right]^{-1}\sigma(0)^{-1}\,R^{-1}\right\}x(1) = 1 \tag{10.71c}$$

$$x(1)^T\,\sigma(1)^{-1}\,x(1) = 1, \tag{10.71d}$$

where

$$\sigma(1) = R\,\sigma(0)\,R^T \tag{10.72}$$

(Just in case it was forgotten, remember that $(AB)^{-1} = B^{-1}A^{-1}$.) Therefore, a powerful result is that the ellipse parameters $\sigma(1)$ at location 1 can be written in terms of the sigma matrix $\sigma(0)$, at location 0, and the transfer matrix R. Also, as promised, no inverses are in this final important expression. Carrying out the matrix multiplication, the elements of $\sigma(1)$ are evaluated as below. The matrix doesn't fit on the portrait page orientation, so the matrix elements are listed.

$$\sigma(1)_{11} = \left[R_{11}{}^2\sigma_{11}(0) + 2R_{11}R_{12}\sigma_{21}(0) + R_{12}{}^2\sigma_{22}(0)\right] \tag{10.73a}$$

$$\sigma(1)_{12} = \left[R_{11}R_{21}\sigma_{11}(0) + (R_{11}R_{22} + R_{12}R_{21})\sigma_{21}(0) + R_{12}R_{22}\sigma_{22}(0)\right] \tag{10.73b}$$

$$\sigma(1)_{21} = \left[R_{11}R_{21}\sigma_{11}(0) + (R_{11}R_{22} + R_{12}R_{21})\sigma_{21}(0) + R_{12}R_{22}\sigma_{22}(0)\right] \tag{10.73c}$$

$$\sigma(1)_{22} = \left[R_{21}{}^2\sigma_{11}(0) + 2R_{21}R_{22}\sigma_{21}(0) + R_{22}{}^2\sigma_{22}(0)\right] \tag{10.73d}$$

Knowing the initial beam parameters (at point 0), and the transfer matrix that characterizes the beam transport system between points 0 and 1, it is possible to predict the beam parameters at point 1.

Sorry to repeat this, but it is very important that the beam parameters at point 0 are known, otherwise there is no way to determine what the beam will look like further downstream even if the transport system is known perfectly. The R matrices alone cannot give this information. Sometimes, measurements of the beam downstream can be used to determine what the beam properties were upstream by transforming backward.

10.12.1 Propagation of a Beam in a Drift Length

Consider a drift length (of length L) that can be described by the transfer matrix

$$R = \begin{pmatrix} 1 & L \\ 0 & 1 \end{pmatrix}$$

The propagation of the phase space ellipse in some simple systems can be understood visually as in Figure 10.28. Assume the ellipse is upright (a waist) at the start. First, a public service announcement: It is sometimes not well understood that a beam 'waist' is not the smallest beam at a location. A waist is defined as a beam characterized by an upright ellipse. Now back to the ellipse.

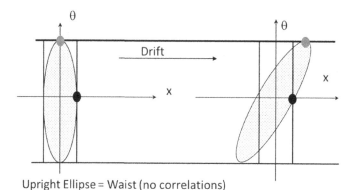

Upright Ellipse = Waist (no correlations)

FIGURE 10.28   Evolution of phase space parameters.

Those particles in the beam with nonzero angles will change position after propagating a distance L and those without angles will not change (e.g. the filled black circle). Therefore, a particle at the top of the ellipse (e.g. the filled gray circle) that was initially at position zero, but with a finite positive angle, will, after the drift, be located at some nonzero distance x, but with the same angle (e.g. follow the top black line in Figure 10.28). This effect is to rotate the ellipse. Propagating the sigma matrix through this drift results in the following expression.

$$\begin{pmatrix} \sigma_{11}(1) & \sigma_{12}(1) \\ \sigma_{21}(1) & \sigma_{22}(1) \end{pmatrix} = \begin{pmatrix} \left[\sigma_{11}(0)+2L\sigma_{21}(0)+L^2\sigma_{22}(0)\right] & \left[\sigma_{21}(0)+L\sigma_{22}(0)\right] \\ \left[\sigma_{21}(0)+L\sigma_{22}(0)\right] & \left[\sigma_{22}\right] \end{pmatrix} \qquad (10.74)$$

The most directly measurable quantity is the beam size, which is related to $\sigma_{11}(1)$ (see Figure 10.27). It is seen that $\sigma_{11}(1)$ is a parabolic function in length L and depends on $\sigma_{22}(0)$, the angular divergence of the beam. This beam size growth is modified by σ_{12} ($= \sigma_{21}$), which is the beam correlation (Figure 10.27). While the extent of the beam is related to the square root of σ_{11}, it is not a strictly linear function; it is unlike the propagation of a single particle through a drift.

Back to the public service announcement: The minimum beam size, as a function of its longitudinal position is, in fact, when it is at a waist, if all focusing upstream is not adjusted. However, the minimum spot size at a specific longitudinal position is not *necessarily* obtained by adjusting upstream focusing to locate a waist at that location. It could be a different condition.

Assume a system is comprised of a lens followed by some optics. The matrix representation from the lens to a target is given by

$$R = \begin{pmatrix} R_{11} & R_{12} \\ R_{21} & R_{22} \end{pmatrix} \begin{pmatrix} 1 & 0 \\ -\frac{1}{f} & 1 \end{pmatrix} = \begin{pmatrix} R_{11} - \frac{1}{f}R_{12} & R_{12} \\ R_{21} - \frac{1}{f}R_{22} & R_{22} \end{pmatrix} \tag{10.75}$$

Therefore, the beam size at the target squared is

$$\sigma_{11}(1) = \left(R_{11} - \frac{1}{f}R_{12}\right)^2 \sigma_{11}(0) + 2\left(R_{11} - \frac{1}{f}R_{12}\right)R_{12}\sigma_{12}(0) + R^2_{12}\sigma_{22}(0) \tag{10.76}$$

Minimizing this function, with respect to the focal length, determines the strength of the lens to obtain the minimum size at the target.

$$\sigma_{11}(1)_{\text{minimum at target}} = R^2_{12} \frac{\det[\sigma(0)]}{\sigma_{11}(0)} \tag{10.77}$$

The distance from this minimum spot size at the target to the waist under these conditions is

$$L_W = \frac{R_{12}R_{22}}{R^2_{22} + \left(\sigma_{11}(0)/\sigma_{11}(1)_{\text{minimum}}\right)} \tag{10.78}$$

Thus the waist position is only the same as the position of the minimum spot size if $R_{12}R_{22} = 0$. The size at the waist is related to the minimum size by

$$\frac{1}{\sigma_{11}(\text{waist})} = \frac{1}{\sigma_{11}(\text{minimum})} + \frac{R^2_{22}}{\sigma_{11}(0)} \tag{10.79}$$

Remember that a waist is a terrible thing to mind. End public service announcement.

Note in Equation 10.77 that since $\det(\sigma) = \varepsilon^2$; a measurement of the minimum size can help to determine the emittance, if the initial size and drift length are known.

10.12.2 REPRESENTATION OF APERTURES

One thing that can definitely affect the beam size is a physical aperture. Beam inside the aperture makes it through, and beam outside the open dimensions of the aperture is blocked (if it is thick enough). It is certainly not a conservative force (Liouville could be disturbed). Put an aperture restriction at a location as in the Figure 10.29(left). Particles with the dashed line trajectories make it through. Some particle trajectories, as shown by the solid arrows, are blocked and some are not. Figure 10.29(right) shows a view head on blocking a Gaussian distribution.

This figure has a collimator with a circular opening, which means that the opening in x gets smaller as the distance y from the midplane increases. Consider only particles in the midplane,

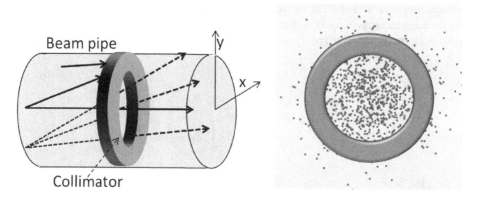

FIGURE 10.29   Side View with collimator (left) and front view of collimator (right).

where the opening is $\pm X_c$. (or an equivalent rectangular aperture), Figure 10.30(left) shows a sampling of the rays that can get through an infinitesimally thin collimator (assuming that the density of the material is infinite – *none shall pass* through that). Rays with arbitrarily large angles can get through as long as the x position of the ray is inside the opening. Figure 10.30(right) is a phase space plot in x and θ. This graph is plotted at the location of the collimator, so the coordinates plotted are effectively in the reference frame of the collimator. The hatched area indicates regions where beam cannot pass, since the coordinates are larger than $\pm X_c$. However particles between that with any angle can pass.

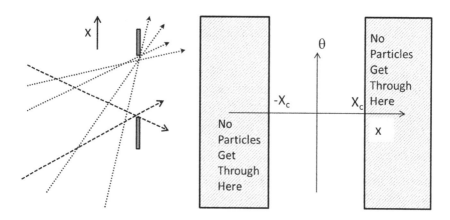

FIGURE 10.30   Collimator rays (left) and collimator phase space (right).

A physical collimator has a finite length, which is required to block particles in the real world. Figure 10.31(left) has a lot of lines; try to stare while reading and they may become clear. Consider particles that are close to the upper edge of the collimator. Most particles coming from below the lower part of the collimator (dotted lines) cannot get through the upper (in this figure) part of the collimator (despite the arrows extending through the material [to guide the eye] which is non-physical). (Also, ignore the reduced length through the material owing to the particle angle – this does have a measurable effect, but not discussed in this book.) When the particle is close to parallel

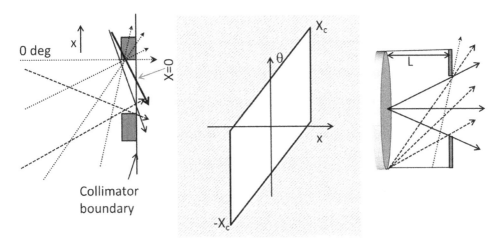

FIGURE 10.31   Thick collimator rays (left), thick collimator phase space (middle) and physical limits (right).

(0 degrees) or above the edge (negative angle), it can get through until the angle is too negative. In that case, the particles are coming from the top and the angle is too steep. The steepest angle possible is that which touches both the inner side of the upper and outer side of the lower (and vice versa), (see the thinner solid black arrow in Figure 10.31(left)). This angle will be determined by the thickness of the collimator. There is a similar constraint for the angle which will pass through the center ($x = 0$) at the collimator boundary (see the thicker black arrow). Thus, the range of angles that can pass are now constrained. The effect of the finite extent of the collimator is to limit the angles that can pass and the admittance in phase space is shown in the white region of Figure 10.31(middle).

These two graphs are actually still nonphysical. They assumed that it was possible to have particles of any angle. However, the particles in the beam actually came from a place in a beam pipe upstream of the aperture. The beam pipe is, in itself, a collimator and there is a limitation of the range of angles that are possible. Figure 10.31(right) shows this additional constraint. All the rays in that figure (except the dotted line which is just a reminder of the previous nonphysical case) are emanating from a distance L, upstream of the collimator. The solid lines come from the center and the dashed lines come from an extreme edge of the pipe (which also wouldn't exist in a real beam). It is useful to represent which of these rays can get through the collimator. This is best represented from the perspective of the location where the rays start. Exploring both this position and the collimator position is instructive and they are plotted in Figure 10.32.

The left side of Figure 10.32 is plotted at the location of the start (upstream of the collimator) and the right side is plotted at the location of the collimator. There are six circles with various filled patterns to focus on. The solid lines of Figure 10.31(right) emanated from the center (or $x = 0$), so the extreme angles are indicated in the circles filled in white and black. The dashed lines originate at the bottom (negative x), so they are represented by the circles filled with diagonal hatches (on the left side of the figure). The lines that would have started at the top (which were not drawn) are represented by the vertical and horizontal hatches (on the right side of the figure). Drawing a line to connect the dots represents the phase space in which particles can pass through the collimator. Those same circular representations of rays, transported to the collimator, are shown in the right side of Figure 10.32. Just follow the horizontal lines realizing that particles with positive angle will move to the right and negative angle will move to the left. The result is a different shape altogether, but

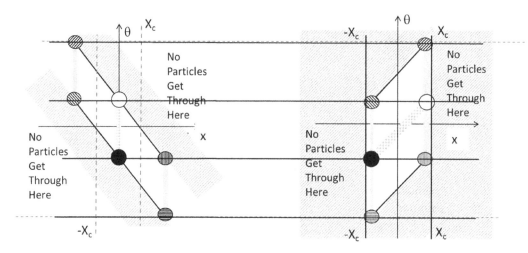

FIGURE 10.32   Phase space representation of collimator at different locations.

with the same phase space area. Of note is that particles with a position greater than x_c at the start can pass through the collimator with aperture x_c. Also of note is that when, at the start, particles were differentiated by angle, not position, at the collimator they are also differentiated by position. This can similarly be carried forward to downstream positions. At the collimator, no particle with position greater than $|x_c|$ gets through.

There are a variety of transport optics conditions that are relevant to the use of a collimator. Beyond just the advantages of the collimator collimating the beam immediately downstream, it might be the case that the sharp edge produced by the collimator could be beneficial further downstream of a collimator. It would mean that each point along the collimator edge should be imaged downstream exactly as it is immediately after the collimator. Therefore, this means that the x position of these points should not depend on θ. By now it should be clear that this implies that $(x/\theta) = 0$. For example, if it is desired to have the effect of a collimator in front of a patient, but it is not convenient to place the aperture there, it might be possible to position it upstream where the effective drift length is zero. Notwithstanding other things, which might affect the sharpness of the edge (which are nontrivial), this could be a powerful tool. Another example is the control of the growth of the beam due to scattering. If the angles of the beam are large and it is desired to limit the phase space area of the beam that will be transported in the rest of the beamline, then it would be necessary to trim those angles. Unfortunately, physical collimators do not directly trim angles, only physical beam positions. There are a variety of ways to do this, some of which will be explored in the exercises.

10.12.3   DISPERSION

A beam with a single momentum has a normal transverse distribution about a mean. This is called a monochromatic beam, which has a monochromatic beam size σ_{011}. (The subscript 0 indicates the monochromatic component of the size.) How would that distribution change if there was a momentum distribution in the beam? If there is no spatial dependence on momentum (dispersion), there is no change. Recall the achromat discussed earlier. Well, now forget it temporarily. What if the beam is measured in a place where the dispersion (x/δ) is nonzero? This produces a spatial dispersion of the beam; particles that have different momenta will be at different x locations. This spread in positions of many particles will have a distribution related to the distribution of momenta in the beam. It shouldn't be surprising to learn that the mechanisms that accelerate and manipulate the beam, most of the time, produce a Gaussian distribution of momenta, σ_p. Thus, this Gaussian momentum distribution will be folded into the nonzero dispersion to give a Gaussian spatial spread of the momenta. Imagine that the coordinates of every particle in the originally monochromatic beam is overlaid with a full spectrum of particles with different momenta at those coordinates. If there is already a Gaussian monchromatic beam (independent of the momenta), then this dispersion will add another Gaussian distribution in quadrature. Therefore, if the total monochromatic beam size squared (σ_{11}) is σ_{011}, then the new beam size will be

$$\sqrt{\sigma_{11}} = \sqrt{\sigma_{011} + \left[(x/\delta)\frac{\sigma_P}{P}\right]^2} \qquad (10.80)$$

Therefore, the beam size is determined, not only by the emittance and focusing conditions (the monochromatic beam size) but also by the amount the momentum distribution is spread transversely due to the nonzero dispersion. This size depends on the dispersion and the range of momenta in the beam, including contributions from the accelerator and range straggling to name a couple.

10.12.4 THE EFFECT OF MULTIPLE SCATTERING

When a charged particle beam goes through a material, it undergoes multiple scattering, and the angles of the particles are modified as was discussed in Section 9.5. Overall, they are increased (although statistically speaking some are reduced) and the scattering results in a Gaussian distribution of the angles with an RMS angle of θ_{MS} above that which already existed. The increased beam phase space is shown in Figure 10.33.

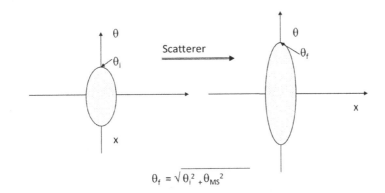

FIGURE 10.33   Phase space growth from scattering.

The additional RMS angular extent and the effect of multiple scattering are determined by the root mean sum of squares

$$\theta_f = \sqrt{\theta_i^2 + \theta_{MS}^2} \qquad (10.81)$$

where θ_i is the incident beam divergence and θ_{MS} is that portion added by multiple scattering. The emittance increases as a result of the increased angle. Given a fixed multiple scattering angle, its effect on the beam would be minimized, if the angles in the beam are maximized so that the RMS addition of the scattering effect has a smaller proportional gain. This could be useful in the design of some beamline situations.

10.13   BEAM MATCHING BEAMLINES

With all of the earlier specific situations discussed, it is still a good idea to be able to fit the beam inside the beam pipe and optimize the cost of the beamline. Well, it might also be said that a beam pipe and/or magnet aperture has to be big enough to fit around the beam. Either way, the beam size becomes another constraint. This is not just about the beam fitting inside the beam pipes. Radiation is produced where the beam interacts with materials in the beam path, pipes, collimators or any equipment that intercepts a portion of the beam. However, this author will still insist that the most important aspect of beam magnetics design is the optical matrix element conditions. This will often give the best beam size solution. As will be discussed, the unmodified beam from the accelerator and beamline system is not normally shaped appropriately for patient treatment. The beam delivery system will modify the beam accordingly, but the beam injected into the beam delivery system will have to be properly matched to be usable. The initial beam conditions are as important as the optics.

10.14   WHAT COULD GO WRONG – 3?

In this chapter, the discussion of charged particle propagation has been extended to beams. While no new components have been added, one very important parameter has been discussed. This is the beam. Much has been discussed about its shape as a Gaussian distribution. But this is not always true. It's helpful that even if the beam starts out non-Gaussian, any material that the beam passes through introduces multiple scattering and the beam can eventually become very close to Gaussian, but the details in any given situation are important to explore. Certainly, any accelerator that uses a septum of some sort to extract a beam will have an edge of the beam that is sharp, as it starts its journey to the patient. Large gantry magnets with higher order fields (wanted or unwanted) can affect the symmetry of the beam. These days, it's all about the journey and what the beam looks like when it gets to where it's going.

One thing that does not get easily cured is if the beam phase space is different in x than in y. In the case of a gantry that rotates, this will present a different projection as a function of gantry angle. The beam parameters that are used for patient treatment are implemented into the treatment planning software. It is not easy to give that software many different beam data sets. This software limitation results in the technical team attempting to find different focusing solutions such that the beam at the target is within tolerance as a function of gantry rotation angle. This is much simpler if the beam is rotationally symmetric at the gantry entrance. If this is not done correctly, then the beam size at the patient could be in error. The author has so very many plots of this type that it was decided not to include any. Just to indicate that it is a lot of work.

Although no new components have been added to the discussion, the fact that the beam phase space has been introduced requires that the effects of accelerator extraction and beam transport are reevaluated with respect to their effect on the beam parameters at the patient target.

10.15   EXERCISES

1. Is the symmetry of the achromat an added constraint? (Ignore L1 and L4.) How does it help, or hurt?
2. Quadrupole doublet
    a. Work out the matrix for the quadrupole doublet.
    b. What is the condition for a double focus (same focal length in x and y)?
3. Quadrupole Triplet (Drift1, Quad1, Drift2, Quad2, Drift2, Quad1, Drift1)
    a. Work out the matrix for the symmetric triplet.
    b. What is the condition for a double focus for the triplet?
        i. In general?
        ii. $L_2 = 0$?
        iii. $L_2 = f_1$?

4. Dipole focusing
   a. What is the direction of the force from B_y in Figure 10.23/10.24?
   b. What direction of pole face rotation is **vertically** focusing?
      i. Entrance rotation (α)
      ii. Exit rotation (β)
5. What is the optical condition that rotates the phase space of an aperture 90 degrees at a place downstream of the aperture?
6. A proton beam from the accelerator has an energy spread of ±0.5%. After going through 30 cm of material (maybe water), at its end of range, what is the new total energy spread?
7. Measurements and conditions:
   a. How can a beam centroid offset be adjusted without changing its angle?
   b. How can the effective drift length be measured between two locations in a beamline? (Assume it is possible to vary the beam centroid θ.)
   c. How can focal length f of the thin lens be determined through measurement? (Assume the position of a ray at a length L from the lens is measurable.)
8. Parallel-to-Point focusing: (a) Write down the form of the transfer matrix of this system (Figure 10.19). (b) What are the conditions? (c) What relationship(s) (e.g. f, Ls) satisfy the conditions? (d) Does the position of a ray at the end depend upon the starting angle? (e) What measurements, can be done to verify that these conditions are met?
9. Derive an equation to determine the required focal length needed to obtain the minimum spot size (at the target) for a system comprised of a lens and a drift of length L to a target.
10. Work out the matrix elements for the achromat in Figure 10.20, except use a thin wedge dipole matrix and ignore L_1 and L_4. What is the focal length of the lens required to produce the achromatic conditions? The final transverse offset and transverse angle do not depend upon the initial momentum offset.
11. A quadrupole with a pole tip field of $B_0 = 3$ kG, a half gap $a = 3$ cm (gradient is 1 kG/cm) and length = 0.3 m, is misaligned by 5 mm. After a 230 MeV proton ray with 0° incident angle passes through the quadrupole on the nominal central trajectory, what is its transverse angle when it exits? (Hint remember Equation 10.28 and that $B' = B_0/a$.)

11 Clinical Perspective of Charged Particle Therapy Beams

The characteristics of a charged particle beam should be determined by the intended therapeutic use. The intrinsic physical properties of particles originating from the accelerators that generate the beam are modified by components to tailor the beam required for treatment. It is required to deposit the prescribed dose to the target volume and to minimize the dose delivered elsewhere. A treatment is usually specified by a prescribed dose to a volume of tissue in the patient, i.e. the target volume. A number of volumes with different definitions are often used to guide the treatment planning and delivery processes. These include gross target volume (GTV), clinical target volume (CTV) and planning target volume (PTV) as shown in Figure 11.1. The GTV is the clinically identified location of the target. It can include the primary tumor, regional and distant metastases. It is identified primarily by imaging techniques. The CTV is a volume that contains the GTV and includes additional tissue with a certain probability of subclinical disease. This is identified with the experience and expertise of the diagnostician. The PTV is a geometrical concept helpful for planning the irradiation field. It starts with the GTV and adds additional margins to compensate for a variety of effects. These could include uncertainties in target location based upon anatomical variations, and the accuracy of beam to target positioning for issues such as patient motion. It is defined so that there is a high probability of the CTV being irradiated, at the expense of irradiation outside the CTV. Sometimes, there is also an ITV (internal target volume) defined. This specifically includes the effect of the target motion during irradiation, to ensure that the target does not pass out of the beam field. The beam must be shaped to accomplish this, even if that target is moving around.

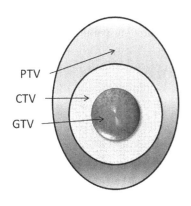

FIGURE 11.1   Target volumes.

When reading about the 'physical' properties of beams, especially in writings that are related to clinical use of beams, the word physical means the physics of the beam, as opposed to the biology or the clinical interactions. The clinical effect of a beam on a cell is the combination of the energy released in the cell (the physics) and the biology of that effect. The latter is not part of this discussion, although it is very important. The physics is also related to the shaping of the beam. There are two main technologies for this, which are briefly mentioned here and covered in the following chapter. While the majority of the proton therapy treatments in the past were shaped by scattered beams, the use of scanning beams has *spread* rapidly in the last few years. The beam delivery system is responsible to direct the beam to the target volume with the appropriate dose distribution.

The dose to the target should conform to the prescription. That prescription could either call for a homogeneous dose or different regions of tissue may require different amounts of radiation doses in consideration of the distribution of different types of cancer cells and other medical issues. Applying a nonuniform distribution of dose could be done over multiple treatments with different target regions defined or, if possible, multiple regions could be treated with different dose levels within one application. This technique, sometimes called simultaneous boosting, is possible with scanning technology.

DOI: 10.1201/9781003123880-11

A number of parameters are needed to characterize the physical properties of the particle beam for clinical use. Different beam delivery systems produce different beam characteristics and therefore their clinical use may be defined in relation to the manner in which the beam is produced. Treatment planning requires a model of the beam and this model may also have a parameterization based on the beam delivery modality. This is then used to specify the desired beam production for the control system, and to specify quality assurance measurements.

During particle treatment, the accelerator and the beam transport system (if the latter exists) send a nearly monoenergetic beam with a small lateral cross section and a sharp longitudinal Bragg peak (BP). As shown in Figure 11.2(left), the longitudinal- or depth-dose direction is along the depth of penetration into the patient. This pencil-like beam is the Bragg curve (BC) introduced in Chapter 9 with a sharp peak in depth (shown below the target in that figure). To obtain the desired depth-dose distribution within the target volume, this sharp peak is modified to a more appropriate distribution. The transverse or lateral direction, as shown in Figure 11.2(right), is transverse to the beam direction. The unmodified lateral distribution of the beam is also relatively small, with a near Gaussian distribution, different than most typical targets. This distribution must also be adjusted to match the shape of the desired target volume. The modified three-dimensional (3D) beam distributions can be characterized by the following parameters in the longitudinal and transverse directions:

- Beam range
- Distal penumbra
- Field size
- Lateral penumbra
- Dose conformance and dose rate
- Beam directions

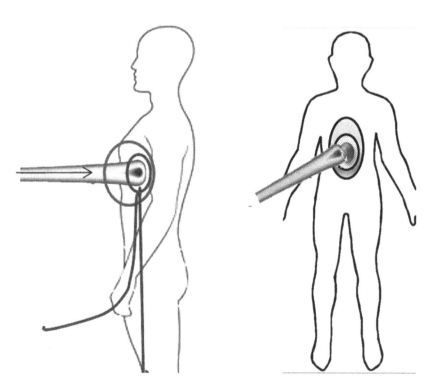

FIGURE 11.2   Longitudinal Tx beam (left) and transverse Tx beam (right).

11.1 LONGITUDINAL DIRECTION

11.1.1 BEAM RANGE

In the direction that the beam is traveling, the longitudinal direction, the beam range must be consistent with depositing dose, as prescribed, from the shallowest (proximal) to the deepest (distal) part of the target volume. The prescription will describe the GTV, CTV, PTV, ITV and any other TVs that the physician identifies. These volumes will include any margins required. The range required and the method of delivering the beam will dictate the beam energy required prior to the beam delivery system. From a clinical point of view, it is generally accepted that for typical clinical sites, the maximum range of the beam that should be available, starting at the surface of the patient, should be about 30 cm. This is not quite enough to reach all parts of the body from all directions, so beam directions must be selected to take this constraint into account. The minimum beam range close to the patient surface is generally accepted to be between 4 and 7 cm, although this range can be further reduced by the introduction of energy-degrading material in front of the patient. The need for a lower value could occur, for example, for pediatric patients or superficial target volumes, like post-mastectomy chest wall. Adding material reduces the beam energy, but it also generates more scattering in the lateral dimension and straggling in the depth-dose direction.

The particular maximum beam range specified is not necessarily the maximum conceivable clinical beam range. However, it was a compromise between the accelerators that could be designed and fabricated for proton radiotherapy and clinical experience. This is a question related to whether the chicken or the egg came first. In an ideal world, the clinicians would specify an ideal set of parameters, independent of the technical capabilities and any compromise would start from there. However, some in the field argue that it's too much work to do that and would rather start with knowledge of the technical capabilities of existing systems and use that information to identify treatment planning methods. This is one reason why, for example, identifying the tolerances of systems is difficult and was long delayed.

Part of this mindset is a result of the fact that, early on, it was very difficult to produce beams suitable for particle radiotherapy. These beams could only exist in the high-energy physics laboratory environment and clinicians with vision did the best they could do with what was available. Some examples include the treatments done at the Gustaf Werner Institute (now the Svedberg labs), the Harvard cyclotron laboratory, Fermilab, Los Alamos and GSI. When it was first considered possible to build a special purpose system for hospital use at Loma Linda University Medical Center, the Particle Therapy CoOperative Group (PTCOG) consortium was formed. The specifications derived therein gave rise to the current values. An interesting extension to this is the beam energy used for heavier ion therapy (such as carbon ions). Essentially the same range, about 30 cm in water, is chosen, even though the accelerator required to produce this is quite large already. Why not have it be a little larger for a deeper range?

Perhaps, to argue the counterpoint, 30 cm beams, from the left and the right directions, for example, results in the equivalent coverage of about 60 cm, which can indeed reach almost any site in most human bodies, so the real question is a cost-benefit ratio and to identify whether the 95th percentile of patients is sufficient to consider. Still, occasionally some deep-seated tumors, like prostate or pancreas, for an exceptionally large patient could benefit from a longer range. Perhaps having 5% of the facilities with a deeper range might be an option, but if the systems constructed are done so by commercial vendors, building a few different types wouldn't be cost-effective. On the other hand, beam scattering, spreading and range straggling through the tissue in the body increases with increasing depth and can perhaps make the beam unsuitable for treatment. These trade-offs are part of the considerations.

Some of the above comments are purely geometric. To date, all, but one particle accelerator for charged particle therapy use have been built to produce particles with a range of about 30 cm. Beam data for a total of 4,033 patients treated by using the spreading technique of scattering, in the two gantry rooms at the Francis H. Burr Proton Therapy Center (FHBPTC) over a ten-year period from

2005 to 2015 has been assembled. Figure 11.3 shows the collection of the 23,603 beams used with the range in a patient and modulation width (to be defined in subsequent sections, but for now represents the depth-dose length in which the longitudinal dose is uniform – or the longitudinal extent of the target). A total of 21,811 fields were used for non-prostate treatment, and only five of them for four patients (1-pancreas, 1-lumbar chordoma, 2-spine metastases) required a beam range greater than 25 cm (group A). Among the 446 prostate patients, 386 had at least one of the lateral treatment fields with beam ranges greater than 25 cm (group B). One can safely say that the beam ranges greater than 25 cm were primarily for prostate treatments. As noted earlier, one goal is to identify the optimal beam direction for the best target coverage and healthy organ sparing. Long beam range for prostate treatment is necessitated by the use of lateral beams (from the left and right). If anterior or anterior oblique beams were used instead, the beam range would be well below 25 cm, as was verified for the ten patients with the longest lateral beam ranges (group C in Figure 11.3). In that case, a maximum beam range in a patient of 25 cm should be sufficient to treat all treatment sites with very rare exceptions.

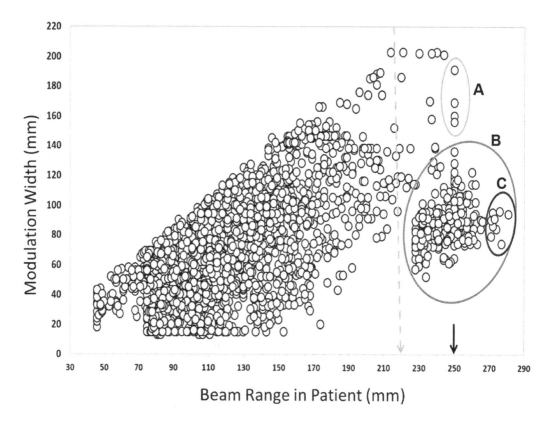

FIGURE 11.3 Clinical energies used.

Regarding the minimum range, it keeps being repeated that if a range less than 7 cm is needed, additional degrading material can be used. One reason for this is because it is harder to produce those ranges with sufficient current in many accelerator systems. So it is a technical limitation, not necessarily a clinical constraint. Well, there is a clinical consideration; a low-energy beam has a very sharp BP and that is actually difficult to use without degrading material to increase the BP width.

These are ranges in the patient, not necessarily the range of the unmodified accelerated beam in water, so the range before the beam spreading system might be larger. The energy of the beam produced from the accelerator is determined not only by the above clinical considerations but also the technique used to deliver the beam. The two types of delivery modes that will be covered in this book are scattering and scanning. Scattering requires the maximum accelerator energy to be higher than that required for scanning since the material used to scatter the beam also reduces its range.

11.1.2 DISTAL PENUMBRA

The distal penumbra of the depth-dose distribution is given by the falloff of the depth dose close to the end of the particle beam range. The distance required for the dose to reduce from its maximum (at the BP) to a desired lower value determines the allowable proximity of healthy tissue. One of the unique aspects of charged particle beams is the BP. This results in little-to-no dose (for protons) (from the primary particles) deeper than the BP. However, as shown in Figure 11.4 (same as Figure 9.13), owing to factors discussed in Section 9.4, the BP distal penumbra is less sharp than that of a mono-chromatic non-scattered beam. The single energy beam (solid line) shows a sharp edge, while the dashed line is the result if there is a Gaussian spread of different energy particles. That spread in energies can result from several factors as has been discussed. However, in many cases, the distal beam penumbra is much sharper than the lateral one. The distal penumbra or falloff is defined as the distance from the point that is 80% of the maximum of the BP (on the distal side obviously) to 20% of that maximum as shown in Figure 11.4. Also, sometimes other values are used, such as the 95%–50%, depending on the situation. Magically, the distal 80% point of the distal penumbra is at a point close to the range of the pristine (single particle) BP. The degree to which external factors (materials in the beam path) affect the distal edge can be factored into the system design. For a typical scattering system, the distal penumbra increases from 3.5 to 5.0 mm over the beam ranges from 4.8 to 25 cm.

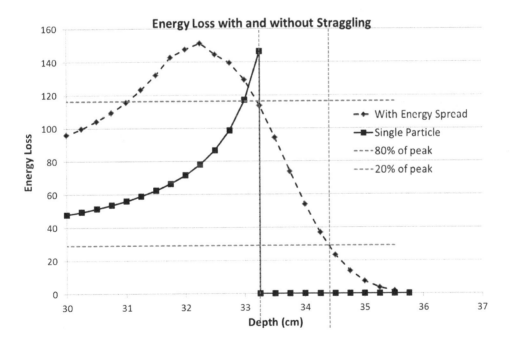

FIGURE 11.4   Distal penumbra.

Human tissue is not perfectly homogeneous. This implies that the effective range of the charged particles will depend upon which parts of the anatomy it is traveling through. This effect is called range mixing. The range of different particles will then be different and the overall distal penumbra can be increased further. It could become larger than the lateral penumbra.

For the most part, what has been discussed earlier still leads to the possible use of the sharp distal edge of the beam as an excellent way to maximize the dose to the target volume, while minimizing the dose to healthy tissue. However, and this is probably one of the biggest howevers in charged particle radiotherapy, the exact knowledge of the range that the beam will travel in a patient is uncertain owing to uncertainties in the anatomy, conversion between X-ray attenuation parameters (conventional CT imaging) and particle beam stopping power as well as other things. If the water-equivalent path length (WEPL) value changes by 1 cm, the location of the distal falloff will change by 1 cm causing either an 'undershoot', missing the distal portion of the target volume by a full 1 cm, or an 'overshoot' delivering full dose to a centimeter of normal tissue behind the target volume. As a result, most treatments do not utilize the distal edge of the beam at interfaces between target and nearby critical structure in the longitudinal direction.

In current practice, the range uncertainty issue is managed by adding an additional amount to the beam range in treatment planning, usually 3.5%, to avoid potentially stopping short of the edge of the target. While this guarantees the coverage of the distal aspect of the target volume, it also increases the dose beyond the target. For example, in the treatment of prostate cancer, as noted earlier, anterior or anterior oblique fields have rarely been used, despite the fact that such fields can utilize the sharp distal penumbra (~4 mm for 95%–50%) to separate the prostate and the rectum behind. Instead, only lateral fields are used, relying solely on the much broader lateral beam penumbra (>10 mm for 95%–50%) to spare the rectum. (This creates the need for the deeper range.) When a water-filled endorectal balloon is used to immobilize the prostate, as widely practiced, the anterior rectal wall is only about 5 mm thick and is situated right next to the posterior side of the prostate in many areas. If an anterior field is used for treatment, the typical beam range is about 15 cm and its 3.5% margin will be 5 mm, just the thickness of the anterior rectal wall. Therefore, the usual method of adding extra range will risk delivering the full dose to the anterior rectal wall. This is unacceptable, given rectal bleeding as the leading treatment toxicity for prostate treatment.

11.2 TRANSVERSE DIRECTION

The beam must be shaped to cover the transverse projection of the target too. Figure 11.5 shows a 3D target and its largest 2D projection on the right. The field size required to cover the projection is indicated by the dashed rectangle. (A tighter fit that more closely matches the target contour is possible with some methods.) Spreading out the beam to match the target shape is a purpose of the beam delivery subsystem. The requirements of this system are intertwined with the clinical needs and the technical constraints.

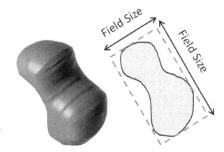

FIGURE 11.5 3D target and transverse projection.

11.2.1 FIELD SIZE

The field size represents the transverse extent of the beam field. For a uniform transverse distribution, it is typically defined as the distance between the locations within the beam that is at a dose level of 50% compared with a uniform maximum (see Chapter 12). The definition is only relevant if there is a uniform field region in between those 50% points. In many cases the clinical treatment field is uniform, even though some beam delivery modalities are more flexible. The field size has a significant impact on most of the parameters of the system design. It can affect the beam production, the beam delivery subsystems and the gantries, to name just a few. Most cases treated by charged particles have relatively small target volumes, with a few exceptions such as medulloblastoma, other central nervous system (CNS) diseases and large sarcomas. A similar balance as discussed about the choice of beam range is to be considered in the case of field size. There is a bifurcation in existing system designs in that some are designed with a field size up to 25 cm in diameter and some are designed with a field size up to a rectangular 30 cm × 40 cm. The target in the case of the entire CNS, including the whole brain and the spinal cavity extending inferiorly nearly to the coccyx could be as long as 80+ centimeters, particularly with young adult patients. There are no systems with a field size of 20 cm × 80 cm, for example, although it might be interesting to have a special purpose treatment room for just such cases. Today such a field requires multiple patient moves. As particle treatments become more accessible, an expanded range of indications are being treated. This includes late-stage diseases often with nodal involvement and thus much larger target volumes, or multiple targets. One heavy consideration, in the case of a scattered beam, is the size and weight of the apertures. Imagine having to handle 25 lb apertures 100 times per day.

It's interesting to take an experiential approach. At MGH up to 2018, close to 70% of treatment fields have been with a radius of 12 cm or less. This includes a limitation of prostate treatments to less than ten per day. The usage of the small field size could be substantially larger at centers with significantly more prostate patients. The treatments with larger field sizes can require more effort than the 70% smaller sizes combined. In the CNS example, using a scattered beam modality, the treatment is often broken into four parts using five fields, two laterals for brain and three abutting spinal fields each covering a portion of the spinal cavity. Care must be taken at these junctions. With repeated setup verifications between different fields due to the patient move, the entire treatment could take up 30–40 minutes. An increase of field size capability would allow for a smaller number of fields and would greatly simplify the effort and shorten treatment time. Some of this complexity is simplified using a scanning beam system. There are still moves required, but the matching and patching tolerances are relaxed.

11.2.2 Lateral Penumbra

The lateral penumbra is that part of the transverse beam that falls off from its maximum in the transverse direction. (See Chapter 12.) As in the distal penumbra definition, the falloff from 80% to 20% is the distance defined to be the lateral penumbra. A sharp lateral penumbra can help to spare critical organs adjacent to the target volume. However, if two beams are to be combined (to obtain an overall larger field size), the dose distribution in the region of overlap is important. If it is too sharp, it is hard to match. The complexities of this situation depend on the spreading modality. If, for example the penumbra were Gaussian shaped, as in beam scanning, that could help the process. The uncertainties in target position are also quite important factors that influence how the lateral penumbra should be shaped. This is discussed in more detail in Chapter 12.

The way that the lateral penumbra is formed can be totally dependent upon the beam delivery mode. In all cases, however, the beam size before the beam delivery system will play a role. For scattering, the lateral beam penumbra is affected by the source size and position, the position of the aperture, the range compensator, the air gap between the compensator and patient's body surface and, naturally, the depth of tissue that the beam must penetrate before reaching the target volume. In scanning, the beam size before the beam delivery system is about the same as the beam size after, until it penetrates the patient.

11.3 DOSE CONFORMANCE

A requirement for charged particle beam delivery is that the target dose must match the prescription to within 2%. This is important not only for the clinical outcome but also to allow for clinical protocols to be accurately followed and compared. The clinical data obtained from accurate beam delivery helps the evolution of treatment plans for subsequent clinical protocols.

Different beam delivery modes have different capabilities regarding the dose distribution of the beam inside and outside the target. A beam delivered using a scattering system can produce a linear dose across most of the target volume. This linear function can also be homogeneous. A beam delivered using scanning can deposit an arbitrary (only limited by the beam size) dose distribution across the target, including a homogeneous dose. It is very important to be clear on this point. The beam described is the beam distribution in a homogeneous target. Taking into account the reality of target volumes that can include different materials and irregular boundaries internally, the dose that is deposited to the ITV will depend on these factors. It is one of the jobs of treatment planning to identify what the incident beam dose distribution should be to achieve the desired dose distribution inside the target. While it may appear that beam scattering has flexibility limitations, the fact that it is essentially a broad beam can make it impervious to small errors.

11.3.1 DOSE AND DOSE RATE

The physical dose rate is essentially the amount of energy deposited in the target per unit mass and time. An external beam radiotherapy system must be able to produce a high enough dose rate so that the treatment time is reasonably short. This is not only in consideration of how many patients can be treated in a day, but also it is directly related to the treatment quality. In most of the treatment procedures today, the patient is first set up by image guidance (2D radiography, cone beam CT, portal imaging etc.), and the treatment is then delivered using one or a few treatment fields, under the assumption that the patient's body configuration does not change from the time of the last imaging to the completion of the dose delivery. While this assumption is more valid for certain types of treatment with appropriate patient immobilization devices than for others, the longer the treatment takes, the more likely the anatomical configuration will change and affect the quality of the treatment.

Studies have found that during prostate treatment, the target volume can stay within a 5 mm margin only for 5 minutes. If an endorectal balloon is used, the margin is reduced to 3 mm but still only for 5 minutes. In this case, the lack of a sufficiently high dose rate would require increasing the planning margins of the target volume and delivering more dose to normal tissues nearby. Patient fatigue is another consideration for reducing the treatment time, particularly for elderly or pediatric patients. However, this does not mean that one should use the highest dose rate possible. Head and Neck immobilization techniques enable submillimeter targeting for extended times.

It's likely the usual dose rate of 2 Gy/min was originally based upon safety considerations. For example, 2% of 1 minute is 1.2 seconds. If human intervention is necessary to ensure that the dose is within 2% of the prescribed dose, then at least 1 second would be required for a human reaction time. Thus, it should not be possible to deliver a clinically significant amount of dose in about 1 second. In 1.2 seconds, 4 cGy is delivered. A standard fractionation regimen is that of 1.8–2.0 Gy per treatment. As technology has evolved, treatments have become more complex, so that 1.2 seconds has evolved into a time frame that can be too long for a reasonable intervention. In this case, the balance between technical safeguards and human backup has to be carefully considered and the question of an appropriate dose rate may be open at the time this was written. In common practice, for example, a dose rate of 10 Gy/min or more is used for prostate treatments. Higher dose rates can also benefit other situations, such as hypofractionated treatment, which deliver a higher dose per fraction. An extreme example is stereotactic treatment. Another type of treatment where a higher dose rate may be appreciated is respiratory gating in which the beam is turned on only for a portion of the respiration cycle, usually 30%, centered on the end of respiration phase. Ideally, the dose rate should be three times more than normal if the treatment is delivered in the usual amount of time. However, the dose rate should not be too high, so that the treatment can be delivered over a sufficient number of respiratory cycles to average out the uncertainties due to breathing irregularities. Fast forward to the possible future use of FLASH radiotherapy; this technique requires a dose rate of 40 Gy/s or maybe more.

The available dose rate depends upon many factors, but in particular it depends upon the beam current that is incident to the beam delivery system. For some accelerator systems, the beam current extracted from the accelerator is not equal to the beam current that is delivered to the beam delivery system. This will in turn affect the design of the accelerator. Also the beam delivery system will affect the designs of the upstream components.

11.3.1.1   Dose

Physical dose is a measure of the energy deposited in a local region. The units of dose are Gray (Gy), where 1 Gy = 1 J/kg. Imagine that a particular dose of 1 Gy is desired in a volume of 1 cm³ of water (with a density of 1 g/cm³). This 1 cm³ of water has a mass of 1×10^{-3} kg and will require 1×10^{-3} J of energy locally deposited. If it's necessary to deliver 1 Gy into 2 cm³ of water, that would require 2×10^{-3} J deposited, but the dose is still 1 Gy. Thus, the same dose (number) was delivered, but given the different masses of material, more energy was required for the larger mass. It may be desired to deliver this dose in a specific time frame that will determine the dose rate needed.

The energy lost in material per unit distance = dE/dx. The stopping power is $S = -dE/dx$. The mass stopping power (dropping the negative sign from here on) is given by

$$\frac{S}{\rho} = \frac{1}{\rho}\frac{dE}{dx}\left(\frac{\text{MeV}}{\text{g/cm}^2}\right) = \text{Mass Stopping Power} \qquad (11.1)$$

Sometimes, the letter S is seen as the mass stopping power with the ρ 'absorbed' into the S. The particle fluence (the number of particles that intersect a unit area) is given by

$$\varnothing = \frac{N}{A}\left(\frac{\text{Particles}}{\text{cm}^2}\right) = \text{Fluence} \qquad (11.2)$$

Particle flux is the rate at which the particles flow through a unit area.

$$\dot{\varnothing} = \frac{\dot{N}}{A}\left(\frac{\text{Particles/s}}{\text{cm}^2}\right) = \text{Flux} \qquad (11.3)$$

where

N = number of particles traveling through a surface area A (N/A is per unit area).

The dot above the letter indicates a time rate of that quantity.

As noted earlier, the physical dose (dose solely measured by the energy released – not including any biological effects) is defined by

$$D(\text{Gy}) \equiv \frac{dE}{dm}(\text{J/kg}) = \text{Gray} = \text{Gy} \qquad (11.4)$$

Therefore,

$$D = \frac{(dE/dx)\times \Delta x \times N}{\rho \times A \times \Delta x}; \text{ each of } N \text{ particles loses } \frac{dE}{dx} \text{ in a } \Delta x \text{ layer} \qquad (11.5)$$

$$= \frac{N}{A}\left(\frac{\text{number of particles}}{\text{m}^2}\right)\frac{dE}{dx}\frac{1}{\rho}\left(\frac{\text{J}}{\text{kg/m}^2}\right) = \varnothing\frac{S}{\rho} \qquad (11.6)$$

where dx is the limit of Δx as $\Delta x \to 0$. Δx is the longitudinal distance into which the dose is deposited. It is important to remember here that the energy loss dE/dx depends upon the speed of the particle, which is slowing down as it penetrates matter. Therefore, this equation formally holds only in a short distance (since v and dE/dx change as the particle penetrates), although dE/dx does not change rapidly in the proximal plateau region of the Bragg curve (BC).

There is some subtlety in this expression. Partly it's associated with the behavior of the energy loss as a function of depth and partly it's associated with the definition of dose. The units of dose are energy/kg. It might be thought that if N charged particles of the same energy were incident on 1 kg of material, that material would receive the same dose, independent of the configuration. This is not the case. The energy deposited should be considered; the sketches in Figure 11.6 help to illustrate that point.

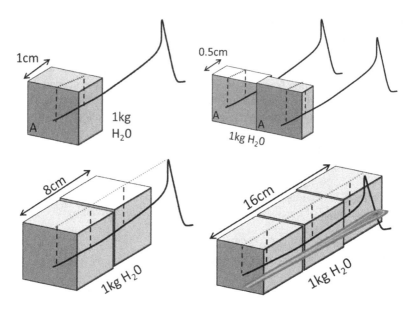

FIGURE 11.6 Bragg peak through 1 kg block (upper left), two blocks side by side (upper right), two blocks from front to back (lower left) and three blocks (lower right).

Assume it is desired to deposit a dose D in 1 kg of water. Figure 11.6 shows a 160 MeV beam coming from the left, that is incident on various configurations of blocks of water. The same number of particles, N, with the same energy is used to irradiate these blocks. Figure 11.6(upper left) has a block of depth $\Delta x = 1$ cm. The beam passes through that block and the high-speed incoming particles deposit some energy from the BC proximal plateau. The 1 kg mass is in an area of 1,000 cm^2 of water (31.6 cm per side if it is square) with a depth of 1 cm (mass length 1 g/cm^2). Therefore, $D_0 = N \, dE/dx \times 1$ cm/1 kg. Since $N \, dE/dx$ will be kept constant, call it equal to S_N. The dose is S_N (MeV/cm) $\times$ 1 cm/1 kg.

In Figure 11.6(upper right), the beam is used to irradiate 1 kg of water but with a block that has twice the area and therefore a depth of 0.5 cm ($\Delta x/2$) (mass length of 0.5 g/cm^2). Now split this block with area $2A$ into two blocks and also split the beam. If just the darker block, to the right, is irradiated, then not only is the energy deposited per particle reduced by half due to the shorter length ($dE/dx \cdot \Delta x/2$), but since the thickness is reduced by half, the mass is also reduced by half. According to Equation 11.5, the $\Delta x/2$ is canceled. It's been assumed that dE/dx is not changing in this Δx distance. The energy per mass is the same per particle. However, the number of particles N had to be split between the two blocks, so the darker block gets one-fourth of the energy loss (half of the particles in half of the thickness) in half of the mass or half of the original dose $D_0/2$ ($D = S_N/2 \times$ 0.5 cm/0.5 kg). Now consider both blocks. If the full 1 kg of water is irradiated, this requires both blocks with the depth of 0.5 cm and twice the area. In this case, the dose is $D = S_N \times 0.5$ cm/1 kg (since the area and, therefore, the mass has doubled and all N protons are used). So it gets half of the energy loss in the full mass and the total dose received is still $D_0/2$. Half the energy has gone into the full volume (mass) and the result is that the total dose in both blocks is half of the dose in Figure 11.6(upper left). In other words, facetiously, 1/2 + 1/2 in this case = (1 + 1)/(2 + 2). The same number of particles has been distributed over both blocks with a mass of 1 kg, with only half the depth. The same number of particles was used with the same initial energy to irradiate the same volume, but the dose delivered is different. That is because the energy loss depends on the linear depth traversed by the charged particles.

Figure 11.6(lower left) shows the case where the same number of charged particles is incident on an 8 cm long block of water (shown as 2×4 cm long blocks with transverse sides 11.2 cm for a square). Note in all cases so far the range of the 160 MeV proton beam (about 16 cm) is longer than the blocks of water. If the stopping power were the same throughout the 8 cm length, then the dose in the water would be $S_N \times 8$ cm/1 kg or $8D_0$, eight times the dose in the case of Figure 11.6(upper left). Once again, the same mass of water has been irradiated with the same energy and number of protons, but the dose is different. This last case (Figure 11.6(lower left)) began to stretch an approximation. Since the energy loss is a function of the particle velocity, it is a function of the particle energy and it increases with decreasing particle energy. There is a long plateau in which the energy loss per unit length doesn't change much. In fact, it changes by about a factor of 2 from 160 MeV to 50 MeV, but that is not the case getting closer to the (BP). Considering the geometry of Figure 11.6(lower right), the beam is stopped in the 16 cm length of water and the full energy is deposited (not uniformly) in the 1 kg of water. The surface area is now 62.5 cm². Therefore, the dose deposited is 160 MeV/1 kg. This is the rest of the energy loss story. Taking the original S_N to be 5.2 MeV/cm (of water), the situation of Figure 11.6(lower left) resulted in an energy loss of 42 MeV/proton (it probably would be closer to 50 MeV). In the situation of Figure 11.6(lower right), the energy loss is not $5.2 \times 16 = 83$ MeV; it is 160 MeV/proton. Therefore, 74% of the dose is in the last half of the range. This is not exact, since the BC starts to rise toward the end of the second third of the range, but it is useful for a back-of-the-envelope estimate of the situation. This is a lesson to pay attention to the geometry.

It is convenient to use units of Giga protons (10^9 protons), and MeV, g, and cm instead of SI units. Converting to these units leads to:

$$D = \left(\frac{Gp}{cm^2}\right)\left(\frac{MeV}{cm}\right)\left(\frac{1}{g/cm^3}\right) = \left(\frac{10^9 \ Protons}{10^{-4} \ m^2}\right)\left(\frac{1.6 \times 10^{-13} \ J}{10^{-2} \ m}\right)\left(\frac{1}{10^{-3} \ kg/10^{-6} m^3}\right) = \cdots \quad (11.7a)$$

$$= 0.16\left(\frac{number \ of \ protons}{m^2}\right)\left(\frac{J}{kg/m^2}\right) = 0.16 \ Gy \quad (11.7b)$$

Thus, the dose with 1 Gp in 1 cm² with an energy loss of 1 MeV in 1 g/cm² of a medium with 1 g/cm³ density is 0.16 Gy and with values not equal to one it is given by

$$D(Gy) = 0.16 \ \frac{N}{A}\left(\frac{number \ of \ Gp}{cm^2}\right)\frac{dE}{dx} \ \frac{1}{\rho}\left(\frac{MeV}{g/cm^2}\right) \quad (11.8)$$

A somewhat cruder, but perhaps more intuitive (from a physics point of view) example of this conversion follows. Start with the assumption of a 150 MeV beam of protons with a beam current of 1 nA. This beam is assumed to be incident directly (without modification or loss) on a target. A proton beam has energy and current, just like the current in house wiring, and therefore the electrical analogies follow:

- The power in the beam is = J/s = energy × current.
  - E.g. 150 MeV × 1 nA = 0.15 W (J/s).
- Recall the definition of dose = J/kg ≡ Gray (Gy).
  - Dose = [power (W) × time (s)]/kg.
  - E.g. assume the energy is delivered in 1 minute: 150 MeV × 1 nA × 60 s = 9 J.

- Assume the dose is delivered in 1 kg of water.
  - The density of water is 1 kg/1,000 cm^3 = 1 kg/liter.
- Therefore, the dose delivered in 1 minute is = 9 J/kg (in a liter) = 9 Gy.
- This assumes the length of the water is about 15 cm, or all the energy is deposited in the column of water.
- 1 nA for 60 s $\Rightarrow$ 60 × 10^{-9} C $\Rightarrow$ 370 × 10^9 protons needed for this 9 Gy.
- Therefore, 41.6 Gp are needed for 1 Gy in this full volume of 1 kg.
- If only the last half of the volume contains the target, the BC does not deposit its entire energy in the target, since it has a finite extent. It's sort of the opposite of the situation depicted in Figure 11.6(lower left). So assume that about three-fourths of the integrated energy loss is in the last half of the target (0.5 kg). Therefore, 1 nA × 1 minute will give 13.5 Gy to that 0.5 liter that is the target volume in 1 minute.
  - Therefore, 27.8 Gp are needed for 1 Gy in the last half of the volume of 0.5 kg.
- For completeness, the first half is receiving one-fourth of the energy or 4.5 Gy with 1 nA in 60 seconds.
  - Therefore, 83.3 Gp are needed for 1 Gy in this first half of the volume of 0.5 kg.

A dose rate constant for the full volume can be defined as (1 Gy/min)/(0.11 nA), or 0.075nA for only the last half. Note that some systems are capable of producing hundreds of nA, or thousands of times this current. The system must be designed for this contingency. This gives an indication of the number of particles needed to treat a target, and the beam current required, depending upon the dose that is prescribed. At least this number must be extracted from an accelerator in the desired time interval. This calculation has considered a rough approximation of the energy distribution in the BP, so to compare like results this must be factored in. Using Equation 11.8 with 42 Gp, 15 cm length and full energy stopping yields about 1 Gy. Well, that's close enough for this back-of-the-envelope book.

The dose rate can be obtained from Equation 11.8. Differentiating this with respect to time

$$\dot{D}(Gy) = \frac{dD}{dt} = 0.16 \frac{\dot{N}}{A} \frac{dE}{dx} \frac{1}{\rho} \tag{11.9}$$

The dot above the N and D indicates a time derivative, so

$$\dot{N}e = i_p \tag{11.10}$$

where e is the charge of a proton and i_p is the proton beam current in amps. A convenient set of units will be current in nA and stopping power in MeV/(g/cm^2) (not MeV/cm).

$$\dot{D}(Gy/s) = \frac{dD}{dt} = \frac{i_p(nA)}{A} \frac{dE}{dx} \frac{1}{\rho} \text{ or } \dot{D}(Gy/min) = 60 \times \frac{i_p(nA)}{A} \frac{S}{\rho} \tag{11.11}$$

The conversion to nA and absorbing the $e = 1.6 \times 10^{-19}$ C/proton gives this result.

These relations are given in terms of the stopping power (S) and/or energy transfer dE/dx. The reader is encouraged to use whichever suits the occasion. (Again, sometimes the ρ is absorbed within the S.)

11.3.1.2 Counting Dose (Ionization Chamber)

A particle beam of a certain current is generated from the accelerator and transported to the beam delivery system. This beam usually goes through an ionization chamber (IC) that generates a signal related to that beam. That beam then enters the target delivering dose to that target. An IC is an enclosed box filled with a gas and has layers of foils, some of which have voltage that create an electric field. Energetic particles passing through the IC ionize the gas in the IC and the charged ions produced are accelerated by the electric field toward a plate that collects the charge. Use of ICs is a regulatory standard (for now).

- There is a relationship between the beam current and the signal rate detected from the IC in the nozzle.
- There is a relationship between the beam current and the dose rate in the target.

A single particle that passes through an IC filled with air at a particular pressure and temperature loses some of its energy in that IC. The amount of energy lost is dependent upon the particle's energy and the stopping power. Some examples for protons are given in Table 11.1.

TABLE 11.1

Some Stopping Power Values for Protons in Air

E = Proton Energy (MeV)	S(E) = Stopping Power in Air (MeV/(g/cm²))
50	10.99
100	6.443
150	4.816
200	3.976
230	3.642
250	3.462

Remember that the units of stopping power are MeV/(g/cm²) (in this table). This indicates that, for example, a proton of 150 MeV will initially lose 4.816 MeV passing through 1 g/cm² of air material. In air, with a density of 0.001225 g/cm³, a proton will lose 0.0059 MeV/cm. A typical (not all) IC is filled with air at ambient pressure and temperature and has a 1 cm gap (or a 0.5 cm gap or some other gap, but 1 cm will be used here). The energy lost by that single proton goes somewhere. This energy is used to create ion pairs (IP) from the gas in the IC. The most common situation is to strip off one electron from the gas atom creating an ion with a charge of 1. A charge of 1 in units of Coulombs is $e = 1.602 \times 10^{-19}$ C. The amount of energy required to do this is $W = 34.3$ eV. Therefore, the number of IPs created per proton of 150 MeV in an IC with a gap (g) is approximately

$$\text{Number of IP} = \frac{S_{IC}(E)[\text{MeV/(g/cm}^2)] \times \rho[\text{g/cm}^3] \times g[\text{cm}]}{W[\text{MeV}]} \quad (11.12)$$

$$\text{Number of IP} = \frac{5 \times 1.20 \times 10^{-3} \times 1}{34.3 \times 10^{-6}} = 175 \text{ Ion Pairs} \quad (11.13)$$

where $S_{IC} = S/\rho$ in the IC, so *the ρ has finally been absorbed into the stopping power*. The density of air, $\rho_{air} = 1.20 \times 10^{-3}$, has been used (at 20°C) because a gain of 175 is what is used in some other examples. The gain would be 178.5 at standard temperature and pressure (STP). If $\rho_{air} = 1.29 \times 10^{-3}$ (at 0°C) then the result becomes 188 IP. The temperature and pressure (remember $PV = nRT$) of the IC play a role in the performance and must be accounted for every day. An average stopping power of 5 MeV/(g/cm²) is assumed in the proximal plateau region (before a significant amount of energy has been lost). Thus, the IC gain has just been calculated. If 1 proton of this energy goes through the

IC, 175 charges are created and the gain factor is 175. The charge released in these number of IPs is $e * 175$. Therefore, the charge created per proton in the IC, Q, is:

$$Q = \frac{(e)S_{IC}(E)(\rho)(g)}{W}\left(\frac{C}{proton}\right) \qquad (11.14)$$

From Equation 11.6

$$D_{IC} = \varnothing S_{IC} \rightarrow S_{IC} = \frac{A}{N}D_{IC} \qquad (11.15)$$

where D_{IC} is the dose released in the IC from N protons (S_{IC} was only from one) in an area A. Therefore,

$$Q(\text{charge per proton}) = \frac{e\rho g}{W}S_{IC} = e \cdot gain = \frac{e\rho g}{W}\frac{A}{N}D_{IC} \qquad (11.16)$$

The total charge created by N protons per cm³ (N/Ag) for a given dose in the IC is

$$Q_T = \frac{N}{Ag}Q = \frac{Ne\rho}{AW}S_{IC} = \frac{e\rho}{W}D_{IC} \qquad (11.17)$$

Define C to be the constant $e\rho/W$. Since dose is defined as J/kg, but Q_T has been defined to be the charge created in a volume of cm³, the units are mixed and the appropriate units for C in this instance can be forced (also note the density used here)

$$C = \frac{e\rho}{w} = \frac{[1.6\times10^{-19}\,C/proton]\left[1.29\times10^{-3}\,g/cm^3 \times 10^3\left(\frac{kg/m^3}{g/cm^3}\right)\right]}{34.3\ eV \times 1.6\times10^{-19}\,J/eV} \qquad (11.18a)$$

$$= 0.03761\left[\frac{(Coul/proton)(kg/m^3)}{J}\right] = 0.03761\left[\frac{C}{Proton}\right]\left[\frac{kg}{J}\right]\left[\frac{1}{m^3}\right] \qquad (11.18b)$$

$$= 0.03761\left[\frac{C}{Proton}\right]\left[\frac{kg}{J}\right]\left[\frac{1}{10^6\,cm^3}\right] = 37.61\left[\frac{nC}{Proton}\right]\left[\frac{kg}{J}\right]\left[\frac{1}{cm^3}\right] \qquad (11.18c)$$

Just a word that may be obvious, the signal from the IC is proportional to the dose deposited to the IC by the incident (high energy) protons. The deposited energy creates the ions in the gas volume, the (low energy) ions created do not create the dose.

While beam is being delivered into the target, the same beam is being counted in the IC. One difference between the target and the IC is the material and, therefore, the stopping power. The stopping power ratio (including ρ) of air to water is about 0.88, very nearly independent of energy (down to at least 5 MeV and lower). However, in *real* units (taking the density into account), the ratio of the energy loss per unit length is about 0.00108 (but also fairly independent of energy). Since the beam loses a small amount of energy in the IC, the energy of the beam at the surface of the target is basically the same and the ratio of the dose delivered in the IC to that at the surface of the target (in the same volume and depth extent) is close to 1 (taking the relative densities into account):

$$\frac{D_{IC}}{D_{Target}} = \frac{S_{IC}}{S_{Target}} \qquad (11.19)$$

Equation 11.17 ($Q_T = C\,D_{IC}$) calculates the charge liberated in the IC. An IC is built so that there is an electric field in the direction of the traveling particle. The ions created are then directed to the plane with the appropriate charge. (A positive ion is attracted toward a negatively charged or grounded plate.) (An IC-design question is whether to count the positive part of the ion or the negative – speed can be a concern.) Sometimes, an IC is built so that the entire plane of the chamber can measure the current generated, and sometimes, it is built with a well-defined region that is sensitive to the current, called a dose pad, with an area that is less than the entire surface area of the IC. Figure 11.7 shows a region where a lighter gray cylinder with radius r_2 intersects a darker gray plane. This area with radius r_2 is the dose pad. The rest of the area is isolated (electrically) from that cylinder. Only that sensitive region captures the ions. If a particle passes through the IC outside the light gray cylinder, for example, if the beam has a radius r_1 (within the dashed cylinder), the part of the beam outside r_2 would not be detected. Therefore it is related to the current within that sensitive volume, or the current density.

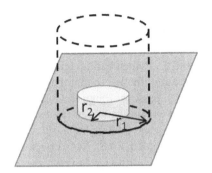

FIGURE 11.7   Dose pad in IC.

This is one example of a selective sensitive region in an IC. Another example is the use of strips.

Figure 11.8 shows a sketch of five strips and a darker plane below them, in a box. Only the strips are sensitive, and they are isolated from each other, so it's known when a particle passes near a particular strip. The ions created are attracted to the nearby strip. If a beam extends over several strips, it is possible to get a measurement of the beam profile which is the transverse particle number distribution in the beam.

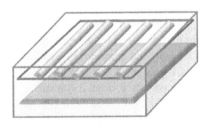

FIGURE 11.8   Strip ionization chamber.

11.3.1.3   Dose Quantities, Dose Rate and Irradiation Time

There are a variety of different types of electronics connected to an IC. These usually are equivalent to charging up a capacitor and sending out a pulse when that capacitor is charged up to a certain voltage. Each of these pulses is called a count. In the old days, each count would increment a mechanical meter that would update with a click and the amount of dose would be counted in clicks (or perhaps that was just in nuclear physics while I was waiting for Sam Penner). These days digital signal processors, programmable gate arrays and other fast smart devices may be used in place of capacitors and pulse counters. The quantum of charge that is implemented is dependent upon the designer, the desired resolution and the electronics noise in the design. One usual example is 10 pC. Therefore, the number of counts measured is

$$\text{Number of counts} = \frac{Q\big(\text{Charge created per particle}\big) * \big(\text{Number of particles}\big)}{10\text{ pC}} = \frac{QN}{10\text{ pC}} \quad (11.20)$$

Assume an IC gap of 1 cm and a dose pad area of 4 cm². The dose delivered to the active volume of the IC that will create 10 pC of charge, using Equation 11.15, is about 70 μGy.

It has previously been estimated (within approximations) that it could take 30 Gp to deliver 1 Gy (in the 0.5 kg target in the last half of the 15 cm range). In this case, 30 Gp fully inside the sensitive IC region would create ~1.0×10^5 counts (depending on the temperature) for a total dose of 1 Gy in the target. This assumes all incoming protons are counted.

It has been assumed that the cross-sectional area of the beam that is detected is 4 cm². This should be replaced with the actual area of the dose pad being used and the fraction of beam inside that area. Alternatively, the active plane of the IC could have an area greater than the beam and thus the full beam area is used.

The IC is used because it is expected that the data obtained from the IC is related to the dose in the target. There is a relationship between the dose near the surface (a small depth distance) of the target and the counts counted in the IC. Remember Equation 11.19. However, the target is downstream of the IC, so that the cross-sectional area of the beam, which passes through the IC, may be different than the area of the beam at the target. For example, if the beam is diverging from a spreading system, there is a $1/r^2$ effect (see Figure 7.20 in Section 7.3.5). This effect is also shown by the dashed circles in Figure 11.9. In that figure, the beam cross sections increase as the beam gets closer to the target. If there is a dose pad smaller than the beam cross section, this effect is also included. Only the beam in the light gray cross-sectional pillbox area (with a radius of r_2) is detected by the IC. The full IC area is given by the darker gray surface. The actual extent of the beam may be given by the dashed line cylinder with a radius of r_2, but only the beam inside radius r_1 is detected (unless the entire dark gray surface is an active area, which is usual for scanning beams). If the dose at the surface of the target spreads to a radius of r_3, this must be taken into account in determining the correspondence between the dose in the IC and the dose at the surface of the target.

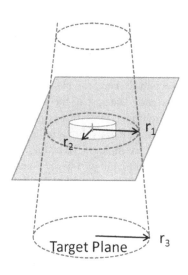

FIGURE 11.9   IC with $1/r^2$ effect.

There is another quantity that is generally used called the number of 'monitor units' (MU). It is just a number scaled down from the counts according to an MU constant. This is facility dependent. It is a calibration constant used to establish a more intuitive number associated with dose.

$$\text{Number of MU} = \frac{\text{Number of counts}}{\text{MU constant}} \qquad (11.21)$$

If the MU constant = 300, 1.5×10^5 counts/300 = 333 MU. Remember that this is close to what would be required for 1 Gy in the target. Therefore, there would be 3 MU per target cGy. Of course, as shown before, the correspondence between the number of protons and the dose in the target is target dependent, so care must be taken. Most of these simplified calculations are to get within an order of magnitude in this book, so factors of a few or several may be different than most facility realities which are quite precise.

Finally, there is a term called cycle. The data from the IC is read in time intervals. Each of these intervals is sometimes called a cycle. (Actually this is quite a good application of Poisson statistics.) A cycle could be the time for a range modulator revolution in the case of scattering (e.g. 100 ms), or the time to check data (10 or 1 ms or 250 μs or faster, in scanning). There are potentially many other cycles in particle therapy equipment, so be careful of the use of that term.

$$\text{Counts per cycle} = \left[\frac{\text{Number of counts}}{s}\right] \times \left[\text{cycle time}\left(\frac{s}{\text{cycle}}\right)\right] \qquad (11.22)$$

Okay, there is more than one particle going through the IC, and these particles are coming at a certain rate (per second); therefore, there is a beam (many particles) with a certain current I_{Nozzle}

(C/s) passing through the IC generating a current I_{IC} in that IC. Going back to the charge created per proton in the IC,

$$Q = \frac{(e)(\rho)(g)S_{IC}(E)}{W}\left(\frac{C}{proton}\right)$$

(11.14 repeated)

The total charge created from N protons (not per volume here) is

$$Q_{Total} = \frac{N(e)(\rho)(g)S_{IC}(E)}{W}\left(\frac{C}{proton}\right)$$

(11.23)

Differentiating with respect to time

$$\frac{dQ_{total}}{dt} = \frac{dN}{dt}\frac{(e)(\rho)(g)S_{IC}(E)}{W}\left(\frac{C}{proton}\right)$$

(11.24)

where dQ_{total}/dt is the current created in the IC and $e\, dN/dt$ is the current in the nozzle passing through the IC (the two are separate)

$$I_{IC}\left[\frac{C}{s}\right] = \frac{(\rho)(g)S_{IC}(E)}{W} \times \left[e\left[\frac{C}{proton}\right] \times \frac{dN}{dt}\left[\frac{\text{number of protons}}{s}\right]\right]$$

(11.25)

Note that the e has been moved inside the parentheses of Equation 11.25 to be put next to the number of protons, so that a current will be obtained. Also note that this equation assumes that all the particles have contributed to the signal from the IC. However, as stated earlier, some ICs have a dose pad of limited size. The ratio of the dose pad area to the beam cross-sectional area in the IC is k. Therefore, the equation can be rewritten as

$$I_{IC} = \frac{(\rho)(g)S_{IC}(E)}{W} \times k \times I_{Nozzle}$$

(11.26)

This is the signal detected by a charge sensitive device connected to the IC. The S_{IC} in Equation 11.26 is the S in the small length in the IC (in whatever gas is inside the IC), not the S in the target (in the material in the patient). With the IC gain, there could be more than 20 nA of current in the chamber. This is going through a power supply delivering 500–1,000 V, or a power of up to 2×10^{-5} W; not very much power.

The counts that will be measured per cycle is

$$\frac{\text{Counts}}{\text{cycle}} = \frac{I_{IC}(\text{C/s}) \times (\text{s/cycle})}{\text{Quantum of charge (C)}}$$

(11.27)

Or

$$\frac{Counts}{cycle} = \frac{[(\rho)(g)S_{IC}(E)]}{W} \times k \times I_{Nozzle} \times \frac{(C/s) \times (s/cycle)}{Quantum\ of\ Charge\ (C)}$$

(11.28)

Or

$$\frac{Counts}{cycle} = \frac{dD}{dt}\frac{[e\rho Ag]}{W} \times \frac{(s/cycle)}{Quantum\ of\ Charge(C)}$$

(11.29)

where dD/dt is nominally the IC dose rate, but that's not much different than the target dose rate (at the surface). Coming back to that 30 Gp that gave rise to 1 Gy; if that was delivered in 1 minute, an empirical dose rate constant of 0.10 nA/Gy represents a useful constant for this example. Using an empirical dose rate constant, Equation 11.29 can also be written as

$$\frac{\text{Counts}}{\text{cycle}} = K \times \dot{D}_{\text{Target}} \left[\text{Gy/min}\right] \times \left(\text{Dose Rate Constant}\right) \left[\frac{\text{nA}}{\text{Gy/min}}\right] \qquad (11.30)$$

The 'constant' capital K will be dependent upon other factors in the beam delivery.

11.3.1.4 Maximum Count Rate
From Equation 11.17

$$Q_T = N \frac{Q}{Ag} = \frac{e\rho}{W} D_{IC} \qquad \text{(11.17 repeated)}$$

A worst-case scenario could be that 300 nA or more passes through the IC. The phenomenon of recombination has not yet been discussed. Depending on the design of the IC, this can occur when the dose rate in the IC gets high. So many IP are created, that some of the ions recombine with the ionized electrons and become neutral again. They have time to do this; the more time they spend uncollected, the more likely they are to recombine. So recombination can be reduced if high enough electric fields are used and if the gap is small enough. Therefore, the charge created in the IC would be less than representative of the actual beam current density in the IC. For example, the signal from the IC could only represent half the actual flux. This would result in the dose delivered in the target not being properly counted. Therefore, it is essential that the system never go into that state, or be detected. This is done through a combination of IC design – ensuring that recombination does not occur at foreseeable currents and that the electronics detects a high enough dose rate before the IC goes into recombination mode. A current of 300 nA contains 1,875 Gp/s. This results in a full-count rate signal of 5.25×10^6 counts/s or a readout frequency of 5.25 MHz (if it is all within the sensitive volume). Thus, the counting electronics must be capable of at least this acquisition speed.

11.3.1.5 Dose Rate Considerations
In the scattering mode, the dose rate depends on the scattering design, particularly the intended field size. The larger the field size, the lower the dose rate for a given beam current at the nozzle entrance. Some scattering systems are designed to give a small field size for a particular type of treatment, (e.g. prostate treatment) where the maximum field size is 12 cm in diameter.

In the case of pencil beam scanning, the dose can be delivered in the transverse direction, spot by spot or line by line (in continuous scanning) and then in the depth direction, energy layer by layer. The overall dose rate is usually specified by the time required to deliver a uniform dose to a 10 cm cube, or 1 liter of tissue-equivalent material. This depends not only on the beam current intensity for each scan layer but also on several other factors to be discussed in Chapter 12. For the same incident beam current, the effective dose rate will also depend significantly on the target size to be treated.

Regular fractionated treatment, i.e. 1.8–2.0 Gy per fraction could mean 0.9–1.0 Gy per field. A dose rate of 1.0–4.0 Gy/min will be reasonable, but current developments in such delivery techniques, such as FLASH, where 80 Gy or more should be delivered in a fraction of second would drastically change the common practice.

The dose rate ultimately depends on the beam current transported to the nozzle entrance where it is to be scattered in scattering mode or to be guided into the patient directly in scanning mode. It is also limited by safety considerations. This largely depends on the capability of the accelerator and the energy selection system, if it exists. For some cyclotrons, the beam current can be continuous and generally has a high operating current, e.g. 300 nA at the cyclotron exit. However, the fixed energy of the beam must be reduced by the energy selection system to the appropriate value before it is sent to patient and this could reduce the beam current significantly. As a result, the smaller the beam range required, the lower is the dose rate available. For synchrotron-based accelerators, dose rate does not depend on the beam energy as much. However, the beam is not delivered continuously but in 'spills' and the overall peak beam current reaching the nozzle must be adjusted accordingly to allow adequate average beam intensity.

Dose rate is a significant parameter. The amount of dose a patient receives is directly related to the safety of the patient. Too little dose and the disease is not properly treated. Too much dose and healthy tissue could be damaged. The accuracy with which this dose can be delivered is related to the accuracy of its detection and the time for a system to react. The tools provided in this section are related to the accuracy of its detection. The timing will be discussed further in Chapter 12.

The highest dose rate available will be determined by the accelerator and beamline. The highest dose rate allowable will be determined by the parameters of the instrumentation and beam control. A limitation is mainly the typical ~100 μs response time of an IC. Such a response time has been reduced in recent years through optimization of the gap and voltage amplitude. Consideration must be given to recombination effects as well. The speed with which one can detect and turn off the beam will determine how much dose is delivered during that time and will therefore limit the dose rate.

Assume a beam current of 1 nA, which is consistent with a healthy dose rate for scanning. Assume that the entire detection and turn-off time takes 100 μs. Then during the turn-off time period, 0.1 pC or $0.625 \times 10(6)$ protons have been transported to the target. Some electronics are designed to *count* the charge in 10 pC units. Don't forget the amplification of the IC, which increases the signal from the beam current through the chamber by on the order of 200, so the electronics will detect 20 pC or 2 counts. At some institutions, the arbitrary parameter called MU is set to be 300 counts. The number of MU detectable during the turn-off time can determine the maximum beam current allowable. In this case, 0.0067 MU were delivered during turnoff. The minimum MU detectable is 1 count, or 0.0033 MU. In a previous example, 3 MU/cGy was obtained so the minimum detectable dose is 1×10^{-5} Gy. The noise in the system should be a fraction of a count and is sometimes in the neighborhood of $1 \times 10(6)$ protons. This is a basis for a minimum deliverable quantity of dose and should be considered for treatment planning. Regions in the target should not be designed to have less dose than this. The plan can be modified for all such regions to deliver either 0 or the minimum dose and the appropriate rounding will have to be devised so as not to compromise the treatment plan.

11.3.2 SOME PARTICLE BEAM TREATMENT-RELATED CONSIDERATIONS

Charged particle treatment does not usually require a large number of treatment fields for each treatment fraction, due to its ability to minimize dose to normal tissues around the target volume. This allows for the use of only one or two fields for each treatment fraction to deliver the prescription dose, although the whole treatment course may use a larger number of treatment fields to spread the dose and further minimize the dose to normal tissues.

When the total prescription dose for a fraction is delivered by only one or two fields, the uncertainties involved in each field become much more important. With more fields, errors can average out. Charged particle treatment requires highly accurate patient setup. Note that since the treatment beam stops in the patient completely, portal imaging during treatment is not possible. However, work is ongoing to be able to use charged particle imaging with high enough energy charged particles that can pass through the body. If the beam energy can be switched quickly enough, it may be possible to take such charged particle scout images during an irradiation. This method gives a direct measure of stopping power and therefore conversion from X-ray attenuations would not be necessary. The dose to the patient is also very low. Some thought has been given to use a combined delivery of carbon and helium ions during treatment. The range of the helium ions would enable them to pass through the body while the carbon beam was delivering most of the dose. The helium ions could then be used for charged particle imaging online, during treatment.

11.3.3 BEAM DIRECTIONS

Ideally, the treatment beam should take the shortest paths to reach the tumor to minimize the volume of normal tissue on the beam path. The entrance dose of the charged particle beam (the plateau) is nonzero and it is sometimes advantageous to spread it out over the normal tissue around the target volume by using a few beam directions, including those with longer beam paths.

The capability of conformal dose delivery has improved considerably with scanned beams. Already it has been noted that charged particle beams can use fewer fields to conformably fill a target. The use of scanned beams can reduce the proximal dose and enable dose modulation. More will be discussed in Chapter 12. It may be possible to deliver the beams with only a fixed beamline, rather than a gantry system. A charged particle gantry system is large and one of the more expensive parts of a system. Whether or not this is possible depends on the specific treatment site and treatment techniques involved. Treatments of ocular melanoma have always used fixed beamlines. Current prostate treatment uses only lateral beams as mentioned earlier and therefore can be treated by only fixed horizontal beamlines as indeed practiced at some centers.

Figure 11.10 shows the number of fields at each gantry angle for all the patients treated in a 12 month period for one gantry treatment room at the FHBPTC. Of all the fields, 23% used only lateral beams, 41% used just the four cardinal angles (0, 90, 180, 270) and 59% used other gantry angles. Interestingly, the angles were only varied in 5 degree increments.

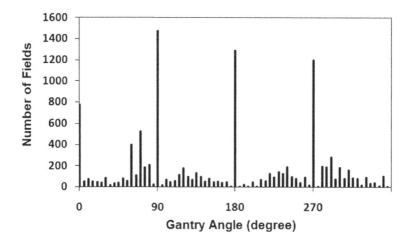

FIGURE 11.10 Gantry angles used in one year.

In principle, the beam orientation relative to the patient provided by all the coplanar beams on the gantry can be achieved equally by a fixed horizontal beamline with the patient lying, sitting in a treatment chair or semi-standing. This would be rotatable around a vertical axis. The fact is that most of the treatment sites require only coplanar beams. Nearly all the noncoplanar beams were used only for head and neck treatments. Treatment sites like nasopharynx and skull base chordoma, were thought to be difficult without a gantry. However, that was before scanned beams became more common. More recent studies show that these could be treated with fixed beam arrangements equally well.

The combination of a chair and standing with a fixed horizontal beam was used in the early days of proton therapy for a broad range of treatment sites, given that it was the only option at the time.

The practice largely stopped as soon as proton gantries became available. With two exceptions it is still used for heavier ion treatments. However, reconsideration of this technique, in general, has appeared recently. While the efforts were mostly aimed at the potential cost reduction of proton therapy equipment, they were also motivated by possibilities offered by new technologies in two main areas. One is patient position control with significant improvements in accuracy and setup speed supported by modern robotics, volumetric imaging, surface imaging, immobilization and monitoring etc. The other is the availability of beam scanning. The power of scanning allows the desired dose distribution to be obtained with fewer beams and more flexibility of beam angles compared to scattered beams. It was found in a recent study that scanned beam plans with a gantry-less system could achieve satisfactory dose distributions for head and neck treatments without the noncoplanar beams that can only be provided by a gantry. See the work of Susu Yan. This is, in fact, not surprising considering the impact of intensity modulation on beam geometry in the photon world. While noncoplanar beams are always needed by 3D conformal treatments for a certain group of treatment sites, IMRT/VMAT can provide equal or improved dose distributions for these sites by coplanar beam directions only. A good example is the Tomotherapy system that simply does not provide any noncoplanar beam directions but can treat all sites, including cranial target volumes. This new effort in re-exploring fixed beamline treatment techniques for proton therapy could come to fruition in the near future.

11.3.4 ADDITIONAL PERSPECTIVE

Optimal treatment partly implies the optimal use of the physical principles of charged particles for adherence to the clinical prescription. Particle therapy has evolved considerably over the past several decades, but the ultimate use of the charged particle has not yet been achieved. Some of the uncertainties that are understood and the margins that are used to compensate have been identified in this chapter. Below is a summary of some of the parameters that will hopefully be improved in the coming years.

- Distal falloff
  - Physical limitation
    - Range straggling $\approx$ 1% of range
      - @ 15 cm = 1.7 mm
  - A typical facility distal falloff used currently – @ 15 cm = 3 mm
  - Clinical margins that are usually added:
    - 3.5% + 1 mm = 6.3 mm @ 15 cm
  - Total = 3 mm + 6.3 mm ~ 1 cm
    - *Compared to an ideal distal penumbra of 1.7 mm*
- Penumbra
  - Physical limitation
    - Multiple scattering $\approx$ 2% of range
      - @ 15 cm = 3.4 mm
  - A typical facility penumbra (Scattering) – @ 15 cm = 6 mm
  - Clinical margins – site dependent ~ 3 mm+
  - Total ~ 1 cm
    - *Compared to an ideal penumbra of ~3 mm*

Although proton therapy has existed for decades, the more widespread use started only in the last 15 years and implementation of the scanning modality started even later. Increased availability of charged particle therapy has enabled the ability to offer treatments for a wider range of clinical sites and improvements in the optimization of beam properties continue. These beams, especially with the scanning modality, are very flexible and improvements in treatment planning with these beams are ongoing. As the numbers of facilities continue to increase, it will perhaps be more economical to consider the use of limited purpose systems instead of one system for all clinical sites. In this way, the clinical and technical compromises can be reduced and costs can be lowered. This might be countered by hospital business considerations. The properties of the beam and the beam delivery subsystem must be matched to the characteristics of the clinical target and understanding how the beam characteristics can be used (or will be used) and modified is an essential part of optimal treatment.

11.4   WHAT COULD GO WRONG – 4?

In this chapter, a somewhat more clinical perspective has been taken. Some of the teachings of my clinical physics guru Dr. Hsiao-Ming Lu have been internalized. There are many sources of errors that are part of the workflow of a radiotherapy department, which are not covered in the scope of this book. However, right from the beginning of the chapter, it is apparent that the wrong definition of the GTV, CTV or any of the XTVs will cause a cascade of potential errors, even if the beam delivery is perfect. Also discussed were some of the beam parameters from the perspective of the clinical sensitivity and margins. This included some definitions of clinical beam parameters. If these parameters are defined differently from one institution to another, there could be differences in the clinical delivery and therefore outcomes, so it is important to maintain communication and accurate protocols. This chapter did introduce some new components, primarily those that measure dose, which should be added to the list of possible errors that can occur.

- The topic of knowing the material, density and range of the beam has been listed before, but its importance is redoubled in the measurement of dose.
- To date, the primary device used to measure dose during treatment is an IC. This has been shown to be sensitive to various parameters.
  - The dimension of the dose pad, for those ICs that have dose pads.
  - The pressure and temperature which will affect the density of the gas in the IC.
  - The energy of the beam will determine the stopping power in the IC.
  - The electronics or processors that read the dose in the IC will have noise and the accuracy of this measurement is only as good as the resolution of these electronics and computer systems.
  - The stability of the beam current will play a role in the accuracy with which the beam dose can be predicted, or whether the beam can be stopped in time before a mistake is made in the dose delivery.
  - Although it has not been discussed, in much the same way that a dose pad works, some ICs are segmented with strips that are used to measure the beam position. The dimension of these strips is important as well as the alignment relative to the beam. (See Chapter 16.)
  - The design of the ICs will determine how well they can measure current, as a function of current. This curve must be known and reproducible. At high currents, IPs that are created can be recombined and the current generated in the IC will result in an incorrect reading of dose.
  - The data rate from an IC can be large, well not large compared to a high-energy physics experiment, but large enough to require some care in the design of the data acquisition system particularly for PBS.

In this chapter, the dose measurement and clinical requirements have been added to the influence of materials in the beam path and magnetic elements.

11.5   EXERCISES

1. IC related
   a. Derive the dose to the active volume of the IC that will create 10 pC of charge.
       i. lowerRoman3. Active volume parameters: Area = 4 cm^2, gap = 1 cm
   b. What if the gap of the IC was 0.5 cm instead of 1 cm?
   c. Derive the number of 200 MeV protons that would be needed to deposit 1 Gy in a full 1 liter volume of water. How many counts would be generated in the IC-sensitive region? (Don't worry about the depth-dose distribution – all of the energy from the single BP goes into the volume.)
   d. If 1 MU = 300 counts, how many MU will be recorded per 1 Gy of 200 MeV protons delivered to the target of 1 kg of water-equivalent material? (Don't worry about the depth-dose distribution – all of the energy from the single BP goes into the volume.) Assume the IC gain is 175.
2. What is the current needed to deliver 2 Gy from 200 MeV protons in 0.5 liter of water in 30 seconds? (Don't worry about the depth-dose distribution – all the single BP energy goes into the volume.)
3. Calculate the data acquisition frequency to handle the counts generated by 500 nA in an IC with an active area of 4 cm^2 and a gap of 0.5 cm. Assume all of the beam is inside the sensitive area.

12 Three-Dimensional Dose Conformation

Effective use of a particle beam for radiotherapy necessitates delivery of that beam in such a way that it covers the target volume with a prescribed dose distribution. Every volumetric element of a target (a voxel) should receive energy from the ions travelling through or up to that voxel as described by the prescription. This requires that the ions are directed to the right place with the correct number of ions at that place. The dimensions of a clinical target are typically different than the dimensions of an unmodified particle beam. The beam extracted and transported from a typical accelerator will have dimensions on the order of millimeters and will have a narrow energy spectrum which results in a narrow spread of ranges (also millimeters) in a target. Therefore, that beam has to be spread out in three dimensions to match the target volume; and, by the way, sometimes to be delivered at the right time, so that's four dimensions. This chapter will discuss aspects of the longitudinal and transverse spreading using scattering and scanning techniques and some of the devices used to implement these techniques.

12.1 LONGITUDINAL BEAM CONFORMANCE

The pristine Bragg curve (BC) is the on-axis depth-dose distribution (dose along the depth direction), in homogeneous material, of an initially quasi monoenergetic proton beam. A BC puts a high percentage of the dose near the end of range of the charged particles in the Bragg peak (BP). Sometimes BC and BP are used interchangeably. By adding multiple pristine BPs of different ranges with the appropriate relative number of ions delivered to the target at any given range, a desired dose at any given depth in the target can be achieved. A spread out BP (SOBP) is a distribution that effectively spreads out the BP into a region of uniform dose in the depth direction.

A carefully measured BC is essential to range spreading design because the shape of the BC will contribute to the depth dose distribution in the patient. The measured shape is determined by, in order of decreasing importance, on

- the energy dependence of the stopping power,
- the transverse size of the beam,
- range straggling,
- initial beam energy spread,
- nuclear interactions,
- low energy contamination,
- effective source distance,
- and the dosimeter used in the measurement (well, this may not be the least important)

The appropriate relative weights can be determined by a linear weighting scheme. Figure 12.1 shows an example of the procedure. If N BCs of different ranges are used, the amplitude of the SOBP at any given depth z_i is determined by the sum of the BCs that contribute at z_i weighted by w_j. In general, the summed dose at any given depth z_i is given by the weighted sums of each of the j BPs used:

$$\text{SOBP}(z_i) = \sum_{j=0}^{N} \text{BP}_j(z_i) \times w_j \tag{12.1}$$

DOI: 10.1201/9781003123880-12

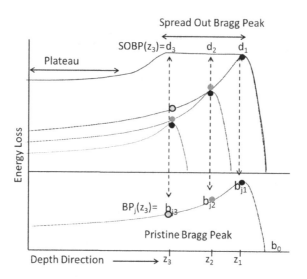

FIGURE 12.1 How to create a spread out Bragg peak.

Figure 12.1 shows the resulting SOBP at three z locations, 1, 2 and 3. In the figure, let $d_i = $ SOBP(z_i) which is the summed dose at z_i, and let b_{ji} equal the contribution of the dose at z_i from BP$_j(z_i)$. For simplicity, if the BPs from depth 1 can be shifted to depth 2 to obtain the depth dose distribution at that range, the shape from just one BP (BP$_1$) can be used. (Let the apostrophe symbol (′) indicate that the BP is shifted.) In the specific example, if the shape of BP$_1$ can be used, j isn't necessary,

$$\text{SOBP}(z_1) = d_1 = \text{BP}_1(z_1) \times w_1 \tag{12.2a}$$

$$\text{SOBP}(z_2) = d_2 = \text{BP}_1(z_1') \times w_2 + \text{BP}_1(z_2) \times w_1 \tag{12.2b}$$

$$\text{SOBP}(z_3) = d_3 = \text{BP}_1(z_1'') \times w_3 + \text{BP}_1(z_2') \times w_2 + \text{BP}_1(z_3) \times w_1 \tag{12.2c}$$

In Equation 12.2b, for example, z_1' means that the amplitude that was at z_1 (BP$_1(z_1) = b_{11}$) is shifted to z_2 to be included in the dose d_2 in addition to the dose from BP$_1$ that is at z_2. The appropriate matrix equations can be solved to achieve the desired weights. For example, for $d_i = 1$, $w_1 = 1/b_1$ and $w_2 = -b_2/b_1^2 + 1/b_1$ are the weights of the BPs to achieve the desired physical dose distribution which can be a uniform region.

Delivering a certain amount of ions of a given energy can be achieved in different ways. Here are two methods:

1. Deliver N1 ions at a range R1. Then change the beam energy extracted from the accelerator and deliver N2 ions at a range R2, and continue until the desired dose distribution is achieved.
2. Deliver N1 ions at a range R1. Then change the material in the beam path keeping constant accelerator energy and deliver N2 ions at a range R2, and continue until the desired dose distribution is achieved.

The distal dose falloff of the deepest peak is a certain dimension. As more peaks are added to build up the uniform region of dose, the distal penumbra can increase because a lower amplitude peak can be added at a depth in the middle of the distal falloff of the deepest peak.

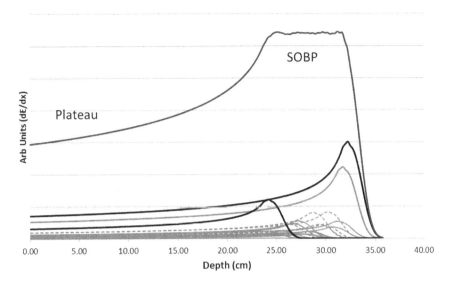

FIGURE 12.2   SOBP using multiple Bragg peaks.

A more semi-realistic example with a larger number of BPs (each BP having a finite energy spread) is shown in Figure 12.2. This is a brute force calculation using the Excel solver and produces a reasonable result. The objective was to optimize the width of a uniform region. Note how the result caused the proximal and distal peaks (black lines) to be emphasized. No attempt was made to minimize the distal penumbra, and there was no thought about deliverability. More meticulous efforts to create a uniform SOBP take these effects into account and may create a different optimized result. Figure 12.3 shows the result with the 12th BP removed. The relative amplitudes of the BPs in Figure 12.2 are shown in the darker gray bars in the bar chart of Figure 12.4. The right side, lighter gray bars shows the distribution for the SOBP in Figure 12.3. The resulting SOBP has the dip shown from the removal of a BP. Also note, in an SOBP, the plateau region is not as flat as that in a pristine BP.

The BP is not a Gaussian and is not quite as magical. If the BP is wide enough, the summation of adjacent BPs can have a smooth result. The width of the BP is related to the energy spread which

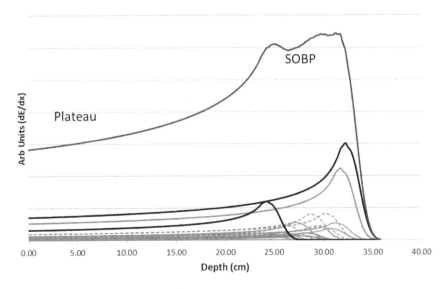

FIGURE 12.3   SOBP with one BP removed.

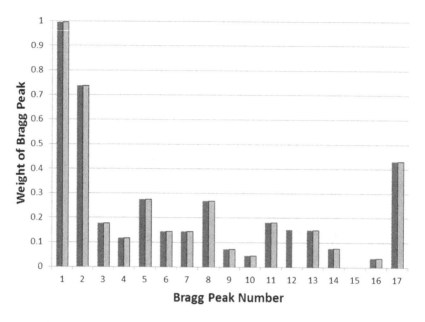

FIGURE 12.4 Relative weights of BPs for an SOBP.

has contributions from various sources as was previously described. At the surface of the target, the distal falloff is related to the energy spread in the beam and any range straggling from interactions in upstream materials. At low energy, the absolute energy spread is smaller (same percentage), and range straggling from a shallower range is also smaller. The result is a narrower range spread and a sharper BP. More such BPs are necessary to create a smooth SOBP because they are so narrow and their sensitivity to the spacing is higher. Alternatively, each of the BPs can be further spread out with additional material to create a wider BP. For deeper targets, first the energy spread in the beam is larger (same percentage of a higher energy), and the range straggling is increased within the patient.

An SOBP exhibits several characteristic features. The shape is conventionally labeled as shown in Figure 12.5. First, it delivers a near-uniform dose distribution in depth along the target volume (flat region). The length of this region is called the modulation width (Mod98). Second, a properly optimized SOBP preserves the sharp distal falloff of the most distal BP. Therefore, the ability of the charged particle beam to spare normal tissue behind the target volume is also preserved. On the proximal side of the target volume, the dose changes gradually (the plateau region or proximal

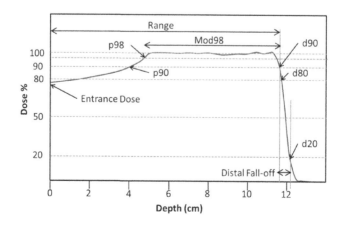

FIGURE 12.5 Parameters of an SOBP.

plateau), i.e. a soft knee. Third, the total entrance dose (at the smallest z position) has increased from about 30% due to the deepest BP to nearly 80% due to the additional shallower peaks (in this particular example). These features are the main clinically relevant properties of the particle beam in the longitudinal direction. In general, the parameters are defined in terms of the positions in depth of the given dose levels, i.e. d20, d80 and d90 at the distal end, with p90 and p98 on the proximal side. The distal penumbra of the SOBP is given by the distance between d80 and d20, corresponding to the 80% and 20% dose levels. The quantity is also called the distal dose falloff. The dose at the entrance is a useful parameter for comparison with ionization chamber (IC) data.

The most clinically relevant parameters in the depth dose direction are the beam range, modulation width of the SOBP and the distal falloff. The beam range is defined as the depth of penetration at d90. Historically, the modulation width was defined as the width of the uniform flat dose plateau in the SOBP at the distance in depth from p90 to d90. Some institutions define the modulation width to be Mod98, as indicated in Figure 12.5, with the following advantages:

1. For SOBP distributions with large modulation width, the magnitude of the upstream (proximal) plateau rises making the proximal 'knee' much less noticeable. As a result, the p90 value becomes overly sensitive to small changes in the proximal BC amplitudes or measurements. The p98 point, on the other hand, is at the steepest part of the knee and is, therefore, more stably defined.
2. For cases where the target volume extends close to the patient body surface, the uniform flat dose of the SOBP must extend close to surface as well to provide full uniform dose coverage. In that case, the p90 point would go outside of the body surface and becomes totally undefined, while the p98 is still valid.
3. Mod98 better reflects the extent of the SOBP high flat dose region with the desired uniformity, to cover the target volume.

12.1.1 DEGRADING METHODS

There are a variety of ways to adjust the beam energy to enable the creation of an SOBP.

12.1.1.1 Ridge Filter

A ridge filter is the only truly passive device for longitudinal beam spreading. This device depends upon an incident uniform transverse distribution. There are multiple ridges across the beam area as shown in Figure 12.6. The labels 1 through 5 indicate different depths of material within one ridge and the thickness at each number is related to the amount of beam energy loss. The beam in

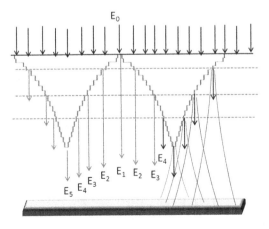 

FIGURE 12.6 How a ridge filter works (left) and photo of ridge filter (right).

Figure 12.6(left) starts at energy E_0. The lowest energy after degrading is E_5, and the closest to E_0 is E_1. The shape of the ridge filter is determined in such a way as to have the appropriate number of ions lose the correct amount of energy, according to the results of the matrix analysis above. In Figure 12.6(left), the slope of the device is less steep at the higher energy section so there is a higher weight in that region. The ridge filter will result in a smoother distribution of BPs than that in Figure 12.4, but it's possible to see how Figure 12.4 could be morphed to the ridge filter shape. The heights of the bars would be translated to width of the steps. The beam is also scattered in this filter (as indicated by the diverging lines in the right of Figure 12.6(left)) so that all the ion energies that entered the ridge filter and those created by the degrading effects of the ridge filter are mixed over the entire field, and, therefore, at any one transverse location, there is a statistical mix of these energies. A photo of a ridge filter is in Figure 12.6(right).

Since this approach is truly passive, all energies are delivered simultaneously, i.e. there is no time dependence. This could be an advantage when considering moving targets and a desired averaging. A disadvantage of this system is that the modulation width of the SOBP is fixed according to the height of the ridge, and each one is designed for a specific energy range. The selection of ranges and SOBP widths is limited by the number of these devices that are fabricated. It is necessary to insert a different one for every class of treatment or have an automated (limited number) library that can insert the required one into the beam path. These days, three-dimensional (3D) printers can manufacture ridge filters.

12.1.1.2 Range Modulator Wheel

The range modulator wheel (RMW) is a dynamic version of longitudinal beam spreading in that something is moving. In any case it is not passive. This wheel has sectors with varying thicknesses. Assuming that the wheel rotates at a constant rate, and that the incident beam flux is constant, the number of ions which are degraded to a given energy will depend upon the amount of time a given thickness of the degrader wheel is in the path of the beam. Thus, in the wheel geometry, the angular extent of various sectors with specific thicknesses will correspond to the appropriate relative weights of the ranges calculated in the matrix formalism noted above. An advantage of this technique, while not passive, is that the wheel revolution frequency can be fairly high. It is typically on the order of ten revolutions per second, and therefore, all the energies that make up the SOBP are delivered in about 0.1 second, which is much faster than typical target motion. One implementation of this, by Ion Beam Applications (IBA), conceived by Miles Wagner, is shown in Figure 12.7. Multiple tracks can be included in one wheel (three

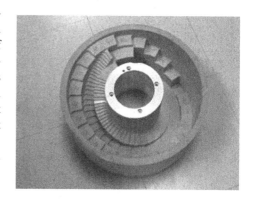

FIGURE 12.7   Multitrack mod wheel.

are shown). By laterally shifting the position of the wheel, a different track can be placed in the path of the beam and be used. This is helpful, in that a wheel is positioned automatically, as needed. However, without any additional techniques, much like the ridge filter, a given RMW is only good for a limited range of incident beam energies and SOBP widths.

12.1.1.3 Beam Current Modulation

In the case of an RMW, the number of SOBPs that can be delivered can be increased if there were another way to adjust the relative weights of the beam. Any subset of a modulation width designed in the wheel can be created if the beam is turned off during the wheel rotation at some proximal range position. Synchronization of the beam turning on and off and the wheel revolution phase is required. This also is not passive.

If it is desired to use a lower distal energy than the RMW was designed to use, the physical sector SOBP weights may not be the distribution that is required. By varying the beam current during the range modulator rotation, the relative weight of the amount of ions delivered at a given range can effectively be changed. An example of beam current modulation is shown in Figure 12.8. Two traces are shown. The upper trace is a signal whose amplitude shows the requested beam current and the lower trace is the signal from an IC showing the delivered beam current. The beam pulse width is less than 70 ms, in this case, which is less than the 100 ms period of the RMW.

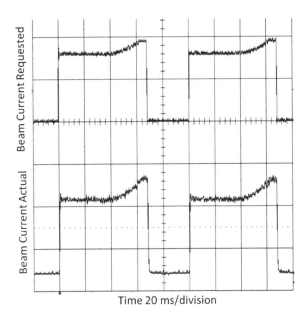

Time 20 ms/division

FIGURE 12.8 Beam current modulation.

The beam is going through different thicknesses of material in an RMW, so it's likely that the energy modulation will result in different scattering divergences for different energies. Therefore, the effective field size can vary or is different for different ranges. In an ideal system, the combined system would result in a constant overall scattering power; thus, the effective ion usage efficiency would be the same over the full SOBP. A combination of materials is used that results in different energy losses but the same scattering divergence (as alluded to in Section 9.7), independent of location on the wheel. This can be done similarly with a binary degrader system.

Much work at the Harvard Cyclotron Laboratory was devoted to an optimized usage of the proton beam, with these scattering systems. Originally, the range modulator used was the most downstream component. In this case, the size of the wheel was large since it was downstream of the beam scattered by the first scatterer. The concept of upstream modulation vs. downstream modulation was developed. Using a compensated upstream scatterer/degrader can result in a smaller device and the most efficient use of ions. (This concept of scatterers soon follows in this chapter.)

12.1.1.4   Other Degrading Methods

Lamination

A different method of delivering the appropriate energy distribution is by putting a certain amount of degrader material into the beam, and letting that material stay in the beam path until the appropriate number of ions have been delivered at that energy. Then a different amount of material is inserted until the appropriate number of ions is delivered etc. This is typically done with a binary degrader system, which includes plastic degraders and lead scatterers in a binary sequence of thicknesses as shown in Figure 12.9. Sometimes these are called lollypops. This method of stacking beam energies is called lamination. This system is not as fast as the ridge filter and RMW for depositing dose over the full set of ranges desired, but it does allow for a high level of flexibility in the depth dose distribution.

Wedge

A wedge degrader system is a variation of the binary lamination scheme. The wedge shown in Figure 12.10 can be continuously moved at a varying speed, so that the amount of time spent through a certain thickness results in the necessary proton fluence at any given energy. Energy changes faster than 0.1 second are possible. A system like this is used at the Paul Scherrer Institute (PSI) to modulate the energy of the cyclotron beam. Also, longer, single-wedge systems are used for other installations by Sumitomo and ProNova. Linear motors are faster than rotary devices.

FIGURE 12.9   Binary degrader system.

FIGURE 12.10   Wedge degrader.

12.1.2   DISCRETE ENERGY CHANGES

The previous depth dose spreading methods are done close to the patient. It is also possible to change the energy of the beam closer to the accelerator, before the beam spreading equipment. In this case, the total dose distribution is constructed by individually delivered BPs, similar to how it's done with the laminations and the wedge but without additional material that scatters the beam. Inaccuracies of the beam energy will create non-ideal distances between adjacent layers in depth and, thus, cause ripples in depth dose that could exceed the tolerance, particularly at the shallower range where the BP is distinctively narrow. The other methods that involve the beam passing through material have the disadvantage and advantage of increasing the distal falloff distance due to range straggling, but the sharp peaks are smoothed out. Increasing the width of the BP in this situation is an advantage. The time it takes to change energies will be the subject of subsequent sections. In the case of discrete energy changes, it might still be necessary to include a range shifter.

12.1.3   RANGE COMPENSATION

Delivering a beam whose range has a fixed longitudinal extent (modulation width) given by the extremes of the target does not provide optimal conformance to the target. The target may not have a fixed depth across the transverse cross section. Also, the actual range to the distal end of the target can be affected by tissue and organs of different densities (e.g. bone) across the target's cross-sectional area. A range compensator is made of material like plastic to mitigate those irregularities as shown in Figure 12.11. It is aligned with the aperture and helps to conform quite precisely to the distal edge across the transverse direction; however, the width of the range modulation is fixed when using some spreading systems, and this contributes sometimes to unwanted proximal dose as shown

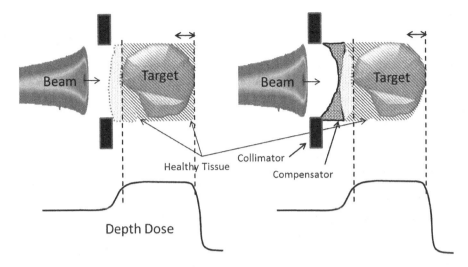

FIGURE 12.11 Range compensator.

by the hatched region on the right side of Figure 12.11. Balancing between the distal and proximal ends is a clinical decision.

Figure 12.11 shows two cases; one without a range compensator and one with. Without the compensator, to the left, the mod width is dictated by the maximum length of the target and the distal edge of the beam field is flat and extends beyond the target above and below the maximum depth. This is to be compared with the figure to the right wherein the range compensator reduces the depth above and below the target, with the result of also shifting the proximal location of the SOBP P98 above and below the target. A range compensator and a patient-specific collimator are shown in Figure 12.12 to the left and right, respectively. The collimator will be discussed in subsequent sections.

FIGURE 12.12 Range compensator (left) and collimator (right).

12.2 TRANSVERSE BEAM CONFORMANCE BY SCATTERING

To spread the beam transversely implies spreading dose over a larger transverse extent than the unmodified Gaussian beam from an accelerator. A beam-scattering system uses the effects of multiple scattering when a beam passes through a material to spread the beam from the unperturbed 'pencil' to a beam size consistent with the target size. In systems that use beam scattering to spread the lateral extent of the beam, the small beam coming to the nozzle is scattered to a larger area, and the scatterers are specially designed such that the beam has a uniform penetration and uniform fluence across the scattered area specified for clinical use. At the same time, the energy of the beam is modulated, as noted in Section 12.1, to spread out the location of the BPs over the target volume in depth. Typically a scattering system is designed in such a way as to very quickly or instantaneously spread the beam range throughout the target. A scattering system is usually configured to produce a homogeneous dose distribution with the same penetration across the beam. For patient treatment, the beam is collimated by an aperture to match the target cross section (see right side of Figure 12.12).

12.2.1 SINGLE SCATTERING

The simplest way to increase the size of the beam is to scatter it by passing the beam through a scattering material. The relative dependencies of scattering and energy loss are different for different materials, so the best material to affect the scattering, without losing too much energy, is lead. An incident Gaussian beam passing through the single scattering system will end up as a Gaussian beam with a larger divergence. After some drift distance, the larger angles result in a larger beam size (tilting the phase space ellipse) since particles with larger angles drift to larger transverse distances (e.g. Figure 10.28). For treatment, it is important to ensure uniform dose coverage across the target. Keeping the overall transverse dose distribution within ±2.5% is a tolerance. If only the top 5% of the Gaussian amplitude is used then this satisfies the requirement. The rest of the beam can be eliminated by collimation. The position on the Gaussian (given here as a function of r), relative to σ, at which the amplitude is 95% is r_{95} is given by:

$$0.95 = e^{-\frac{1}{2}\left(\frac{r_{95}}{\sigma}\right)^2} \tag{12.3}$$

$$r_{95} = \sigma\sqrt{-2(\ln(0.95))} = \sigma \times 0.32 \tag{12.4}$$

This is plotted in Figure 12.13 for a Gaussian with $\sigma = 1$. The upper gray horizontal line is the 95% value, and the lower one is at $r = \sigma$. The vertical dashed line intersects at the 95% amplitude. Integration of the Gaussian over $\pm r_{95}$ results in using about 25% of the incident beam. (This involves using the error function, $\text{erf}(0.32\sigma/\sqrt{2}\sigma)$.) This is not a very efficient use of beam, but it uses a lot more than 5% and it works.

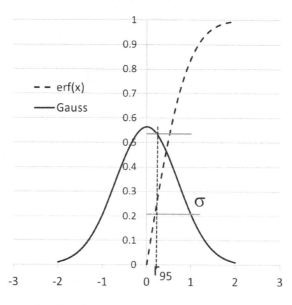

FIGURE 12.13 Single scattering Gaussian.

12.2.2   DOUBLE SCATTERING

Previously used for electron beam therapy, the idea of combining two scattering devices to optimize the percentage of ions in a uniform field area was developed for ion beam delivery, as well. An example of a scheme is shown in Figure 12.14 with each step numbered. In this figure, where beam is represented, the left-right axis is the beam particle number and the up-down axis is a transverse position. So these are particle number distributions. A view in the phase space plane (x,θ), throughout the process, is shown in Figure 12.15. Each phase space diagram has one or two projections in the x plane and/or in the θ plane below and/or to the right of the each plot. The step numbers shown in Figure 12.14 also refer to the plots in Figure 12.15. (1) The incoming beam enters the first scatterer. This may be similar to that in a single scattering system or may actually be the longitudinal spreading device, acting to spread out the range and create a Gaussian beam profile with a larger divergence (e.g. Figure 10.33). (2) This translates to larger width after a short drift. The phase space has also rotated. (3) The second element of this system provides a differential scattering

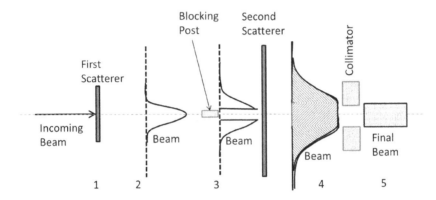

FIGURE 12.14   Double scattering system components example.

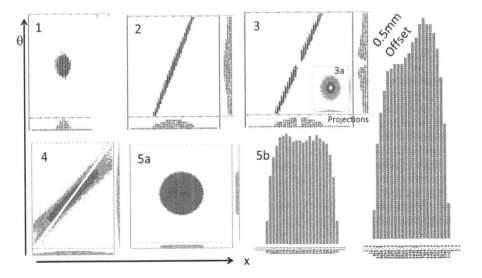

FIGURE 12.15   Double scattering phase space example.

power along the transverse dimension. One extreme such example of this is a post (can't get much more differential than that) in the center of the beam which occludes a portion of the beam and produces a hole. The xy phase space of the beam looks like the inset in plot 3a of Figure 12.15. The $x\theta$ phase space shows the occlusion in the 2D plot and the projections. (4) After a second scatterer and more drift, the angular spread in each half (in the $x\theta$ plane) of the remaining beam grows and then the beam phase space rotates further. The occlusion has also rotated in phase space (see the rotated gap in Figure 12.15 (4)); however, the projection in x, at that location, does not show any sign of the occlusion. No physical measurement can detect this, even though it is visible in phase space. Only the positions of the particles are measurable and, therefore, for the treatment, only the spatial xy plane plot of 5a is relevant. The resulting projection in 5b has a uniform region, and the circle of 5a also looks uniform with uniform projections. More modern techniques use a contoured scatterer system together with a scattering power compensated first scatterer, developed by Bernie Gottschalk et al.

The double scattering system is sensitive to beam steering and a slight change in the beam position relative to the blocking post can affect the field flatness and symmetry. If the beam is misaligned by 0.5 mm with respect to the blocking post, the profile of 5b becomes skewed as shown in the rightmost plot of Figure 12.15. This requires an effective monitoring system and also frequent quality assurance verifications by independent means. More details of these sensitivities are discussed in Chapter 16.

12.2.2.1 Beam Properties

A uniform transverse dose distribution can be obtained. Some important features of the transverse dose distribution are shown in Figure 12.16. In the lateral direction, the most clinically relevant parameters include the field size; defined at the 50% level and the lateral penumbra or the lateral falloff at either end; both 80%–20% and the 95%–50% values are used for a similar specification. While the former is used traditionally to reflect the general quality of the lateral penumbra, the latter is particularly needed for determining the margins of an aperture for a given treatment beam.

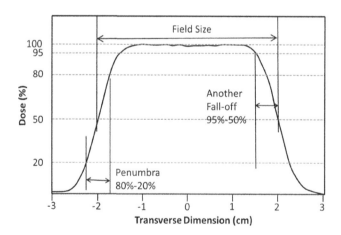

FIGURE 12.16   Transverse double scattered beam parameters.

12.2.2.2 Patient-Specific Hardware

The scattering system, thus far discussed, has created a transverse distribution to roughly match the projection of the patient target. More can be done to better conform that distribution. The target is generally irregularly shaped. The scattered beam is circularly symmetric and has a size dependent upon the scattering power of the scattering system. To improve the conformation, an aperture is interposed between the scattered beam and the target. This aperture can be any material thick enough to stop the transport of ions before they can enter the patient. Typically, this aperture is brass. The aperture is machined with an opening that conforms to the transverse projection of the target in the direction of the beam field as shown on the right side of Figure 11.5. Only ions that will enter the target are allowed to pass. The aperture is designed specifically for the patient. An example of one is shown in Figure 12.12. This aperture, also known as a collimator, absorbs the ion beam outside of the treatment field. However, it also contributes to a sort of secondary lateral penumbra.

A scattered beam appears to come from a finite-sized effective source. Figure 9.17 showed how a scattered beam can be back projected to approximate the effective source position (but that example started with a zero-sized beam). A larger scattered beam will have a distribution of back projections, resulting in a finite effective source size. Due to physical constraints of the collimator mounting hardware and the patient surface outline, there is usually space between the collimator and the patient skin. This is shown geometrically in Figure 12.17(left), as well as a different view via the phase space representation of the beam in Figure 12.17(right).

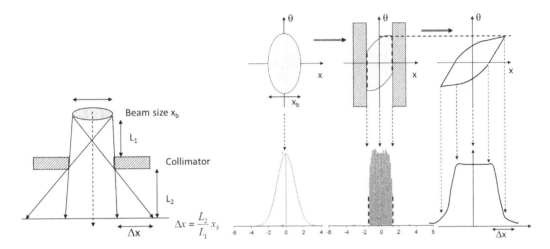

FIGURE 12.17   Collimation components (left) and phase space representation of collimation (right).

Since the effective scattered beam source has a finite size and the beam particles have a distribution of angles, particles from any part of the source can pass through the collimator with a range of angles (remember Figure 10.31). For example, in Figure 12.17(left), a particle on the extreme left of the source, with an angle to the right can end up on the right side of the target along with a particle starting from the right side of the source that has a smaller angle. Thus, beam particles have reached the target outside the limits of the collimator. The number of those particles depends on the size of the source and the angular distribution in the beam. The angular distribution in the beam is Gaussian, so there are fewer particles with larger angles and larger offsets than particles with smaller angles and offsets. Therefore, there are fewer particles at the target on the extreme right, beyond the limits of the collimator. This reduction in the number of particles from the edge of the collimator to beyond the collimator (in the distance Δx) is the lateral falloff. This distance is given by

$$\Delta x = \frac{L_2}{L_1} x_b \tag{12.5}$$

where x_b is the beam size and the distances L_1 and L_2 are defined in Figure 12.17(left). The smaller the source size and the closer the collimator is to the target, the smaller the penumbra.

Figure 12.17(right) shows the initial elliptical beam phase space area (distribution in position and angle for one dimension) on the left, the collimator effect in the middle, and the image on the right is after a finite drift distance during which the ellipse rotated. The collimator cuts off the edges of the beam (from Figure 12.15) and creates a sharp edge (see the projections in the lower half of the figure). However, the cartoon on the right-hand side indicates that after an additional drift distance, and owing to the angles in the beam, the cut-off beam, which started out to have a sharp edge in x at the collimator, grows to a triangle shape. Therefore, looking at the profile of the projection along the transverse dimension after the collimator, as shown in the lower graphs, a flat central distribution is retained. However, as the triangular shape pinches at the ends, there is a reduction of particle number as a function of distance and this becomes the penumbra.

12.2.2.3 Scattering System Components

The components of such a double-scattering system might look like those sketched in Figure 12.18. A combined range modulator and first scatterer are followed by a contour second scatterer. The scatterers and the modulator determine the source size, and they are positioned upstream, as far as possible from the aperture, resulting in a much longer source-to-axis distance SAD (e.g. >200 cm). The collimator should be as close to the patient as possible to reduce the effect of source size. This also reduces the air gap that further increases the penumbra.

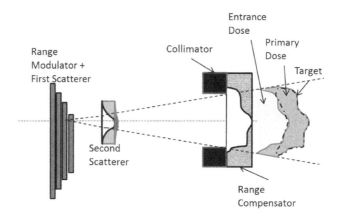

FIGURE 12.18   Double scattering system components.

12.3 TRANSVERSE BEAM CONFORMANCE BY SCANNING

Another method to spread the transverse distribution of the beam is called scanning. Beam scanning, in theory, is quite a general technique, and it has acquired many acronyms such as PBS (pencil beam scanning), IMPT (intensity modulated particle therapy) and SS (spot scanning), among others. The author prefers to simply use particle beam scanning (PBS). A non-scattered and tightly focused beam is sometimes called a 'pencil beam', although sometimes this term is also used for a beam that has a dimension on the order of a few millimeters or less. It is also possible to obtain an unmodified (unscattered and uncollimated) beam which is on the order of several millimeters or even a centimeter, in which case, the author uses the term 'crayon beam', owing to the larger size. In any case, it is more important to define the terms and understand the regime being considered rather than rely on unclear acronyms which through misunderstanding of the technology could minimize the perception of the power and generality of this beam spreading approach.

This is not a new technique; its advantages were recognized early, having first been implemented in Japan in 1980 using a novel system including an RMW while scanning the beam with magnetic dipoles. It was developed in 1989 at Lawrence Berkeley Laboratory (LBL) and used for routine clinical treatments at the PSI and the Gesellschaft für Schwerionenforschung (GSI) in the 1990s. However, widespread adoption has only begun in the past decade. Implementation outside the research environment began with commercial and academic hospital collaborations at MD Anderson and Massachusetts General Hospital in 2008. Currently, many particle therapy facilities are using beam scanning as the primary beam delivery modality.

12.3.1 SCANNING METHODS

In the transverse dimension, there are a variety of ways of moving the beam across the target. Some of these methods include:

- Scanning by mechanical motions
  - Physically moving the patient with respect to a fixed beam position
  - Mechanically moving a bending magnet to change the position/angle of the transported beam
  - Using an adjustable collimator to effectively adjust the location/edges of the beam with moving apertures
- Scanning by magnetic field variation to bend the beam trajectory
  - Scanning an unmodified beam (pencil or crayon beam)
  - Scanning a slightly scattered beam, so that the beam scanned on the target is a larger dimension. This is called 'wobbling' or a version of this can be called uniform scanning (US).
- Combinations of the above
  - One-dimensional 'ribbon' scanning of a beam wide in the dimension perpendicular to the direction of motion; the wide beam extent is adjusted by a variable collimator
  - Scanning the beam magnetically in one dimension and moving the target mechanically in the other dimension
- Other combinations are possible

12.3.2 GENERAL DESCRIPTION OF SCANNING

Beam scanning can be defined as the act of moving a charged particle beam of particular properties and perhaps changing one or more of the properties of that beam. They are all adjusted in such a way as to deposit the appropriate dose at the correct location and time and minimize the dose outside of the target. Physical equipment in the system is used to control these properties.

Pictorially, Figure 12.19 describes the scanning process. It is important to understand that true beam scanning could involve the variation of many parameters of the beam while it is being scanned. A beam at position A can be characterized by a variety of parameters including the transverse coordinates (x_A, y_A) its energy E_A, which determines its depth (the longitudinal dimension), the beam current I_A, the beam size (σ_A which may be different in x and y), the time it stays in a given location, t_A and others. The beam deposits a dose D_A in the voxel around A. After that

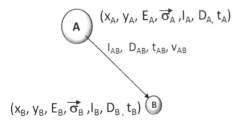

FIGURE 12.19 Scanning process parameters.

dose is deposited, having stayed at A for a time t_A, the beam is moved to location B. The time it takes to move from location A to location B is t_{AB}. The beam current during that movement is I_{AB} which could be a function of position (or time). The average velocity that the beam moves from position A to B (in the x direction) is $v_{xAB} = (x_B - x_A)/t_{AB}$, but this could vary as a function of time and/or position, and the average current change rate between A and B is $dI/dt = (I_B - I_A)/t_{AB}$ (which could be a time dependent function also). The charge delivered (related to the dose absorbed) is determined by the integral of the beam current and the time spent in a given location. In this way, all the terms that are necessary in the delivery of beam scanning have been defined. No assumptions are made about what happens when and for how long, yet.

The above description appears intrinsically discrete or digital. There are usually, two interpretations of this description related to one extreme or the other. This has unfortunately caused confusion in the specification of the system and in the terminology. In the limit when $t_A = 0$, the beam does not stop at a particular location, and its motion is characterized by $\mathbf{v}_{AB}$. Also, in such a case, the concept of I_A is undefined, but rather the quantity I_{AB} is defined. This extreme has been called 'continuous or raster or line scanning'. It is equivalent to the beam motion used in the old CRT (cathode ray tube) televisions. (So beam scanning has a long commercial history.) In the case when this extreme limit is not valid, it has been called 'spot scanning' or even sometimes a form of 'raster scanning' (different than the previously identified continuous raster scanning). However, the distinction between these extremes in the case when t_A is sufficiently small and/or $\mathbf{x}_B - \mathbf{x}_A \ll \boldsymbol{\sigma}_A$ (where these terms are vectors) is not relevant. Whether or not the $t_A \to 0$ limit is reached, it is an implementation decision or based upon physically realizable quantities. Thus, the level of discreteness vs. continuousness will all lead to the measurements and controls to be used. These may all be determined by the capabilities of the hardware and software and requirements of the clinical treatment.

In the case when the beam fully stops at a given point, it may be necessary to measure D_A, if the integrated current is not very predictable; however, even when the beam does not 'stop' at a given point, it is necessary to measure some form of the quantity D_A at places on a 3D grid in the target volume to compare with the prescribed dose distribution.

It is desired to put dose where the target is and not put dose outside the target. It is also desired to conform as closely as possible to the prescription both inside and outside the target. In the extreme, the dose would be nonzero, and equal to the prescription at the edge of the target, and then zero an infinitesimal distance outside the target. This would be possible only in the extreme limit of an infinitesimally small beam sigma, $\boldsymbol{\sigma}_A \to 0$, and an infinitely fast beam current change, $dI/dt = \infty$. The change in beam current over a distance of a moving beam, dI/dx, can be related to the gradient which is essentially dD(dose)$/dx$(position) across the beam. When this is not infinite, optimization is done to obtain the best distribution possible. Of course, clinically, sometimes a sharp gradient may not be warranted.

12.3.3 Technical Scanning Delivery Techniques

12.3.3.1 Time- or Dose-Driven

The control of these quantities, whether controlled by open loop, or closed loop techniques is an important subject. In an ideal world, all quantities would be perfect and no control is needed. In another idealistic world, but not ideal, all parameters would be controlled in a closed loop mode and verified with very fast response time. In the material world, wherein the measurements made require finite periods of time and money, open loop delivery and subsequent correction or feed forward processes are used. In the case when $t_A \neq 0$, this just means that the dose is delivered in quantized doses, and there might be a correction for the next quantum if the previous quantum was not as expected. This allows the beam delivery to be possible even with a fluctuating beam current (assuming its maximum value is limited). The beam is stopped (or moved) when the dose is reached independent of its time dependence. However, if $t_A = 0$, the position of the beam at any given small time interval is not known, (similar to the Heisenberg uncertainty principle ☺), owing to the time it takes to make the beam measurement, and therefore, an analysis and then a dose distribution correction after the fact may be necessary. Of course, this assumes that any fluctuations are of a magnitude that does not lead to an unsafe dose delivery. In any case, the time scale will have to be identified for measurement and control of each parameter and/or device. The 'control' strategy to obtain the appropriate dose distribution also has to be defined.

Various methods of scanning the beam are possible. Sometimes the style of the delivery is clinically based, and sometimes it is a style that industry would like to market. In the case that the beam is magnetically scanned transversely across the target, this process has different options. It is convenient to divide the options into two categories.

1. How the beam current and total charge delivered are controlled
2. How the beam is moved across the target

The dose deposited in any given region can be controlled either by measuring the dose as it is delivered and controlling the beam accordingly (turning it on or off, or moving it faster or slower [see next section], changing the beam current), or it can be controlled by a timer, assuming that the beam current and transverse velocity is appropriately controlled. The former can be called *dose-driven* and the latter *time-driven*. This is represented in Figure 12.20. There are three traces in these figures. The top trace shows the beam current as a function of time, the middle trace shows the dose accumulated during that time and the third trace shows the beam position. Dose-driven refers to making decisions based on dose. For example, positioning the beam at a given location and waiting while the beam deposits charge/energy into the target, counting the dose deposited and ending the irradiation of that voxel when the required dose is reached. In this way, up to a point, the time dependence of the fluence may not matter. Of course if the fluctuations are too large and unpredictable, and recognizing that it takes time to measure and react to the dose, it would be necessary to limit the beam so that at the highest peak of the fluctuations the charge being deposited is consistent with the ability of the system to stop the beam in time if the dose exceeds a threshold. This could have the effect of significantly lowering the dose rate. Note how the first and second pulses in Figure 12.20 are well controlled and have a higher and lower current resulting in different beam-on times. The pulse width of the first pulse is shorter compared to the second pulse for the desired dose to be reached. A less controlled pulse is shown in the third pulse and the integrated dose is not smooth, but the integrated dose is correct.

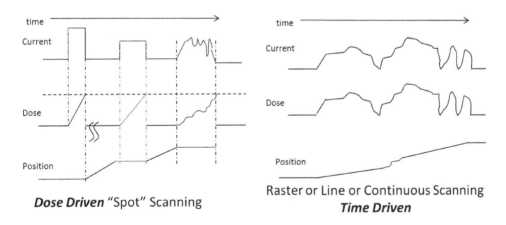

Dose Driven "Spot" Scanning

Raster or Line or Continuous Scanning
Time Driven

FIGURE 12.20 Dose- and time-driven scanning.

The delivery of dose can be discrete or continuous. In the case of a discrete delivery, such as 'spot scanning' the beam is off when the scanning magnet fields are adjusted to position the beam to a desired location. Between spots, the magnetic fields are adjusted as fast as possible by applying the maximum power supply voltage to the deflection magnet. When the magnetic field has settled, the beam can be turned on. Most of the implementations use the dose-driven mode, as depicted in the left side of Figure 12.20, but some are starting to use a time-driven, with corrections, mode.

Time-driven refers to making decisions based on time. One method of time-driven SS could have the 'pulses' of current on for a set time interval, not depending on the dose collected. The dose per voxel is controlled by the amount of time the beam spends at any given location, moving or not. This works if the beam current is well controlled, e.g. held constant. The case when the beam is not stopped (moving continuously) can be approximated by a series of small moves and can be implemented both in the time and dose-driven modes if the dose readout is fast enough. The dose trace on the right of Figure 12.20 shows how it tracks the current with time. The lower trace shows that the beam is moving continuously and therefore the dose at any location depends on the current and speed at that time. One important aspect of the choice between these two implementations is that of the ability of the accelerator to deliver a well-controlled intensity beam, and whether the beam is continuous or pulsed. This will be addressed in some detail in subsequent sections.

An intermediate method is used whereby the beam is spot scanned but is not turned off between spots. For example, the dose-driven SS technique was employed in Gantry 1 at the Paul Scherer Institute. A variation of this in which the beam is not turned off between spots was used at the Gesellschaft fur Schweisstechnik, GSI Helmholtzzentrum für Schwerionenforschung und ist used by Varian. Continuous scanning, in recent years, has been developed both at research institutions such as PSI and NIRS as well as commercially for clinical use by Sumitomo cyclotron-based machines. There is the general feeling that the main advantage of a continuous raster technique would be that of speed, not necessarily speed in delivering the entire dose in a field, but speed in covering a dose layer with a fraction of the dose and the possibility to repeat that several times. This could be an advantage to mitigate organ motion, but the details of the timing are not really as favorable as may be thought at first glance. For example, if it were too fast, it would not average out the organ motion aspect. Some timing situations can be advantageous. The dose delivered could be monitored and, if not as desired, could be corrected on a subsequent scan, if there are multiple scans.

12.3.3.2 Variation of Speed and/or Current

The continuous scanning method can be characterized by a continuous motion of the beam across the target. Figure 12.21 contains two graphs in which the upper curve is dose rate (not dose) and the lower curve is the scan speed as a function of a transverse direction. The one on the left represents a constant velocity scan and the one on the right shows a constant dose rate scan. The dose to the target as a function of position is the quotient of the upper and lower curves. Both deliver the same modulated dose to the target. It is instructive to compare these two cases.

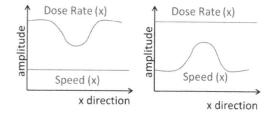

FIGURE 12.21   Dose rate and speed parameters.

In the case of *constant speed*, one can examine several scenarios. Assume the time to deliver this scan is, say, $t1$ with a total integrated dose of $D1$. If the speed is increased by a factor of 2, but not the dose rate, since the time it now takes to move across the line is reduced by 1/2, the dose delivered is also 1/2 or $D1/2$. Therefore, one would have to rescan twice and the overall time is the same if there is no overhead time to prepare each scan. The scan speed was increased without changing the dose rate and that didn't gain anything – unless it's advantageous for other reasons, to rescan (e.g. to make corrections), but if that scan is fast compared to any human motion, again, it's not helpful for organ motion without tracking. If the dose rate is doubled, the scan speed must be increased by a factor of 2, or the dose delivered during time $t1$ would be $2 \times D1$, twice the required dose. In this way, the time to scan the beam is $t1/2$, reduced from the original time by 1/2. Increasing the scan speed without increasing the dose rate does not gain any time. However, increasing the dose rate requires instrumentation that can safely detect the beam properties with commensurate speed. In all cases, the dose delivered in the center of the scan is less than the dose at the edges as required by the graph.

In the case of constant dose rate (the right side of Figure 12.21), one can examine similar scenarios. A goal, implied above, is that the dose in the center of the scan is reduced compared to that at the ends – this is dose modulation. Using a constant scan speed of 1 is not good, since, this would not result in reducing the dose in the center of the scan. Therefore, the speed has to be increased in the center of the scan, so that less time will be spent in the center (with the same dose rate). If the ratio of maximum to minimum dose required during the scan is a factor of 2, one solution to consider would be to start the speed at 1 and rise to 2 in the middle of the scan. If each segment has been approximated to be 1/3 of the length, this will take approximately 5/6 times the original time. The time can be further reduced by increasing the dose rate. While it may seem trivial, a dose rate limitation is the main issue. Given a dose rate, faster speed requires repainting and does not reduce time. However, a higher dose rate, if safe, can reduce the time if there is the capability of a higher scan speed and detection speed to use it.

Assuming the motion is similar (same average speed) in both situations, it is clear that the shortest time to deliver the dose is by a method that can deliver the beam with a maximum dose rate. In fact, the velocity of the scan can be used to control the dose deposited during continuous scanning, instead of modulating the beam current. This could allow the use of a maximum dose rate. Alternatively, a spot by spot approach can also allow the maximum dose rate. There are different limitations to each of these approaches.

This brings up one of the safety issues. In the process of radiotherapy, an overdose cannot be proactively prevented without predicting the future. An overdose is stopped by reacting in sufficient time to stop it from becoming clinically relevant. Therefore, depending upon the system parameters, including dose readout time, and beam control time, the dose rate is limited by instrumentation time constants. This will be reviewed in more detail in subsequent sections. Recently systems with exceptionally fast scanning speeds have been developed.

12.3.3.3 Dimensional Priority

Another choice for the beam delivery technology involves the direction that the beam is scanned. There are three physical directions, two are transverse and one is longitudinal, in the direction of the depth dose. There are choices to the direction the beam can be moved. Two cases are shown in Figure 12.22. The beam is coming from the left side of the figure. The upper situation is one in which the beam is moved transversely in a given fixed depth. This is called a layer. After a layer is 'painted', the beam range is changed and the next range layer is painted. Alternatively, the lower figure shows the situation in which the beam is positioned within a plane, including the longitudinal direction and the beam is scanned both transversely within the plane and also the range is changed to keep within that plane. Each layer (horizontal in the figure) is painted and then the next horizontal layer is addressed. Perhaps the most significant issue is the time difference in the application of these two techniques which will be discussed below.

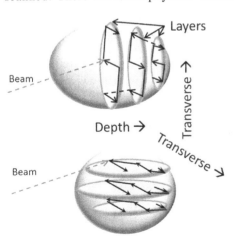

FIGURE 12.22 Directional choices.

12.3.4 CLINICAL DELIVERY STYLES

Three main categories of dose delivery that have been used in beam scanning include US, single field uniform dose (SFUD) and multi-field uniform dose (MFUD).

12.3.4.1   Uniform Scanning

US employs a fixed scanning pattern with constant beam current for each layer. The relative dose of the layers is set to produce a flat dose plateau longitudinally (an SOBP) for a homogeneous medium. It also generally uses a larger, lightly scattered beam and uses an aperture for beam collimation and a range compensator for distal conformity, just as for beams produced by scattering. However, the size of the lightly scattered beam is smaller than the aperture, and therefore, a smaller portion of the beam is collimated and can result in lower secondary radiation when compared to a scattered beam field. The dose distributions produced by US are largely the same as those by double scattering, except that the maximum field size is no longer limited by the scattering system. Because of this, the beam is treated in the same manner as scattering in treatment planning, as well as in delivery. That is, the beam is specified by the range and modulation width for the overall dose distribution rather than the energy and amplitude of each individual BP. Also use of collimators and compensators are standard.

12.3.4.2   Single Field Uniform Dose

The scanning pattern is customized when SFUD is used for a treatment field, but the resultant dose distribution at the end of each field is still uniform over the target volume.

As seen in Figure 12.23, the SFUD technique has interesting consequences. Assume each depth dose layer is painted individually as in Figure 12.22. Imagine a target distal edge has the concave (bullet) shape shown in the left side of the figure. It is desired to conform the dose as much as possible to the target; in other words, to not paint it with a large uniform brush as in the US and scattering techniques. The first layer will paint the smaller transverse volume at the distal end of the target. The transverse dose distribution is shown as the dotted single square distribution at the right of the figure in the

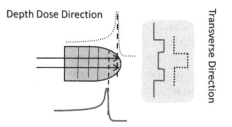

FIGURE 12.23   Single field uniform dose.

shaded rectangle. This has been painted with the dotted BC shown above the target (deeper than the one below). This BC has deposited some dose on its way to the distal end. Therefore, the next layer to be painted must take this dose into account and the transverse dose distribution will be given by the solid line in the right side of the figure, which has a drop in dose in the middle region where the previous deeper BC was (now it is seen how Figure 12.21 can be useful). Depending upon the shape of the target, this can have interesting patterns.

In general, one has the central dose at the distal end as shown in the left side of Figure 12.24 and 'islands' of dose around this in shallower layers as shown in the right side of Figure 12.24. Note the darker, reduced dose region in the center.

A lesson is that the SFUD dose delivery technique requires dose modulation across the target, even for simple geometries. Therefore, dose modulation is a required capability for any scanning system and is not reserved only for what is sometimes thought to be the more complicated technique of multi field delivery.

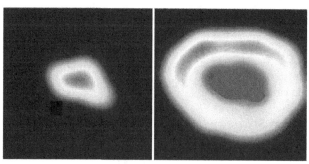

FIGURE 12.24   Dose pattern at different layers for SFUD scanning.

SFUD might be considered another analog of double scattering in the sense that a uniform field is delivered, but is not required. Figure 12.25 indicates the point, but also shows a disadvantage. Two fields are shown one from the lower right and one from the lower left. The transverse dose distributions are shown from each field direction in the hatched rectangles whose height indicates the dose level as a function of position. They are flat and the SOBP's flat portion is inside the target. Each single field produces their own uniform dose distribution in the target. The dashed lines outline the transverse edges of the two fields. But the black circle is a healthy organ at risk (OAR) and in this mode it is fully irradiated. It is beneficial to not irradiation that organ. It would be difficult in a SFUD scenario to reduce the dose to the OAR without additional complicated field configurations.

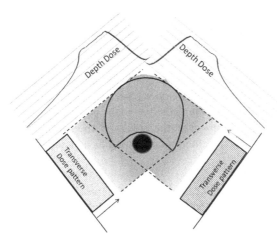

FIGURE 12.25   Two field SFUD example (only one is needed for a single uniform field).

12.3.4.3 Multi Field Delivery

It is possible to get closer to that goal by adjusting the transverse dose pattern as shown in Figure 12.26. The dose pattern is now modified so that the dose is lower in the direction of the OAR. The grayscale image, when compared with the previous figure shows a bit of lighter gray surrounding the OAR. However, owing to the dose pattern, each field does not fully cover the target with uniform dose. Assume the right pattern is delivered first, then there is already some dose deposited in the target. The second field, the left pattern amplitude is lower than the right pattern to account for that already delivered. However, the total of the two fields delivers a uniform dose

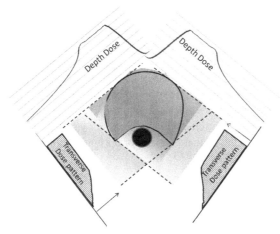

FIGURE 12.26   Two field MFUD example.

in the target while lowering the dose to the OAR. Thus, it took multiple fields to deliver a uniform dose and is called MFUD in keeping with SFUD. It is also sometimes called MFO, for multi-field optimization, explaining the need to optimize the multiple fields together to achieve the desired result. It is also called IMPT, using the photon dose therapy analogy of IMRT. However, it is not IMRT. MFUD/MFO does not necessarily require intensity modulation, but is applying dose appropriately across the target. It includes the conformation properties of charged particles with little or no dose beyond the end of range and requires fewer fields to cover the target volume. The low dose beyond the distal edge and the lower dose from the SOBP plateau combined with the transverse edge shaping possible with scanning, result in a highly conformal dose with much less integral dose (the dose outside the target integrated within the body). With beam scanning, abutting multiple fields is simplified and a tapered dose distribution across the 'match line' combines multiple beams nicely. It eliminates hot/cold spots caused by geometric uncertainties from matching with a scattered beam.

For both SFUD and MFUD, the treatment planning system considers each pencil beam BC explicitly, rather than their combinations as for US or scattering, thus being consistent with the beam parameter characterizations previously defined. The concept of modulation width can become irrelevant. The specification of a treatment beam is basically a list of BPs, each with the range of the particle, the lateral location of the beam projected on to the target plane, and the number of particles (in units of giga-protons (10^9)). In this case, the quality of the beam is determined largely by the quality of the individual pencil beams, their relative weights and how they are combined.

12.3.4.4 Distal Edge Tracking

Another approach for beam delivery is called distal edge tracking (DET). It is interesting and perhaps non-intuitive to learn that simply pointing BPs toward the edge of the target from a variety of angles (e.g. every 30 degrees, or less depending on the case) as in the left side of Figure 12.27, results in a highly uniform target dose as shown on the right side.

FIGURE 12.27   Distal edge tracking.

This method relies upon the accurate placement of BPs. Clinically one of the topics of present research is the accurate determination of the end of range. While the range can very accurately be predicted in homogeneous water, it is not the case in human anatomy. Also, currently particle therapy systems take time to move from beam angle to another beam angle. In addition, charged particle beams are known for the fact that fewer angles are required to cover a target volume in more conventional modes of delivery. For these reasons, this mode has not yet been widely adopted. Rotating the patient can actually be quicker if the patient is seated or upright and this possibility is being explored.

12.3.5 Beam Motion (Not Patient Motion) Effects

It is useful to note that the dose delivered to the target can be affected by motion. This includes the motion of the beam and the motion of the target. Of course the two interact with each other. The clinical aspects of organ motion and the 'interplay' effect between them are well covered in the literature and not discussed in this book. Other aspects will be explored here. In the case of dose-driven SS, the dose delivered is what one would expect from a static beam. Indeed, in the extreme example, the beam is only turned on when it is pointed at the correct location. If, however, the beam is not turned off before moving to the next spot, the dose delivered during the beam motion must be accounted for. For example, the beam can be moved before the full dose is delivered expecting the remainder as it is moving. The final dose will depend on the stability and predictability of the current delivered.

In the other extreme example, wherein the beam is continuously moving there are a variety of effects to consider. One may desire to change the dose from one location to another and this change can be as extreme as turning it off. If the beam isn't turned off (which is possible, but that also takes time), the distance the beam travels while the beam current is changing is important. The left side of Figure 12.28 contains several Gaussians, each one displaced a given distance (as time evolves and the beam moves to the right) and each one with a smaller amplitude as the beam current is being reduced.

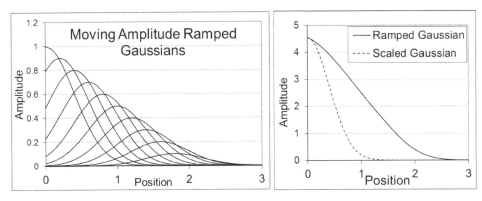

FIGURE 12.28   Moving Gaussian with ramping amplitude.

Take, for example a moderate scanning dipole with an effective 30 Hz frequency that sweeps over 30 cm. This results in an effective sweep speed of 18 m/s. To control the desired current change before the beam moves 1 mm, this requires that change to be done within 55 μs. As the beam is moving, it is effectively integrating at each position. The right side of Figure 12.28 shows an example of the effective increase in beam size owing to the motion. At 20 m/s, with a 3 mm sigma, beam off in 500 μs produces an effective penumbra growth, from the undisturbed Gaussian of the dashed curve to the ramped beam (solid curve), of 70% (from 3 to 5 mm sigma). The effect is only 10% if starting with a beam that has a 1 cm sigma, since the absolute distance traveled does not change and makes a relatively smaller effect on the initially larger beam.

In the case of SS, the beam on-time is on the order of milliseconds, and this is a very short time compared to human organ motion. In principle, given the ability to modify the beam parameters for each spot, it should be possible to track the motion in the transverse direction. However, a difficulty arises if the target density and/or organs in the beam path change. This brings up the question whether the beam energy should be changed quickly to compensate for a different effective range, and whether the accelerator and beam line system can be adjusted to change quickly enough. Thus, the treatment planning process is more complicated. The subject of this book is technology and safety, so only those aspects relative to motion and timing are discussed.

12.3.6   SCANNING IRRADIATION TIME

The events that take place during a scanning sequence will determine the time required to deliver the treatment plan and, depending upon the motion effects, will actually determine the dose distribution delivered. How a treatment map is delivered might have nothing to do with the treatment plan, other than the final dose distribution, but rather with the system implementation. For example, it could be faster to move in one direction compared to another. It is helpful to consider, within safety limits, how to best deliver an efficiently scanned beam. A treatment plan is a 3D map (it will probably grow into four-dimensional maps as adaptive therapy evolves). Many of the contributions to the scanning time can be listed. As noted earlier, it is useful to separate the timing information into that required to deposit the dose, the time needed by the scanning magnets to move the beam between adjacent locations and the time taken by other equipment such as the dosimetry system. Three main categories are in the bulleted list below and these will be referred to in subsequent sections by the italicized *xxxTime* words.

- Beam control (*doseTime*)
  - Time to deposit dose in a location (dose rate)
  - Time to change the beam intensity or to turn it on and off.
  - Safety considerations that might affect the dose delivery time.
    - As discussed earlier with respect to the dose rate – the faster the better, but for safety reasons the maximum dose rate will be inversely proportional the time to measure and stop the beam.
- Beam positioning (*magnetTime*)
  - Time to move from one location to another in the transverse plane
  - The time to change the magnetic field is a balance between speed and accuracy. Some practical limitations such as the power supply voltage available come into play. Also important is the time to detect that the desired magnet current has been reached; this can sometimes be longer than the time to move the beam. Figure 12.29 is an example of a 5 ms settling time of a scanning magnet current (measured by voltage).
  - Time to change the beam energy
- Instrumentation (*equipTime*)
  - As in all the above contributions, instrumentation plays a crucial role. For example, the time to measure the dose is determined by a number of factors including the ion drift time in the ICs and the speed of the electronics readout. The graph in Figure 12.30 shows multiple signal rise times as the resistors of an IC electronics were modified to achieve faster response times. (It's all about Rs and Cs.)

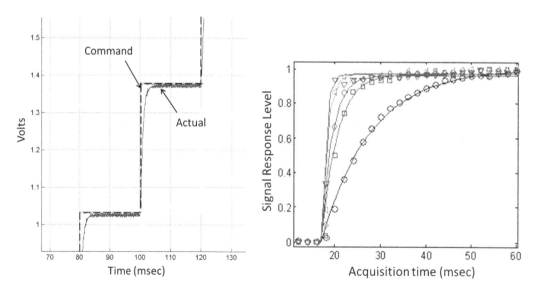

FIGURE 12.29   Power supply response.   **FIGURE 12.30**   IC response time.

This is a very high level start. Digging a bit deeper, tasks before and after the irradiation must be done, which also take time. This leads to Table 12.1.

TABLE 12.1
Some Tasks before and after Irradiation

Step	Function
1	Data is loaded that defines the scan
2	The plan trajectory is converted into hardware control points[1]
3	The converted trajectory is downloaded into the scanning controls
4	Beam magnetics are set to the first layer values
5	Fine beam steering correction is done (first layer)
6	Scan starts – Scanning controls send out control settings and read back data
7	Scan proceeds – Irradiation
8	Data is analyzed and control is adjusted[2]
9	Scan ends normally
10	Go to step 2 for the next layer
11	If this is the last layer, end
	If things don't go normally – go to a much, much longer table[3]

Notes:
1. A trajectory (not the same as a beam transport trajectory) is a set of consecutive, in time, control points, containing position, dose and/or fluence, beam size, beam velocity and maybe other information.
2. Some of the data needed by one or more of the scanning control modules may be the result of a required analysis of beam parameters.
3. The control system should accept the usual pause or resume. When resumed, it must use the information from before the last pause to continue accurately. Also, the pause or emergency stop signal should be sent directly to the hardware, but also to the scanning controls.

Sometimes, the details of the irradiation timing depend upon the mode used. Table 12.2 explores the steps that take time for SS and continuous scanning. It is assumed that the energy dimension takes the longest time and is, therefore, done the fewest number of times.

TABLE 12.2
Time Elements of Spot and Continuous Scanning

	Spot Scanning	Continuous Scanning
1	Set system so beam would appears at A	Set system so beam would appear at location A
2	Turn on beam at spot A	Turn on beam AND start moving beam
3	Deliver dose to spot A	Adjust beam current and or scanning speed
4	Read the dose and beam info during step 3*	Read the dose and beam info during step 3*
5	Turn off beam	Keep going
6	Move from spot A to spot B	Keep going
7	Loop back to step 2 until layer is complete	Keep going until layer is completed
8	Change beam energy	Change the beam energy
9	Loop back to step 1	Loop back to step 1

* Actually the dose and beam info should always be read.

During the beam delivery, other things may also happen. For example, something analyzes the data, during and/or after a layer, maintains the correction vectors (to correct the beam position if needed) for each setpoint and determines the following:

1. Beam is not positioned properly – make a beam position vector correction.
2. Information may be sent from an external system that a patient has moved – make a beam position vector correction (if that is implemented), or pause
3. The beam size has changed; make a correction or pause
4. Other corrections

Some of these times can be encapsulated. For SS, with the two transverse dimensions being x and y, the time to deliver a layer (*layerTime*) includes the three main categories as follows:

$$layerTime_{Spot} = N_x \times \left(doseTime + magnetTime_x + equipTime\right) + N_y$$
$$\times \left(doseTime + magnetTime_y + equipTime\right) \tag{12.6}$$

where N_x and N_y are the number of x and y spots.

For continuous scanning, it's a little trickier since the dose is delivered as the magnet is moving the beam. Let's say they are synchronized so that these times are the same and consider only the *magnetTime* (which could be an upper limit to the magnet time, not necessarily as fast as it can go).

$$LayerTime_{Continuous} = L_x \times \left(magnetTime_x + magnetTime_y + equipTime\right) \tag{12.7}$$

where L_x is the number of x lines and the time includes the time to move in the y direction between each x line. This can be interchanged. The *magnetTimes* are separated into x and y values since most systems have different parameters in each direction (but some do not). Remember that in time-driven scanning mode (e.g. continuous scanning), the maximum dose rate depends upon maximum scan speed and the instrumentation time constants. The maximum scanning speed dependencies include the beam intensity, time it takes for a beam current change and the desired effective penumbra. The hardware contribution to *magnetTime* is covered in Section 12.3.10.1.

The number of spots and the spacing between lines depends on the field size, the beam size and the spacing between the beams.

$$N_{x \, or \, y} = \frac{\text{Field size} \times \text{overlap factor}}{\text{Beam sigma size}} \tag{12.8}$$

where the overlap factor is related to the spacing between Gaussian beams required to achieve the desired field conformance (or flatness if that's what is desired). Numerically it is the number of spots per sigma. If, for example, the spot spacing was 1 sigma, the overlap factor would be 1. If the spacing allowed for 2 sigma, the overlap factor would be 1/2. This could be different for x and y.

The total time to deliver the 3D irradiation is

$$\text{Total time} = \text{Layer time} \times \text{\# Bragg peaks} \times \text{time to change Bragg peaks} \tag{12.9}$$

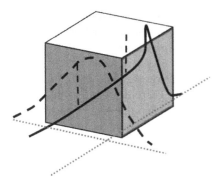

FIGURE 12.31   Spot voxel.

Consider the SS mode first. The beam is used to deposit dose in the voxel depicted in Figure 12.31. The transverse beam is Gaussian (dashed curve) and the longitudinal beam is a BP (solid curve). Assume, for this first crude approximation, that the size of the voxel is chosen so that all the energy from the Gaussian and Braggian peaks is deposited in the voxel. The volume of that voxel is $V \cong \sigma_x \sigma_y d$, where the σ is from the Gaussian and d is the 'thickness' of the BP. The linear dimension is σ instead of 2 sigma because the Gaussian is being moved 1 sigma between spots. The amount of time that the beam is on, while stationary, will determine the total dose and dose from neighboring spots will add to the dose in this voxel, but ignore that here.

Now consider the continuous scanning mode. In this case, the beam is moving transversely as it is passing across the voxel. Looking at the moving beam, the amount of energy that will be deposited is the integral of each part of the transverse distribution. The big rectangle, in Figure 12.32

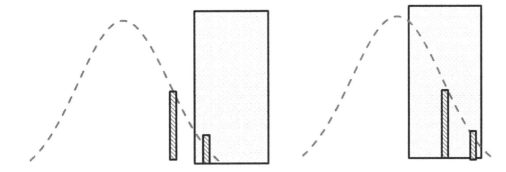

FIGURE 12.32   Charge deposited during moving beam.

represents the front face of the voxel. The small hatched rectangles are small portions of the integrated intensity at two regions on the Gaussian. The Gaussian (dashed line) and hatched rectangles are moving to the right relative to the big rectangle. The time it takes for each of these to cross the big rectangle is the same. Therefore, the amount of energy deposited will be related to the current in the beam distribution times that time or basically the integral of the Gaussian current and the time it took for the beam to cross.

$$\# \text{ Coulombs in region} = I(\text{amps}) \times \frac{\text{distance traversed}}{\text{speed of beam}} \qquad (12.10)$$

Therefore, the amount of dose deposited will depend on the speed of the beam movement as implied several times earlier.

There is yet another component of the total time which depends upon the accelerator used. In the above equations, it was assumed that the beam was always instantly available when required. Some accelerators, such as a synchrotron, are characterized by a stored beam which is slowly extracted until it is used up and then refilled. The time it takes to refill and accelerate, to be ready again must be added to the above.

Some accelerators produce a pulsed beam. The pulse is at some given frequency to coincide with the accelerator parameters (see Chapter 13). Consider this situation in the case of SS. Imagine a target that is 30 cm × 30 cm × 15 cm. The beam sigma is 10 mm and the BP width is 5 mm. Thus, there are about 1,000 spots/layer and 30 layers or a total of 30,000 spots. In the old days, a synchrocyclotron (for example) had a pulse frequency of about 60 Hz. Assuming that there is enough charge in each pulse to fill a voxel, it would take 30,000 pulses of beam or 500 seconds (8.33 minutes) to irradi-ate this target. Now, what if a safety analysis concluded that there was a probability that the dose delivered in one pulse fluctuates more than an amount to ensure the dose delivered would be within an accuracy of, say, 2%. One way to deal with that is to configure the system so that three pulses are delivered to each voxel. Then it might take over 25 minutes to deliver the dose. Well, this isn't sufficient, if, for example, the possible error in each pulse is up to 3%. Therefore, imagine that after the first pulse, there is an accurate measurement of the dose delivered and the dose in the second pulse is corrected for the error in the first pulse. This is covered in more detail in Section 12.3.9.

12.3.7   TIME SENSITIVITIES TO SCANNING HARDWARE

It is useful to explore the sensitivities of the irradiation time to various parameters. This can help to identify some safety criteria and to set constraints on design parameters. Stephen Dowdell was a talented student (no longer a student) loaned to me by Harald Paganetti. We investigated some of these parameters.

One painting is used to irradiate the 1-liter volume of a cube of water with a beam current of 2 nA. The spacing between the Gaussians and the BPs was chosen to assure a conformation to a uni-form dose distribution of 5%. This spacing and the number of curves can change depending upon the tolerances and the irradiation parameters.

12.3.7.1   Time vs. Dose

The time required to deposit the dose as a function of the prescription dose is shown in Figure 12.33. Once the time to deposit the dose exceeds the hardware waiting times, the time it takes to deliver the required dose increases linearly with the prescription dose in a single layer. This is not necessarily the case when considering a total irradiation. The inclusion of dose tolerance in the optimization could lead to a reduced spacing of Bragg and Gaussian peaks required to achieve the required flat-ness when increasing the overall prescription dose. This leads to an increase in the total time needed to move the beam and wait for the equipment as the number of spots in the irradiation is higher. The time required to deposit the dose increases linearly with the number of paintings.

12.3.7.2   Time vs. Current

Increasing the beam current reduces the time required to deposit the required dose. Figure 12.34 shows the effect of changing the beam current on the time required to deposit dose during the irra-diation. There is a time benefit when increasing the beam current, but the benefit is reduced with

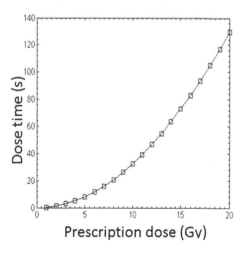

FIGURE 12.33   Scanning time vs. dose.

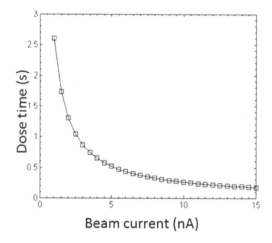

FIGURE 12.34   Scanning time vs. current.

increased beam current. For example, increasing from 10 to 11 nA only decreases the required time by 9% compared to the 50% improvement achieved moving from 1 to 2 nA because the hardware time is limiting.

The lateral field size determines the number of spots, affecting the total irradiation time. Increasing the field size in the depth direction increases the time required to change energies. This is also true for the dose tolerance parameter. A tighter tolerance causes an increase in the number of BPs required, leading to a longer irradiation time.

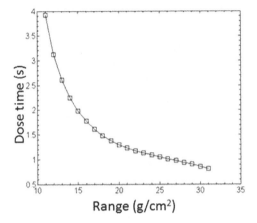

FIGURE 12.35   Scanning time vs. range.

12.3.7.3   Time vs. Range

The range of the field influences the irradiation time as seen in Figure 12.35. This effect is due to the dependency of the time taken to deposit the dose on the power in the beam which changes with energy and, thus, range. A higher energy beam has more power. However, increasing the range also affects the time to move the beam between adjacent spots because the scanning magnet current increases with range which can result in a longer settling time when trying to move the beam.

12.3.7.4   Time vs. SAD

Altering the SADs of the scanning magnets produces little effect in irradiation time. Larger SADs lead to shorter irradiation times since the magnet bend angle is smaller and it requires less current; however, the time change is almost insignificant when counted over the course of an entire irradiation. Increasing the SAD of one of the scanning magnets from 1 to 4 m decreases the irradiation time by approximately 0.1 second.

12.3.7.5   Time vs. Magnet Current Ramp Rate

The time for the irradiation increases linearly with the rate of current change required to move the beam across the maximum field size. As shown in Figure 12.36, increasing the rate at which the current can be supplied to the scanning magnets decreases the time to move the beam throughout

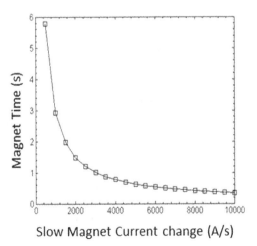
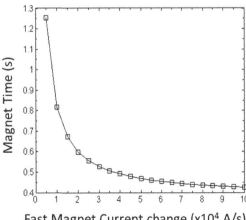

FIGURE 12.36   Time vs. dI/dt: slow magnet (left) and fast magnet (right).

the course of an irradiation. However, this takes either a reduced inductance or higher power supply voltage. Figure 12.36(left) is for current rate supplied to a slower dipole and Figure 12.36(right) is for a faster scanning magnet.

The irradiation time scales linearly with each of the quantities that comprise the *equipmentTime*. The values should all be of the order of 10^{-3} second, or smaller, to ensure the irradiations are completed within a reasonable time frame (<1 minute). These parameters do not determine the total time alone; however, since for SS they affect the time for each spot, they can potentially have a large effect on the overall time required for an irradiation.

12.3.7.6 Time vs. Beam Off Time

Figure 12.37 demonstrates the effect on the *equipmentTime* if the time required turning the beam off is increased from ~1 to ~100 ms. The increase in turn off time leads to an increase of approximately 2 seconds to the total irradiation time. (This time is very dependent upon the accelerator, or subsequent equipment used to help [e.g. fast kicker dipole].)

12.3.7.7 Time vs. Energy Change Time

The variables examined above are predominantly focused on single layer irradiations or the time during a single layer of a multiple layer irradiation. By far the largest contributor to the total irradiation time is still the time required to change energy. This requires changing the set points of all magnets along the beamline from the beam source to the nozzle. Figure 12.38 shows the effect on the total irradiation time when the time to change the energy is adjusted. Assuming the time to change energy is 1 second, this irradiation would take 18.6 seconds to complete. If the time to change energy is as high as 5 or 10 seconds, the irradiation time becomes 82.6 or 162.6 seconds, respectively. The irradiation time data shown in Figure 12.38 assumes a cyclotron beam source such as the one at Massachusetts General Hospital (MGH).

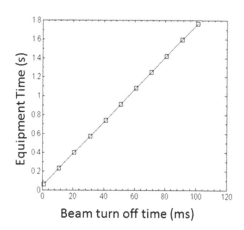

FIGURE 12.37   Contribution of beam off time.

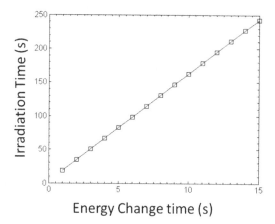

FIGURE 12.38   Contribution of energy change.

Combining all the effects there are a variety of parameters combinations that can be explored. Among these are the relative time of the extreme of SS compared to continuous scanning, as a function of dose rate (or beam current) as shown in Figure 12.39. For the very specific conditions explored here (e.g. 2 Gy delivered to 1 liter of water using dipoles capable of 30 and 3 Hz, located about 2 m from isocenter), and for one energy layer, the highest curve in Figure 12.39 includes five repaintings of SS, the middle is SS for one painting and the lower is continuous scanning which, due

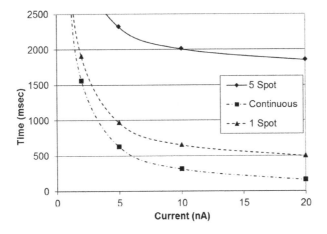

FIGURE 12.39 Layer time for spot vs. continuous scan.

to the speed of the magnetic field ramp used, requires multiple repaintings anyway (as discussed in Section 12.3.3.2). It is important to be sure not to use a dose rate that is so high that an error could happen as will be discussed in the next section. Owing to the extra steps in a SS method, compared to continuous scanning, in which the beam is turned on and off (not all dose-driven methods require this!), there is indeed some increase in irradiation time. However, for some realistic parameters, the difference within a given layer may only be a fraction of a second.

When considering the overall time of the irradiation (including the time to change the energy) the difference can be less than half a minute assuming repainting with both methods as shown in Figure 12.40. This plot includes six cases. The lowest solid black curve is continuous scanning with 0.5 second energy change time. The higher solid black curve is continuous scanning with a 2-second energy change. The dashed lines are for SS with 0.5 second energy change time. The lower one is one painting per layer and the upper one is five repaintings per layer. The dotted curves are for 2 second energy changes. Note how the continuous and one time SS are similar in times despite the turn on and turn off times, owing to the repainting required in continuous scanning. Again, safety considerations vis-à-vis dose rate are critical.

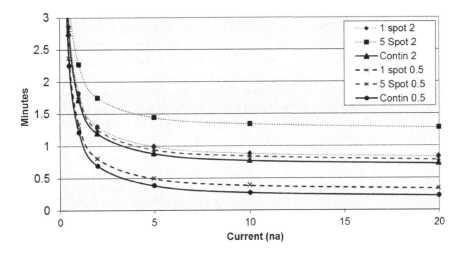

FIGURE 12.40 Full irradiation time for various combinations of spot size, repainting and ΔE time.

Some typical timing numbers (three possible sets) that contribute to a SS irradiation time are summarized in Table 12.3. Bolded numbers highlight the relative change compared to other columns.

TABLE 12.3

Examples of Some Numbers Contributing to Irradiation Time

Parameter	Units	Times 1	Times 2	Times 3	Reason for Parameter
Beam off	μs	50	50	50	
Move beam	μs	5,000	5,000	**1,000**	Reduce settling time of scanning dipoles
Beam on	μs	100	100	100	Ion source and beam path length travel time
Measure beam	μs	150	150	150	Ionization chambers and electronics
Field size	cm	20	20	20	User defined
Spot size	cm	0.5	0.5	0.5	System capability and user defined
Number spots		1,600	1,600	1,600	Dependent
Time per layer	s	8.48	8.48	2.08	Calculated
Change energy	s	**0.2**	1	1	Time to change the energy
# Energies		25	25	25	Number of Bragg peaks
Total time	s	42.4	212	52	Calculated
	min	**0.71**	**3.53**	**0.87**	

Some of the calculations outlined above cannot be used, in their present form, to calculate the irradiation time for any arbitrary scanning field. The calculation of the number of spots assumes that the field is rectangular. This is not usually the case in the clinic, where fields are shaped to match the tumor volume. It also assumes that the spot size does not change throughout the irradiation. The beam can change during a layer or as a function of layer and the number of spots in the layer can change based on the energy of the current layer. Only evenly spaced BPs were used in the simulations to generate the SOBP. This is again not necessarily what could be reflected in a clinical situation. It was assumed that the scanning system used for the calculation uses two magnets to scan the beam. Other systems include a combined function scanning dipole with much faster time constants. Using a different type of system will lead to different conclusions regarding which elements of the system can to be optimized to reduce treatment time. Much of the formalism presented herein holds, but the parameters have to be identified for any given system.

The time to change the beam energy is accelerator specific and will be covered in the Chapter 13 addressing accelerators. Until then, here are a few words. Unlike the transverse position of the beam, today the range is difficult to change quickly for the types of accelerators and beamlines that are generally in use. This time depends very much on the accelerator and beamline implementations. Some energy changes can be as fast as 50 ms for mechanical energy changing systems while other implementations can take a second or more. Therefore, in practice the energy change is done separately in between each layer of transverse painting. While some accelerators have certain beam delivery constraints with respect to the time-optimized order of the steps described above, new accelerators being developed may not, and may be, able to vary the beam energy, spot size and position and current more quickly. A beamline can also contribute to the time in an irradiation since it takes time to change the beamline magnetic fields. Fast feed-forward systems have been implemented to reduce this time considerably. Perhaps this can be done at the same time as the accelerator is changing energy, but it still takes time. A system without a beamline, like an accelerator on a gantry (e.g. Harper hospital and MeVIon) does not have this restriction. These days, multiple synchrotron accelerators now have the capability to change the extracted beam energy within one extraction period. Therefore, it is possible to vary the energy from the accelerator in tens of milliseconds instead of what used to take seconds just a few years ago.

12.3.7.8 Other Considerations

If a synchrotron is used as the beam source, the number of cycles required during each layer comes into play because only a finite number of protons can be in the synchrotron ring at any given time. Once the ring is populated, these protons are used after which it can be repopulated. This introduces the time required to accelerate and decelerate the synchrotron and the number of protons which can be in the ring into the time calculations. (See Chapter 13.) The acceleration and deceleration times are inherent properties of the individual synchrotron and vary throughout an irradiation as they are dependent on the momentum of the protons being accelerated. There is benefit in increasing the number of protons which can be in the synchrotron ring as this reduces the number of cycles required per layer, but this increases cost. Short of containing enough particles for an entire irradiation (which could also minimize the time for energy changes using multi-energy extraction), a reasonable scenario would be if the synchrotron could store enough particles to deliver an entire layer of the maximum field size at the maximum prescription dose required. This would reduce the number of cycles per layer to 1, thereby minimizing the irradiation time using a synchrotron as the beam source compared to other methods.

12.3.8 DOSE RATE TOLERANCES

One of the elements of time referred to above (*equipTime*) is the time it takes for an instrument to detect, measure, calculate, send messages and react to an event. In Figure 12.41, the upper curve shows the actual dose rate as a function of time. The lower curve shows what an instrument might see. The solid curve is beam on time. In the ideal world the beam would be shut off, at the time the vertical dashed line shows. Thus, the dashed gray dashed line, on the curve shows the beam turning off (taking some time). How is it known, however, to turn off the beam at that time? It is not known, unless one is counting time (the time-driven mode) with a well-controlled beam. Otherwise, the lower curve shows what is known by the instrument. Waiting longer, until the lower curve shows that the desired dose has been reached (gray dotted vertical line), would then allow more dose than desired. In other words, if the trigger to turn the beam off was the gray vertical dotted line, the beam would not have been turned off at the right time and the upper dotted beam dose rate would characterize the extra dose delivered. The gray dotted ramped line shows the beam turning off in this case.

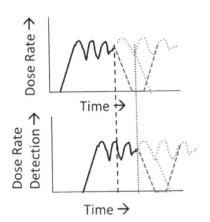

FIGURE 12.41 Dose rate detection time lag.

Therefore, it is normal to consider this time difference, factor in the dose already measured and project what the dose will be in the next interval of time, thereby beginning to turn off the beam before the measurement shows that the full beam has been delivered. That is indeed proactive, but not necessarily foolproof. Any deviation can be measured and possibly accounted for in the next fraction, assuming there is a next fraction.

The dose rate is shown with ripples. What if the ratio from high point to valley of the ripples were a factor of 10, and what if the ripples were not regular or predicable? In that case one would not know the actual dose rate to predict and the best scenario is to assume the worst case. In both this case and a less ripply case, it is necessary to consider how much beam is likely to be delivered in the dark period (after the dashed vertical line).

This is not just the situation for ending a 'normal' beam delivery. This measurement is continuously happening and must detect if there is an anomalous delivery. All of these scenarios require that the dose rate delivered take into account the worst case scenario, such that the dose delivered during this dark period is clinically insignificant. This is the dose delivered during the time that the beam can be stopped which contributes to less than, for example, 1% of the desired dose. There is no standard, other than the total (2%–5%), but local staff must determine the allowances.

There are several related questions. Is it necessary to:

1. Wait until beam is read out before continuing in SS,
2. Continue, analyze and readout in the meantime and stop if there is an error, or
3. For continuous scanning, is the dose rate/magnet speed limited for a 1% safe reaction time?

A straightforward example of this effect does not include beam current fluctuations, just the finite reaction time. To obtain an order of magnitude of the effect consider that the IC drift time and the electronics processing time and the reaction time is about 100 μs. A typical irradiation of a target on the larger side may require 100 Gp to deliver an appropriate dose. If the target is 10 cm $\times$ 10 cm and the beam spot sigma is 5 mm, the equivalent SS mode would require 400 spots/layer. Assume also that the BP width is 5 mm and this irradiation will require 20 BPs, so there are a total of 8,000 spots in this irradiation. The 100 Gp are divided equally into these spots (in real life they are not – since the depth dose distribution is not equally distributed and the distal layers would be more highly weighted and overlap with previous BPs delivered is ignored). Then each voxel receives 1.25×10^7 protons. If an accuracy of 2% is required then the accuracy of the number of delivered protons is 2.5×10^5 protons/voxel. Now, the total reaction time has been taken to be 100 μs. Therefore, the maximum rate at which the protons should be delivered in this dark period is $(2.5 \times 10^5 \text{ protons})/(1 \times 10^{-4} \text{ seconds}) = 2.5 \times 10^9$ protons/s. This converts to 4.0×10^{-10} C/s (A) or 0.4 nA. Thus, in this case, assuming a well-controlled beam, but perhaps not perfectly controlled, to be safe the beam current should be no higher than 0.4 nA. If there were fluctuations in the beam current on the order of $\pm100\%$, it would be necessary to err on the side of the highest possible beam current and the maximum current would be a factor of two lower, as would the dose rate. Now use this value with Figures 12.39 and 12.40 and it is seen what a severe restriction this dose rate is to the irradiation time.

Earlier in Section 12.3.5 it was shown that it is desirable to be able to turn off the moving beam within 55 μs. If the dark time is 100 μs, the beam will have moved 1.8 mm at 30 Hz. If the scanning speed were 100 Hz, 17 μs time resolution would be needed to meet the same position accuracy requirement.

12.3.9 PULSED BEAMS

The previous sections have assumed that a beam is available when it is needed and that the only concern was the stability of the current in that beam. However, some accelerators will produce short pulses (as was briefly mentioned in Section 12.3.6) and while the average intensity is high enough there are some considerations that should be analyzed. Two of these, in particular involve the pulse rate, which will directly relate to the time for an irradiation and the accuracy of dose delivered for each pulse. The two are intertwined with respect to the overall treatment time. If the dose delivered per pulse were perfectly predictable and reproducible, only one pulse would be needed. However, if this is not the case, and it is truly not the case, a strategy for delivering the dose within the desired tolerances is necessary.

A pulsed beam, by its nature is consistent with a time-driven, SS delivery approach. One strategy is to consider delivering multiple pulses of beam, each one with fewer particles than is needed for a full dose and to try to deliver on each subsequent pulse an amount to make up for any deviation from what was desired in that previous pulse. For example, 75% of the total desired dose can be delivered on the first pulse (assuming the stability of the beam was better than 33%). Then 75% of the remaining dose (accounting for what was delivered in the first pulse) is delivered on the second pulse, followed by the remaining dose in the third pulse. The question is how many pulses are necessary to ensure the total dose per voxel is sufficiently accurate. Figure 12.42 addresses this question. The horizontal axis is the number of pulses delivered. It is assumed that the dose delivery error is a percentage of the desired dose and not an absolute dose error per pulse, but that error can be evaluated as needed. The vertical axis is the total dose delivered. The desired dose, in this case

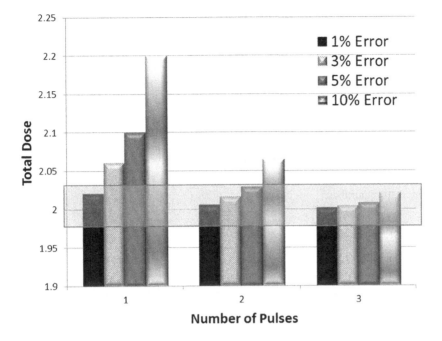

FIGURE 12.42 Pulsed beam dose accuracy.

is 2.0. The errors introduced are systematic, so this is a worst case analysis. There are four bars per pulse number. Each of the bars represents a different error level from 1% to 10% per pulse.

The graph shows the cases for 1, 2 and 3 pulses according to the strategy just described. If the desired tolerance is 2% (the shaded horizontal bar), of course, a 1 pulse delivery would have to be within that 2%. According to the graph, a 2 pulse delivery can tolerate up to 6% error and a 3 pulse delivery can tolerate beyond a 10% error per pulse. Therefore, it is the case that most pulsed beam delivery is done with multiple pulses, depending upon the particular constraints of the specific system.

These pulses will depend on the nature of the accelerator and the equipment. Given this information it is possible to estimate how long a treatment can take. For example, the graph of Figure 12.43 shows the number of minutes it would take to deliver 8,000 spots. (This comes from the example in

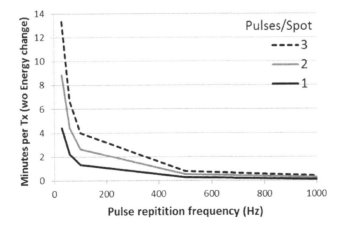

FIGURE 12.43 Pulsed beam irradiation time.

the previous section.) The time to change energies is not included in the estimate. The horizontal axis is the pulse repetition frequency. While in the old days a repetition rate for a synchrocyclotron was 30–60 Hz, these days those systems use solid state components with a frequency of about 1 kHz.

The technology of particle therapy is evolving as is the biological and clinical application. One of the developmental goals is that of mitigating organ motion. If the target volume were irradiated in 1 second, that could mitigate the issue. Various time factors discussed in this chapter would have to be investigated to see what design configuration could achieve that. Another development (at the time of this writing) is revisiting the possibility of FLASH irradiation or dose rates of 40–100 or more Gy/s. Pulsed beams have been shown to be capable of over 50 pC in a 20 μs pulse or 2.5 μA instantaneous current. Averaged over 1 kHz, the average current is then about 50 nA, an interesting number.

12.3.10   SCANNING HARDWARE

The hardware required to spread out the beam in the scanning modality includes equipment to control the beam properties and instrumentation to measure the beam properties. Most beam-scanning systems use magnetic deflection to move the unperturbed charged particle beam across the target cross-section, thus spreading out the dose delivered by the beam. The unperturbed beam is usually characterized by a Gaussian profile. Longitudinally the beam profile is that of a BC whose energy is adjusted in such a way as to obtain the desired longitudinal dose distribution. A general representation of most of the systems currently built is shown in Figure 12.44. This system can be treated as being decoupled from everything upstream. The beam that enters this hardware has some incoming properties. The beam range is given by the accelerated beam energy or the beam energy resulting from a degrader system. The beam size is determined by the intrinsic emittance of the beam and any beam focusing elements in the beamline.

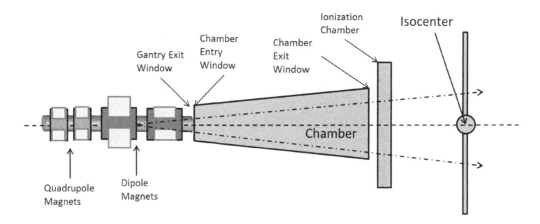

FIGURE 12.44   Scanning hardware components.

Fine tuning of beam focusing can be done with quadrupoles in this subsystem, shown as a quadrupole magnet doublet. The position on the target is controlled by the beam angle bent by the scanning magnet(s), shown here as a pair of separated dipoles. The overall system of scanning magnets and power supply will determine the speed of the scan (limited, of course, by the speed of the instrumentation used to measure the beam properties). Instrumentation to measure the fluence, position and beam profile is generally included in the beam path just before the patient, shown here as an IC.

An example of an actual scanning system with these components (except the quadrupoles) is shown in Figure 12.45 (courtesy of Pyramid Technical Consultants and Hefei ion Medical center). From left to right are the two scanning dipoles, a helium chamber followed by a pair of ICs (yes – a pair).

FIGURE 12.45 Photo of a scanning system.

12.3.10.1 Scanning Dipoles

A room temperature electromagnetic scanning dipole, sketched in Figure 12.46, comprises an iron core (diagonally hatched region) and current carrying coils (the squares with the X inside). The current generates a magnetic field which is contained within the iron and the gap. The iron, however, saturates at a magnetic field of from 16 to 18 kG, depending upon the makeup of the material (e.g. iron and carbon). There are many variations to this. Higher magnetic fields can be generated through the use of higher excitation currents. Even higher fields can be generated with superconducting current carrying coil material. Magnetic fields can be shaped by the coils or through a combination of the coils and the iron. In the case of scanning dipoles, it is desired to change the magnetic field on a timescale consistent with a scanning irradiation modality. This is best done with room temperature coils and a laminated iron and/or ferrite core.

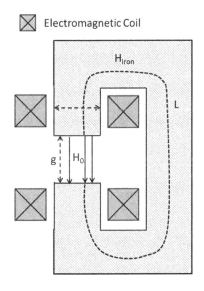

It may be helpful to start this section with an apology because there are three L terms that will be used. Finding other alternatives was difficult, so please pay attention.

Consider the C shaped dipole (as opposed to an H shaped dipole) shown in Figure 12.46. The loop represented by the dashed line of length L and the solid line of

FIGURE 12.46 Electromagnetic dipole.

length g surrounds an electric coil pair of N turns carrying a current I in each turn. Ampere's law teaches us that

$$\oint \vec{H} \cdot \vec{dl} = NI \tag{12.11}$$

where $H = B/\mu$, B is the magnetic field, dl is the incremental length along the loop, μ is the permeability of the material, $\mu_0 = \mu_A$ = permeability of free space = $4\pi \times 10^{-7}$ Tm/A (Tesla × meters/Amperes), I is the current inside the loop.

To evaluate this integral, the loop can be divided into two sections,

$$\frac{B_0}{\mu_A}g + \frac{B_0}{\mu_{iron}}L = NI \tag{12.12}$$

Since, the permeability of iron is over 1,000 times greater than that of air the equation reduces to

$$\frac{B_0}{\mu_A}g = NI \tag{12.13}$$

(The two permeabilities are $\mu_A = 4\pi \times 10^{-7}$ H/m and for pure iron $\mu_{iron} = 6.3 \times 10^{-3}$ H/m.) (Henry = (kg m²)/(s² A²).) From this, a very useful design equation is obtained:

$$B\,(\text{gauss}) = \frac{0.4\pi\ NI\,(\text{Amperes})}{g\,(\text{cm})} \tag{12.14}$$

From a practical point of view, the electric coil can get hot (ask me how hot, I can show you the scar), depending upon its engineering, cooling capability and the amount of current it contains. Current densities of between 1,000 and 3,000 A/cm² in a water cooled conductor can work. An additional practical consideration is that there must be a power supply to energize the coils. The power generated by the power supply, dissipated in the conductor is $P = I^2R$. Where I is the current in and R is the resistance of the coil. So the power is proportional to the current squared. It is possible to consider the cost of the copper coil per unit volume and the cost of the power supply (and running costs) per watt of power generated. There will be an optimum over-all cost for any given current density.

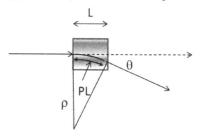

In scanning, the beam is spread transversely by this magnetic dipole. The angle of bend with respect to the beam path through the dipole in Figure 12.47 is given by Equation 12.15.

FIGURE 12.47   Particle bent in a dipole.

$$PL = 2\pi\rho\frac{\theta}{360} = \rho A sin(L/\rho)\,\left(\text{this is the particle length in the magnetic field}\right) \tag{12.15}$$

Using our favorite formula (Equation 10.5) provides a relationship between the bend angle and field:

$$B(kG)\rho(m) = 33.356P\left(\frac{GeV}{c}\right) \tag{12.16a}$$

$$\int B\,dl \sim B(kG)\cdot PL(m) = (33.356P\ GeV/c)\ \theta\,(\text{Radians}) \tag{12.16b}$$

For a 230 MeV proton beam, a 5 m bending radius in a dipole can be achieved with a magnetic field of 5 kG. Remember that this is a scanning magnet and the magnetic field will be varied as fast of possible, so it is not advisable to have a high magnetic field. Assume that a treatment field size of 30 cm at a distance of 2.5 m from the dipole is required. This leads to a bend angle of ±60 mrad. The 5 m bending radius translates to a dipole of length of about 0.3 m. The total Ampere-turns will be about 6,000 (both coils). With a reasonable current of about 100 A, this will result in 60 turns of the coil.

The inductance of such a magnet and coil, with a pole face area A, is given by

$$L\,(\text{Henry}) = \frac{N^2 A}{\left(\dfrac{g}{\mu_A} + \dfrac{L\,(\text{Length in iron})}{\mu_{iron}}\right)} \tag{12.17}$$

So the value will depend upon several engineering choices, and this must be added to the design optimization. In this example, the result is an inductance of 13 mH. (If 200 A were used, the induc-tance would be $L\sim 3$ mH.)

12.3.10.2 Dipole Contribution to Scanning Irradiation Time

The magnet bends the beam through an angle θ. This is an electromagnet which has an inductance and resistance. The electric current in the coils of the dipole will generate a magnetic field of strength proportional (most of the time) to that current. Using a power supply with a given voltage capability, the time to change that current (dI/dt) is determined by

$$V(\text{Volts}) = L(\text{Henries}) \frac{dI(\text{Amps})}{dt(\text{seconds})}; \frac{dI}{dt} = \frac{V}{L} \qquad (12.18)$$

where V is the voltage (above that required to maintain the desired magnetic field), L is the inductance, dI is the change of energizing current in a given time interval (dt), dt is the time interval.

The target is at a distance SAD away from the center of the dipole in the plane of motion. Then the speed with which the beam is scanned at isocenter is given by

$$\text{SAD}(m) \cdot \frac{\theta_{\text{max}}}{(L/V)I_{\text{max}}} = \frac{dx}{dt} \qquad (12.19)$$

If it is desired to scan a 30 cm distance at 30 Hz, or roughly 18 m/s, the time to scan 1/2 that distance is 1/4 of the cycle and with that dI/dt, the additional voltage required would be about 160 V.

A typical scanning frequency can be between 3 and 100 Hz. In the case where the field size is 30 cm this results in a transverse speed from 1.8 to 60 m/s. The inductance of a magnet is very dependent upon the gap of the magnet. In the case where there are two dipoles (one for each plane) in series, the gap of the second dipole will be larger than the gap of the first to accommodate the swept beam out of the first. Therefore, the second dipole will be slower than the first. It's possible to combine both planes of bend in one dipole with a single gap. A couple of examples of combined dipoles are shown in Figure 12.48. The same speed can be achieved in both planes in this configuration. This can come in the form of octupole (Figure 12.48(left)) (courtesy of Pyramid Technical Consultants) or as a combined dipole (Figure 12.48(right)), developed in Indiana University (and installed at MGH). The octupole is actually quite creative. It can also act as the equivalent of a quadrupole. A system with this component is that much closer to being able to perform the more general scanning functions described in Section 12.3.2. There are even potential uses of the octupole to shape the penumbra of the beam.

FIGURE 12.48   Octupole (left) and combined function dipole (right).

12.3.11 Scanning Beam Parameters

Once again, be reminded that the goal of radiotherapy is to deliver the prescribed dose to the target with the prescribed dose distribution. In general, the dose fidelity (conformality to the prescription) is given by properly controlling the beam position, size, range, current and gradient at any given time or integrated over any given time interval. Each of the devices affecting or measuring the beam properties can be controlled and/or measured in a finite time period or continuously depending upon the measuring device (e.g. power supply current vs. beam position), which should become part of the specification. The scanning system can control equipment parameters, read back instrumentation parameters and make decisions about the settings of the equipment parameters based upon the instrumentation parameters.

It is useful to separate the discussion of beam parameters into the static and motion regimes. The former is the unperturbed property of the beam when it is not in motion and the latter are effects arising from the fact of beam motion. The regions of dosimetric interest for a static scanned beam (perhaps an oxymoron unless it is beam to be scanned) are different from those of scattered beams. They are represented in Figure 12.49.

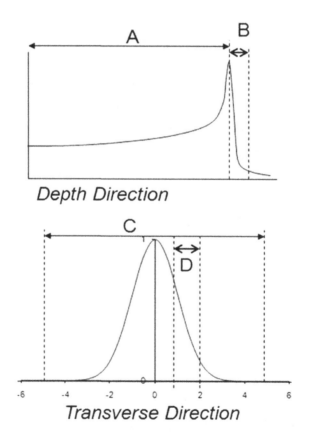

FIGURE 12.49   Dosimetric regions of scanning.

- Region A + B represents the BC. Since the beam has not been modified, the only relevant longitudinal parameter is given by the unmodified BC. Multiple BPs can be added to distribute the dose over the depth of the target region, but this is very dependent upon the scanning technique and the treatment plan. An SOBP is possible but not necessary from a technical perspective. The range is given by A at the d90 amplitude. The only other relevant quantity is the distal falloff in region B given by the distance between the d80 and d20 heights, since this will determine the distal penumbra.

- Region C represents the Gaussian transverse beam distribution. Since the beam has not been modified, the only relevant transverse parameter is given by the raw beam transverse distribution, or basically the σ of the Gaussian distribution.
- The transverse penumbra of the Gaussian in region D, also the 80%–20% distance is indicative of the overall minimum extent of the penumbra.
- Multiple beams can be added together to form virtually any transverse distribution. Technically the concept of a uniform field region is not relevant, since the transverse distribution can be uniform or not, depending upon the treatment plan.

12.3.11.1   Static Beam Parameters

12.3.11.1.1   Depth Dose Distribution

The beam range is normally given by the accelerated beam energy or the beam energy resulting from a degrader system and perhaps slightly further modified by a minimum of material in the beam path. Any material in the beam path will scatter the beam and increase the beam emittance and distal falloff.

The depth dose distribution will be determined by the superposition of the BCs used in the delivery of the dose volume. Unlike an SOBP as used in the scattering technique, the delivery of a scanned beam can be general. Two such examples are shown in Figure 12.50. However, without any degraders (material) in the beam path, the width and distal falloff of a BP for a low energy (shallow) beam is smaller and creating a smooth dose distribution by superposition of these sharp peaked depth doses is difficult as shown in lower plot of Figure 12.50. A range shifter or ridge filter is generally used to artificially increase the width of the distal falloff in this situation. Sometimes even the scanning beam may need to be modified. Sometimes sharp is too sharp. It should be added that while scanning is considered beam delivery without modifying devices, use of a range compensator can provide advantages in some situations and the use of a collimator can provide certain advantages sometimes, as well. The compensator would allow one beam range to be used to irradiate a distal irregular shape with an extended width, sort of a mini-SOBP and, thus, reduce the number of energy layers required for the dose delivery, considerably shortening the time of the irradiation.

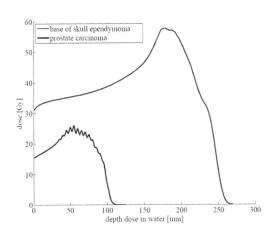

FIGURE 12.50   Non-SOBP depth doses.

12.3.11.2   Transverse Dose Distribution, Penumbra and Modulation

The beam size is determined by the intrinsic emittance of the beam as modified by the beam focusing elements in the beam line. Sometimes a set of final quadrupoles can be used to fine tune the beam size in the nozzle. The penumbra of the beam size can be enlarged by adding Gaussians and modulated dose distributions are necessary to win back the smaller penumbra.

The transverse dose distribution is given by a superposition of the transverse raw beam profiles. Much depends upon the shape of that raw beam. The distribution of particles in a beam is typically statistical and owing to the physics of the source of ions and the accelerator. This results in a Gaussian distribution.

$$\text{Amplitude} \approx e^{-\frac{1}{2}\left(\frac{x}{\sigma}\right)^2} \tag{12.20}$$

The sharpness of the unmodified Gaussian beam falloff will determine the sharpest dose falloff possible with a clinical beam. The steepness of this falloff determines how much dose will be deposited into healthy tissue outside the target, as shown in Figure 12.51. The beam sigma (σ) is the single parameter characterizing the ideal Gaussian shape. The full-width-half-max (FWHM) is given by $2.35*\sigma$. It is desirable to separate the target from an OAR. One example is to ensure that the target dose is within 2.5% of the nominal dose. Therefore, the edge of the target must be contained in the left shaded box, similar to the situation with single scattering. The difference here is that, as will be shown in the next section, the extent of the target can be much larger (to the left of the plot), without scattering, and the dose distribution across the target can be much more uniform, if desired. Suppose that the OAR occurs in the right shaded box; assume, for example, its edge is 5 mm from the edge of the target. Suppose also that the physician has determined that any part of this OAR cannot receive more than a maximum of 50% of the target dose. In this case, the distance between the target and this critical structure has to be at least 0.85 σ. This sets the scale of the beam size needed for different treatment sites. For example, if the target and OAR are separated by 5 mm then the beam sigma should be smaller than approximately 6 mm.

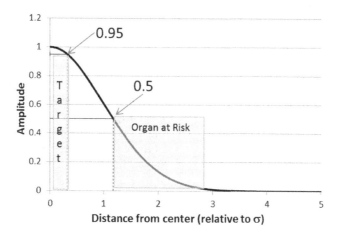

FIGURE 12.51   Gaussian edge for target and OAR.

There are a few other interesting observations in this case. If, in the worst case, the target were small enough to be covered by the 100%–95% top of the Gaussian, as in the single scattering case, the target would receive 25% of the total charge in the Gaussian. Using the erf (or just looking at the graph), the integrated relative charge to the OAR is about 1/3 of that. That only goes down as the target size increases.

Alternatively one might modify the beam, or apply an aperture to sharpen an edge. Apertures are generally thought of as an inconvenience and expense when confused with the type used in scattering systems. However, it's possible to conceive of a beam optics solution which allows an edge sharpener to be placed upstream of the target, and the beam on target is imaged from this aperture so that the beam that is scanned has a sharper edge at the target as was discussed in Section 10.12.2. It was first proposed by the author at a PTCOG in 2002 and subsequently implemented independently by Eros Pedroni in the PSI Gantry 2 design. Of course, even a scanned beam passes through material on the way to the target and the multiple scattering in this material may broaden the penumbra again unless care is taken to prevent this.

At the time of this writing, there is no one single clinical requirement for the pencil beam size. There are competing factors. Some treatment planning studies have been published to evaluate the dose volume histograms and dose to critical structures as a function of the size of the beam. Deeper targets are subjected to multiple Coulomb scattering and result in a larger beam size at depth anyway (see Section 12.3.14.6). Shallower targets may benefit from a smaller penumbra. At the suggestion of the author, A. Trofimov, T. Bortfeld (and the author) studied the issue. Treatment planning for a head and neck case was studied using a range of spot sizes. It was found that reducing the spot size from 8 to 5 mm lead to a marked improvement in dose conformality for the target volume, whereas the difference was not as dramatic from 5 to 3 mm. It was concluded that for most clinical cases, pencil beams of widths $\sigma = 5$ mm will be sufficient for delivery of the planned dose with a high precision. Reducing the beam spot size below 5 mm does not lead substantial improvement in the target coverage or sparing of healthy tissue in most cases. On the other hand, slightly larger beam sizes are more forgiving for organ motion mitigation and can be delivered more quickly. Smaller beams cost more. This is discussed in Section 14.7.

If a goal is to achieve the smallest transverse beam penumbra it is possible to only do almost as good as achieved with a collimator if the unmodified beam sigma is very small. Very small beam sigma in a scanning beam system is expensive to achieve, commands very tight tolerances on beam delivery and can add significantly to the beam delivery time for several reasons. The use of a collimator, on the other hand invokes several perceived (and real) disadvantages such as; air gap minimization requirements; bulky equipment; therapist involvement installing and removing the equipment; and the time involved for manual work, are some examples. The use of a Multi Leaf Collimator (MLC) as normally envisioned to improve the workflow and reduce the time for manual handling has been regarded as not a great advantage for particle therapy owing to the large size of a typical particle MLC and the air gap requirements. However, the use of a scanning beam does not require a full size MLC. First, the scanned beam motion can stop on the collimator, and therefore, the collimator only needs to have a transverse extent of a few beam sigmas, instead of the full length of the maximum delivered field size. Also, it is possible to conceive of a collimator bar which is synchronized with the beam motion and dynamically collimates the beam where needed. Such a system has been designed by the Iowa group and an equivalent system is included in the new scanning system used by MeVIon.

The penumbra of a scanned beam is not the same as the penumbra of the pristine Gaussian. The addition of Gaussians across the field results in a growth of the falloff distance. However, in much the same way that one sharpens the edge of an SOBP by emphasizing the BP at the distal edge, it is possible to achieve similar results by modifying the distribution of the number of protons across the field as shown in Figure 12.52. To achieve a flat distribution with optimal edges it is necessary to modulate the dose delivered across the target even for a single field delivery. The left side of Figure 12.52 shows a number of gray Gaussians equally spaced which add up to the black curve. The figure to the right shows an attempt to create a profile with sharper penumbra and a wider flat top. The gray Gaussians are summed, and different proportions can be noted. The edge Gaussians are more highly weighted than the others. This results in a sharper penumbra when compared to the equally weighted Gaussians. The dashed curve in that figure is the same as the solid black line in the figure to the left. The top of the summed curve is not perfectly flat, but the number of Gaussians used in this example was only 17 and even with 17 it is possible to have an acceptable optimization.

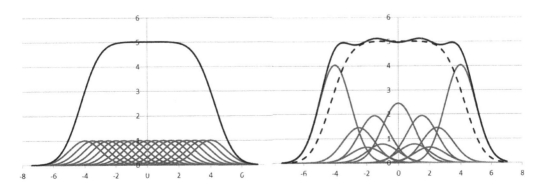

FIGURE 12.52   Transverse dose modulation.

This technique was first developed at Berkeley and then by Eros Pedroni at PSI. The example shown here was simply the result of an excel solver optimization.

What is the lesson here? Once again the lesson is that even for an SFUD delivery, to sharpen the edges, dose modulation across the target is necessary. So a standard tool in any scanning beam delivery is dose modulation, except for US. Therefore, the beam delivery methods are NOT different and are just PBS. Understanding these distinctions may help to refrain from using a term such as IMPT. It can also be helpful in properly specifying a scanning beam system.

If the shape of the beam is not close enough to that of a Gaussian, much of the above discussion becomes invalid and specific calculations are required to determine the adequacy of the beam for treatment. Note that in some synchrotron extraction schemes the beam has a sharp edge in the plane of the extraction (this may occur if there is a septum in the synchrotron). These are usually smoothed out when the beam passes through material in the beam path and/or a patient.

12.3.12 SENSITIVITIES (SCANNING)

A Gaussian shape is particularly well suited for scanned beams since Gaussians can combine well and produce a uniform or otherwise conformal pattern. Figure 12.53 shows how multiple Gaussians can be combined to form a flat top (without the dose modulation described in the previous section). The upper graphs are the individual distributions used. The left hand side uses symmetric Gaussian

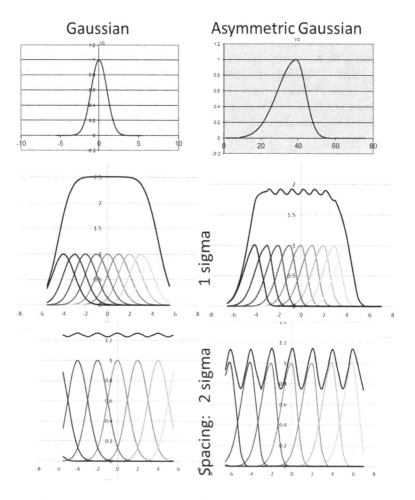

FIGURE 12.53 Effects of spacing and asymmetric shape.

and the right hand side uses asymmetric semi-Gaussian shapes. Consider first only the left side of the figure. The middle graph indicates an acceptable spacing of the symmetric Gaussians. The lower set shows that as the symmetric Gaussians are separated, the combined dose eventually shows the beam structure. However, not all beams are exactly Gaussian shaped. Depending upon the accelerator and other factors the Gaussian may not be fully symmetric and that can have important consequences. Once they are not symmetric, they lose some of their magic. Consider now the right side of the figure. The middle plot shows asymmetric beams, with the same spacing is not as smooth. The same number of Gaussians, spaced the same distance apart, is compared. It gets even worse in the lower figure. Indeed this has been seen at some particle centers and it is not usually an easy condition to diagnose.

Another possible error would be that of mispositioning the beam. While Gaussians are magical, they are not all powerful. The left plot in Figure 12.54 shows the ideal condition. Pay attention to the dashed Gaussian, which is in the correct position. The Gaussians are spaced at 1 sigma intervals. The middle curve shows the dashed curve moved to the left and the right graph shows it moved to the right by 0.3 sigma. A movement of this beam by 0.1 sigma results in a 2% perturbation of the flattop. Moving this beam by 0.2 sigma results in a 4% perturbation. A tolerance of 0.15 sigma has generally been accepted. This is still quite a generous tolerance, owing to the magic of the Gaussian. More of this is discussed in Chapter 17.

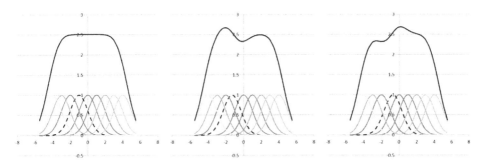

FIGURE 12.54 Misalignment of one of the summed Gaussians.

While nothing has been mentioned about where, in depth, these observations are made, it is the case that these are mostly valid at any depth. However, an asymmetric beam will become less asymmetric at a deeper penetration.

In the scattering modality, the beam penumbra is sharpened with a collimator. This is necessary since the beam divergence is large due to the scattering mechanism. When comparing how this collimated beam propagates with respect to the pristine Gaussian peak some observations can be made. Even a collimated beam still has a greater angular divergence and its growth may be faster than the growth of a Gaussian beam with smaller angles. Relatively speaking the scattering in the medium is proportionally larger for the Gaussian beam, but the larger initial angles of the collimated beam dominate at depth. This is one situation which Murphy's law has ignored. It is generally advantageous to have a sharper penumbra for shallow targets and collimators for shallow targets are smaller and lighter and can be used with larger sized scanned Gaussian beams.

12.3.13 SCAN PATTERNS

The 'field area' is the cross sectional area in which the beam is to be delivered according to a particular pattern of scan lines, or positions with a certain distance between them. This pattern can be given by a path from voxel to voxel, or it can smoother. In continuous scanning, a pattern with constant frequencies along X and Y axes (f_x and f_y, respectively) can be chosen. This would result in a Lissajous pattern with spacing between the painted lines. To reach the required line spacing (δ), which depends upon the beam size and desired overlap, the frequency ratios could be adjusted.

Changing the frequency ratio $\alpha = f_y/f_x$ while keeping nearly the same size for the scanning area, can theoretically adapt the pattern to reach any required line spacing with a repeating pattern. A couple of examples are shown in Figure 12.55. Do not adjust the picture, it is being controlled. However, the limitations on frequencies and speeds along both axes due to the hardware inherent in any power supply will limit the lowest reachable spacing between scan lines. One way to solve this is to work around those hardware limitations and modify the relative phasing of the frequencies. This will reduce the distance between the scanning lines since the scan will not repeat for a to-be-defined number of cycles.

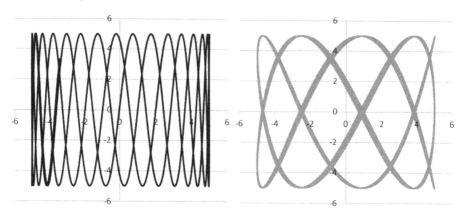

FIGURE 12.55   Lissajous scanning patterns.

In the ideal case, the beam trajectory is arbitrary and isn't dependent upon a set of fixed frequencies. In most cases the beam is essentially moved digitally to any location; however, this used to be used for US. In addition, there will always be a dependence upon the hardware constraints of the system. Such patterns can be useful for US and CS.

12.3.14   The Effects of the Scanning Nozzle on the Beam Size

The nozzle contains the components that comprise the beam delivery system as was shown in Figures 12.44 and 12.45. The penumbra of the beam at the target surface is, as discussed above, related to the beam size there. The nozzle layout greatly affects that beam diameter and can be optimized. In this section, the nozzle components inside of which the beam travels are considered, not including the magnetics. This includes the components sketched in Figure 12.56. It starts with the end of the gantry (if it exists – if not then the last dipole when it exists) which has a vacuum

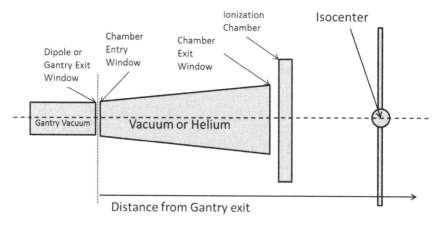

FIGURE 12.56   Scanning nozzle components for size effects calculations.

chamber that is evacuated. At the end of the vacuum chamber is material (called a vacuum window) that serves to maintain the vacuum, but also to try to minimize the scattering as the beam passes through it. This is followed by what will be called a 'chamber'. This is a general purpose container, which can also be no container and represent air. Or it can be a physical chamber that is either evacuated or filled with helium – a gas that has a lower density than air (0.16 vs. 1.29 g/l). In the latter two cases, the chamber has entrance and exit windows. There is usually a longish distance (1–3m) from the end of the vacuum chamber to the target. The scattering which will increase the beam size is minimized if the material and material density it passes through is minimized as well as the distance beyond the windows. However, the degree to which different options matter can depend on the implementation since there are trade-offs.

Even with a scanning system, there are downstream materials in the beam path such as vacuum windows and ICs. Much has been said about the advantages of an unmodified beam, so it is helpful to see how reality modifies those advantages. Some of the scanning efficacy is dependent on the beam size, and understanding the effects of these materials helps to better define the capabilities of a scanned beam. The author was fortunate to have the opportunity to work with Stefan Schmidt studying the influence of materials in the scanning beam path. It is noted that the results described herein are gantry and scanning system dependent and were performed for a scanning system mounted downstream of the last dipole of a gantry (but could just as well be without a gantry).

12.3.14.1 Chamber and Windows

The chamber must be designed with an exit window that provides enough transverse width for a beam scanned to the desired field size e.g. 30–40 cm at the target. Since the chamber ends before the target, the window width is about 20–30 cm. The thickness of the window depends on the strength required to withstand a pressure differential. Consider a plate with a uniform load (force) as in Figure 12.57. This force could be a uniform pressure differential as in between an evacuated volume on one side and air at standard pressure on the other side. The result, if the ends are fixed would be a deflection, y, as in the right side of the figure (deflection is only shown in one dimension). The actual deflection would be in the two dimensions and all four edges would be constrained. The maximum deflection allowed y_m is given by an amount that would permanently deform or break the foil. The Timoshenko formulae for thin plates provide an expression for the stress σ (not beam size).

$$\sigma = K\left[E\left(\frac{pa}{t}\right)^2\right]^{1/3} \tag{12.21}$$

where a is the short dimension of plate (m, in), p is the uniform pressure load (Pascals or lb/in^2), E is the Young's modulus of the material (N/m^2 or lb/in^2), t is the plate thickness (m, in), $K\sim0.4$.

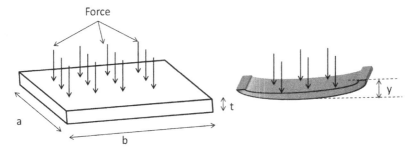

FIGURE 12.57   Force and deflection.

The goal is to ensure that the stress on the foil does not exceed the Elastic Modulus. For Kapton® that is $E = 231$ MPa. For example, a foil with $a = 30$ cm to withstand standard temperature and pressure (STP) on one side and vacuum on the other (a difference of 1 atmosphere) requires a thickness of 33 μm. This result does not include effects like bending the foil over chamber edges or non-circular windows. In addition, the effect of the large window breaking could result in destroying the ion chamber and injuring the hearing of patient and personnel nearby. This risk must be reduced (see Chapter 15) to a low probability which leads to a thicker window. The closest standard thickness with a safety margin is 125 μm.

A thicker window means more scattering. If the chamber were filled with helium instead of being evacuated this would avoid the problems associated with a large vacuum window while only adding a small additional scattering owing to the light He gas. Because the helium gas in the chamber can be kept under standard pressure, requirements on the window strength are low. For these calculations, Mylar® windows with thickness 2.5 μm are possible for the entrance and exit of the chamber.

Moving upstream, in the case of an evacuated chamber, there are no windows between the exit of the last dipole (if it exists) and the entrance of the chamber. However, when making use of a He chamber, a dipole exit window separates the dipole vacuum from the open space under normal air pressure or from the helium chamber. Again, a standard material used for such purposes is Kapton®. In this case, the minimum window thickness needed would be about 25 μm for a window diameter of 5 cm since this is before the scanning system. The window is thinner because it's smaller. On the other hand, this is further from the target, with increased leverage for the scattering angle to grow the beam size. The helium chamber could be directly attached to the gantry exit window, omitting the chamber entrance window. For service and maintenance purposes it is, however, advisable to keep the helium chamber as a separate device.

The window thicknesses and any gas in the chamber will lead to multiple scattering of the beam. The root mean square (RMS) scattering angle is proportional to the square root of the thickness of the plate. Therefore, as the field size is increased, so the beam size will be increased. This is illustrated by the plots in Figure 12.58 showing the relative dependence on field size. The dotted line is related to the goodness of the field size – it's better (in principle) when it's bigger. The dashed line is related to the inverse of the scattering angle. It's better when the curve value is larger (or when the scattering angle is smaller). The solid black line is the sum of these showing how quickly a minimum is passed. Alternatively perhaps the appropriate comparison is the product (gray line)? Of course the exact parameters will determine the appropriate optimum and the subjective 'goodness' of the relative terms. It is important to consider this in the overall system design.

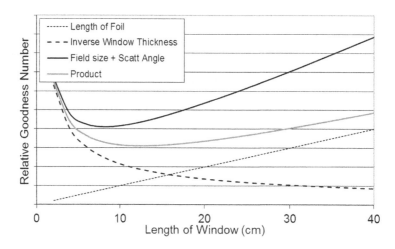

FIGURE 12.58   Effects of window size.

12.3.14.2 Downstream Materials

The beam monitor (IC) is generally installed downstream of the chamber. These chambers are constructed with windows, foils and usually have air at atmospheric pressure. The IC is followed by an air gap extending from the IC exit window to the isocenter. The length of the air gap depends on the position of the IC on the beam axis. Typically air gaps of 600–800 mm are used.

The targeted tissue to be irradiated is usually placed at the isocenter position. That is actually the definition of isocenter in this book – the location at which the target is placed. To investigate the effects of scattering in the tissue, the tissue can be simulated by various thickness water layers in front of the target.

Simulations to determine the effects of these materials were done starting with a beam width of zero transverse extent so that only the effect of the material is identified. To obtain the total width of the scattered beam, the profile of the incident beam is convoluted with the Gaussian profile from the result of these calculations, i.e. for a Gaussian incident beam the widths of the profiles at the evaluated point have to be added in quadrature. This method delivers correct results as long as any beam effects by active elements in the nozzle does not depend on the lateral beam position.

12.3.14.3 Influence of the Dipole Exit Window on the Beam

In this example, the dipole exit window is placed 2.4 m in front of the isocenter. Entrance windows are generally not used if the chamber is evacuated as it could be connected to the upstream vacuum system (although sometimes it is used for ease of maintenance). However, a window is required to separate vacuum from He, and the case considered here involves a chamber filled with He. The beam width at isocenter is sensitive to the thickness of this window. The beam width was evaluated for varying thicknesses of this window (material: Kapton®). Results as a function of the thickness of the window in micrometers are displayed in Figure 12.59. All calculated beam widths are in air at isocenter position. In this case a nominal beam at isocenter, without that window is taken to be 2.5 mm. The calculations were conducted for the minimum beam energy of 70 MeV where the scattering effect is strongest. There is a strong contribution of the dipole exit window to the total beam width, amounting to up to ~2 mm for a window thickness of 50 μm. The need for this window is a disadvantage of the He chamber. If a window thickness of 25 μm is used for a 5 cm diameter window, would the 0.36 mm additional beam size (from the addition in quadrature) be important clinically? An answer is probably not for the large majority of clinical sites.

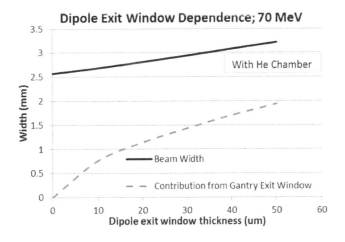

FIGURE 12.59   Scattering effect from dipole exit window (needed for He chamber).

12.3.14.4 Influence of the Large Vacuum Window on the Beam Width

If vacuum is continued through the chamber, there would be large, thick vacuum window a meter from the isocenter. A system with extended vacuum typically results in the smallest beam size, so this calculation starts with a beam size, without that window of 1.9 mm. The effect of this window (material: Kapton® 125 μm) at a position of 1.05 m in front of the isocenter is displayed in Figure 12.60. Due to the thickness (at a minimum) needed for strength, the contribution of this window to the total scattering of the nozzle is a larger than the contribution of the previously described gantry exit window, even though it is much closer to the target. The need of this window is a disadvantage when implementing a vacuum chamber. So, the effect on the beam size is larger for the downstream vacuum window than the upstream window needed to separate the dipole from the chamber. Added to that, the possible dangers described earlier, of the downstream vacuum window, can provide helpful information for a safety analysis.

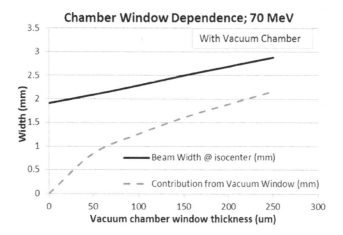

FIGURE 12.60   Scattering effect of downstream chamber window.

12.3.14.5   Effect of the Air Gap on the Beam Width

There is a gap between then end of the IC and the target. This is an air gap and the distance will have an effect on the beam width. Different nozzle configurations can be considered as well as different beam energies. Energies of 70 and 160 MeV were chosen. Resulting data are displayed in Figure 12.61.

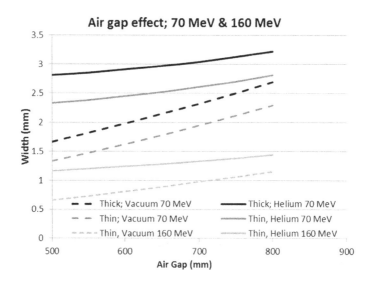

FIGURE 12.61   Air gap.

The principle effect of an air gap variation was studied for a thick window configuration (black curves) and for a thin window configuration (medium gray curves) at the lowest energy of $E = 70$ MeV where the scattering effect is largest. All show the beam widening as the air gap increases. The vacuum chamber situation results in a larger rate of increase, of about 1 mm, but it started with a smaller beam size. The helium chamber starts with a larger beam size but the growth is smaller, on the order of about 0.5 mm. The appropriate comparison is the thin window plus He chamber (medium gray solid line) and the thick window plus vacuum (dark dashed line). The He chamber configuration is less sensitive to the air gap. Since the beam is a bit smaller with the vacuum chamber, the effect is exaggerated compared with the helium chamber. Figure 12.61 also shows the results for this nozzle configuration at an intermediate energy of $E = 160$ MeV (lighter gray and lowest curves). Here, the effect of the airgap on the beam width is only about half compared to the 70 MeV results.

12.3.14.6 Beam Widths with Air Gaps Inside Target

The previous calculation stopped at the surface of the patient. Although the effect is relatively small already at intermediate energies, the previous calculation results suggest reducing the air gap. However, the lower end of the air gap range is limited by the need to keep some space between the nozzle and the patient open for the medical personnel to assist the patient and to work with other medical equipment. A minimum air gap of 600 mm is typically reasonable. The maximum is limited by the space taken by the scanning magnets and instrumentation for downstream scanning systems (these studies are gantry and scanning system dependent) and may be around 800 mm. For these two values, a more detailed investigation of the scattering effects, including inside the patient may give some insight into the priorities. Figure 12.62 shows some of the results.

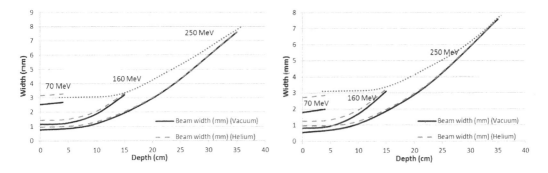

FIGURE 12.62 Scatter in water 800 mm gap (left) and 600 mm gap (right).

There is a wealth of information contained within these plots. Most notable is the reminder that the beam continues to scatter as it penetrates the patient. Therefore, the beam size at the target is not the same as the beam size in air just before the patient. In fact, for all cases, while the lowest beam energies have the largest beam sizes in air just before isocenter (because of scattering in the nozzle), the beam sizes at the end of range for all the other energies are in fact, larger. It is also the case, that except for the lowest energies, the difference, at depth, of the beam size is small when comparing the vacuum with the helium chambers. The light gray dotted line is really the most relevant curve, unless the beam size on the way to the target is important (which could be if the beam passed between critical structures, but that isn't the way it's done).

At the surface, there is definitely a difference between a vacuum/helium system, on the order of 2.5/3.1 mm (800 mm gap) and 1.9/2.9 mm (600 mm gap). The worst case, (low energy at 4 cm depth), the beam size difference between vacuum and He is 0.6 mm and 0.87 mm for an 80 cm and 60 cm air gap respectively. Is that clinically significant?

The results show that the choice of an air gap between 60 and 80 cm only matters at very low energies. At higher energies

- the air gap can be chosen nearly freely,
- the difference between using a helium or a vacuum chamber diminishes, and
- scattering in the tissue starts dominating the scattering by the nozzle elements at depths larger than 5 cm.

12.3.14.7 Effect of Range Shifter on Beam Width

Now, here's a clever and perhaps, counterintuitive, trick. Remember that the beam size is growing in the nozzle as it passes gasses and windows and the fact that this effect is reduced at higher energy. Wouldn't it be nice if it were possible to take advantage of that while also ensuring that the beam range was appropriate for treatment? A way to do this is to place a range shifter upstream of the isocenter, so that a higher energy beam can be transported through the nozzle, and thus, the beam size growth in the nozzle is reduced through all the upstream windows and gasses. This range shifter then shifts the range just before the target to obtain the correct range. The energy of the beam incident on the range shifter is adjusted so that the energy after the range shifter is the desired range. The graphs of Figure 12.63 indicate the results of this technique.

Also included in these graphs is a negative side-effect of this trick. Since the beam has passed through more material, the amount of range straggling will increase and this increases the extent of the distal falloff at the end of range. Whether or not this is significant when compared to the distal falloff in the patient depends upon the energy as seen in the figures.

Another opportunity for a range shifter is for fixed energy accelerators that use a degrader/energy selection system for energy selection. Since the dose rate drops steeply with energy, transporting a higher energy through the beamline can help to increase the dose rate at the patient. Finally, it still takes some time to change the energy of a magnet system. To date, it is still possible to shift the energy more quickly with a mechanical system, such as a moving range shifter.

The graphs of Figure 12.63(top) (70 MeV) and Figure 12.63(middle) (100 MeV) show how the beam size changes as a function of the thickness of the range shifter. Curves are shown for the beam size at the surface and the beam size at the final depth. (For 70 MeV, this is about 4 cm.) These results include both the smaller beam through the nozzle and the increased scattering in the range shifter. With increasing range shifter thickness, the incident beam size at the end of range is reduced and then increases. For the lowest energy, at both the surface and at depth, there is a clear minimum beam size achieved with range shifter thicknesses between 3 and 6 cm. In such a case, the beam energy transported through the Nozzle is closer to 120 MeV instead of 70 MeV. A similar result is obtained for the 100 MeV equivalent, however the beam size at depth is not susceptible to optimization. For 160 MeV, shown in Figure 12.63(bottom), there does not seem to be an optimization possible at all, since the beam energy is already high enough. Note that for all these cases, the distal falloff grows with Lexan thickness. At the lowest energy it grows approximately an additional 1 mm at the optimum.

These results, for realistic scanning nozzle configurations indicate that in the worst case, the difference of the beam widths between a vacuum and a helium chamber is about 0.4 mm even at 4 cm depth. At 160 MeV, this difference already drops to 0.4 mm at the surface and only 0.1 mm at depth. The beam width at the low energies can be further reduced with a range shifter. The design question of the cost, potential dangers of vacuum window breaks must be compared with the sub millimeter difference in the two configurations. These are elements of a safety analysis.

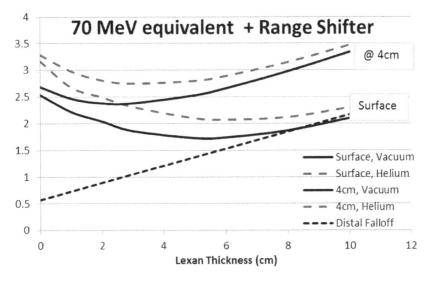

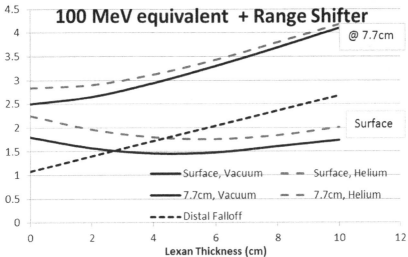

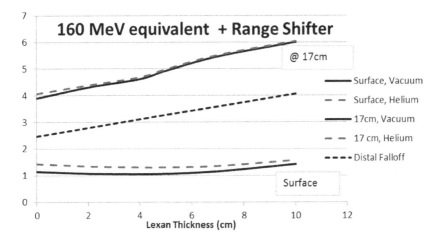

FIGURE 12.63 Range shifter: (top) 70 MeV, (middle) 100 MeV and (bottom) 160 MeV.

Overall, it is shown that helium is a lower cost, effective and possibly safer alternative to a vacuum system. It works – as shown in Figure 12.64 that are photos of a scan pattern in air (left) and the same pattern after the beam has gone through helium (right). Helium is less dense than air. However, many vacuum systems are in operation and are also performing safely. Alternatively if the helium chamber were extended upstream and the IC were replaced with a thin scintillator or equivalent the beam size could improve at the expense of additional straggling and regulatory challenges.

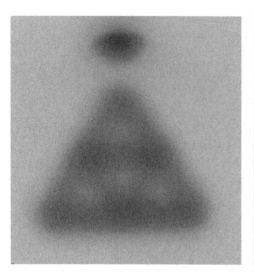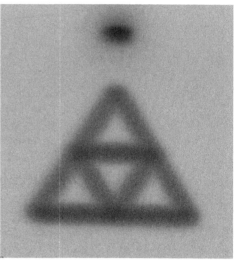

FIGURE 12.64   Scan pattern in air (left) and after He (right).

12.3.15   SCANNING CONTROL AND REQUIREMENT CONSIDERATIONS

The highest importance must be given to a safe and accurate delivery of dose to the patient. The scanning system must ensure that the correct dose fidelity is achieved, which includes giving the right dose to the right place, and not giving the wrong dose to anyplace. This could be a voxel by voxel consideration or a more global view such as an on-line gamma index (see Chapter 17). The method to use to ensure this, represents severe constraints on the implementation of the scanning system. This will impact the control strategy given the time constants built into the system including the time required for measurements and beam current limitations (i.e. dose rate). Also, a safety plan should contain the definitions of the required types of redundancy and sensors to be used and help define the interfaces with which the scanning system will interact both for measurements and beam control (see Chapter 15).

12.3.15.1   More About Parameters

Quantities Required to Control the Beam Dose Delivery

The scanning system controls equipment parameters, reads back instrumentation and makes decisions about the settings of the equipment parameters based upon the instrumentation parameters. Table 12.4 represents a traceability of some important beam quantities to those devices used to set those properties and thereby a first look at the mitigations that could be used to address errors. In addition, the instrumentation used to verify that the settings of the equipment are correct, as well as the instrumentation used to measure the beam properties as a 'functionally' redundant measurement, are listed. This is all to be considered a first pass at data for scanning controls and safety analyses, useful for Chapter 15. Table 12.4 is for a system with a beamline and an energy selection system. Some systems do not have these subsystems.

TABLE 12.4

Scanning Setting and Measurement Parameters

Beam Properties	Setting This Subsystem	Setting This Subsystem	Equipment Instrumentation	Beam Instrumentation
Energy	Beam transport ESS Degrader	Beam delivery	PS current Mag field probe	Range check
Position	Input beam trajectory	Scanning magnet current	PS Reg Ok SM current SM voltage Flux coil Hall probe	ICs
Transverse velocity		Scanning magnet current and voltage	PS Reg Ok SM current SM voltage Flux coil Hall probe	ICs
Size	Quadrupole settings	Quadrupole settings	Quad current Quad voltage BPM reading	ICs
Beam current	Ion source ESS Beam transport system		Ion source Slit settings BPMs maybe	IC cyclotron ICs IC2/3 dose pad IC2/3 QC
Dose		Scan controls	Scan controls	ICs Redundant dose counter

Of course all the above parameters will need a time scale associated for the specification. Some acronyms that were used above include:

ESS = energy selection system
SM = scanning magnets
PS = power supply
IC = ionization chamber
BPM = beam position monitor
QC = ionization chamber quality channels

High Level Beam Scanning Parameters and Tolerances

There are high level requirements that are necessary, independent of the implementation, but they might be used to determine the implementation. The design will define the limitations of the equipment. For example, the speed with which the dose can be acquired and the rate at which the scanning magnet current can be set are important quantities. These might come from higher level desires (e.g. treatment time) or not.

In general, one or more of the following conditions should be met (some are alternatives):

• *Dose delivered should be accurate to within 2% at any point in the target volume.* This results in a requirement that the scanning controller react sufficiently quickly to a signal to terminate the beam. It is likely that there should be redundant methods to accomplish this.

• *Dose delivered should be within 2.5% of the prescription at any point in the target volume or within 1–3 mm of the desired location.* (Gamma index analysis.) This means that at any given time, it is necessary that the correspondence between the beam position and beam current be sufficiently accurate and the beam position be sufficiently accurate.

- *The spatial distance over which the beam intensity should be changed, is small (e.g. 20% of the beam sigma).* (See Chapter 17.) The absolute requirement of this in terms of the time to modify the beam current, depends on the speed of the scan and the size of the beam.
  - If the speed of the scan is v, the time to vary the current (from full intensity to 0 or vice-versa or anything in between) $= 0.2*(\sigma/v)$.
- *The entire 3D dose distribution should be delivered in a time period between 10 seconds and 2 minutes.* Shorter than that is usually considered too quick (based upon old-fashioned beam delivery methods during which a therapist could possibly intervene) and longer than that would result in target motion or patient discomfort issues. There are many contributions to the overall delivery time, such as time to change the beam range, and whether or not each 2D scan is delivered once or repeatedly. This is a softer spec.
- In the near future it could be that much of this is modified and it will be important to deliver the entire volume, with high 100 Gy in a fraction of a second, if, for example, FLASH becomes a required treatment modality. This is quite challenging.

12.3.16 FASTER SCANNING SYSTEMS

Within the last few years, what some may consider the next generation of scanning systems are starting to be used clinically. Actually, these systems were designed several (up to 10) years ago, but that started about the time that some commercial systems were already being introduced for clinical use. The primary attribute of these systems is speed (with commensurate accuracy). While it may be that the time it takes to irradiate an average volume in a hospital setting these days is from 30 seconds to a couple of minutes, the newer systems seem to be able to irradiate an entire volume in a somewhat reduced time (e.g. several seconds). While, clearly, if a full target volume were to be able to be irradiated in about 1 second, this would, change the game with respect to the issue of organ motion. In any case, with a fast delivery and appropriately synchronized timing spread over the course of a minute, the effects of target motion can be averaged out. The evolution to a few seconds of time to irradiate a volume is going in a useful direction, but it has not reached that point yet unless, for example respiration holding is used and cardiac motion is not included.

How have these systems accomplished this feat when scores of commercial systems currently installed are slower? These systems, at laboratory clinics like PSI and the National Institute of Radiological Sciences, QST (NIRS) use a cyclotron and synchrotron respectively, and therefore, that is not the issue, although they have implemented some unique aspects to both to achieve the goals. It should also be understood that each of the system parameters identified in this section have been addressed from the perspective of time and safety, so systems that provide safety have been improved. To be fair, commercial firms are also coming out with faster scanning systems now.

Energy

The time to change the beam energy has been reduced from seconds to fractions of a second, as low as, or lower than 100 ms. In the case of the NIRS system it is done through multi-energy extraction (see Figure 13.29) and fast beamline magnetic field changes. In the case of PSI, it has been done through the use of a fast linear motion mechanical degrader and closed loop, feed forward, beamline magnetic field changes. These fast magnetic field changes are not simple, requiring predictive algorithms and very fast correction magnets. Note that for 25 energy layers, a 0.1 second energy change time contributes 2.5 seconds for just the energy change time, so the energy change time is still a dominant factor and too long for simple respiration mitigation.

Dose Rate

Until recently the parameters used in clinical ICs have been relatively standard, but now these values are better optimized for speed, linearity and accuracy. The NIRS instrumentation can operate at 200 kHz (5 μs), and the instrumentation used at PSI is focused on precision.

Scanning Magnet Speed

While the speed with which the current in an electromagnet can be changed is related to the amount of voltage applied $V = L \times dI/dt$; an optimized design of the scanning dipole can reduce the inductance (L) and, thus, increase the current change speed for the same voltage applied. In addition the effects of magnetic field variation such as eddy currents are controlled by care and detailed design of the laminations and electric coil designs of these systems. Taking care of such parameters has resulted in these systems being capable of increasing scanning magnet field change speed from 4–40 Hz to 100–200 Hz. Table 12.5 illustrates the performance of these new scanning systems.

TABLE 12.5
Parameters of Recent Faster Scanning Systems

Parameter	NIRS System	PSI System
Beam intensity	3×10^7 to 1×10^9 particles/s	>0.5 nA (3×10^9 protons/s)
Scan speed (at maximum energy)	v_x > 100 mm/ms	20 mm/ms (shorter direction)
	v_y > 50 mm/ms	5 mm/ms (larger direction)
Energy change time	<300 ms	<100 ms
Minimum time/spot	25 μs	
Average time/spot	200 μs	3–5 ms
Time to delivery one layer	~40 ms (10 cm $\times$ 10 cm)	~80 ms (12 cm $\times$ 20 cm)

12.3.17 SPREADING BEAMS SUMMARY

Particle beam scanning represents a technical improvement in obtaining the most conformal dose distributions possible with particles. Additionally it can allow for dose delivery without patient-specific hardware. This enables the most efficient beam utilization. Since, in many situations no patient-specific hardware is required, multiple field irradiations can be delivered without entering the treatment room between fields and this has indeed started. The depth dose distribution together with the conformality of this approach leads to a minimization of the number of beam fields necessary for an excellent dose distribution, although there may be clinical considerations in these numbers. The number of fields can be so reduced that consideration of gantry-less solutions are being considered.

The overall system is in fact simpler than some scattering systems; however, the implementation of the system with respect to timing and tolerances is more challenging. In almost all scattering implementations (with the exception of those using ridge filters), the term 'passive' scattering has been misused. Most modern scattering systems require synchronization of the beam with a moving device, either a wheel or paddles, for example. In some cases, the beam current is further modulated. In fact, the implementation of an 'active' beam scanning system can, in some ways, be considered less complex than some so-called passive scattering systems that are used.

Single scattering provides the smallest penumbra (given the small effective source size), but is very inefficient in terms of beam usage and can result in secondary radiation produced. Double scattering is more efficient, in some cases up to 40%, and uses patient-specific apertures. Beam scanning is essentially 100% efficient and minimizes the secondary radiation delivered to the patient. However, it may not provide the best penumbra depending upon the size of the unmodified beam.

Creating a very small beam is very difficult and expensive and can lead to longer treatment times (under some conditions), and it may be worth considering some cases for which lightweight apertures can produce an advantage.

A scanned beam delivery may provide the most efficient treatment scenario. This can open the possibility to deliver multiple fields without entering the room. On the other hand, there may be some situations in which the use of a range compensator is useful to minimize the beam-on delivery time (by reducing the number of energy layers required), when complex 3D concave shapes are involved. Hundreds of thousands of patients have been successfully treated using these techniques. While beam scanning can produce a more conformal field distribution, quite complex cases can be safely treated using any of the techniques described.

Given the fact that it is not normally necessary to prepare hardware that will control the beam delivery, in advance, for a scanning system, it is ideally suited for adaptive radiotherapy. Adaptation is not only relevant for interfraction effects, but can also be important for intrafraction effects. Indeed the largest obstacle to widespread use of scanning in all body sites is still due to organ motion, although it is used more and more. In general, the scanning technique contains knobs to compensate for all types of organ motion effects, if the treatment planning can be adapted to account for these effects and the accelerator system is nimble enough to compensate.

On the whole, the cost of delivering a beam using scanning can be lower than other alternatives and the treatment can be more efficient, thus positively contributing to the cost effectiveness of particle therapy. Scanning has been integrated into hospital-based clinical treatment facilities since 2008 and is not just in the laboratory environment anymore. This represents a new phase in the evolution of particle beam therapy and research.

12.3.18 How to Build a Scanning System

The development of a scanning system for a hospital environment requires careful attention to detail and inclusion of robust safety and manufacturing practices along with sophisticated engineering. It has to be a system that will be safe, foolproof and easy to use. Documentation and extensive testing require the utmost concentration. A team is needed who can withstand the long hard working hours and tower over the terror of riding the ups and downs with hands raised together in unison like the team in the photo in Figure 12.65 representing some of the key contributors who built the scanning system for MGH in 2008. Left to right: Didier Leyman, Damien Prieels, Ben Clasie, Frank Franssen, Alberto Cruz, Philippe Thirionet, Yves Claereboudt and in front of them, to the right, this author.

FIGURE 12.65 It takes a scanning team.

12.4   WHAT COULD GO WRONG – 5?

A lot has been added in this chapter, so hold onto your hats. This part isn't easy. The beam delivery systems are the workhorses of the treatment. They carve the beam into the size and shape that is required for the patient treatment and determine the dose that is delivered. Depending upon the beam delivery modality there are different components, parameters and potential errors.

It's probably best to start with the beam delivery parameters that characterize these beam spreading modes (from Figures 12.5, 12.16 and 12.49) and reproduced in Figures 12.66 and 12.67.

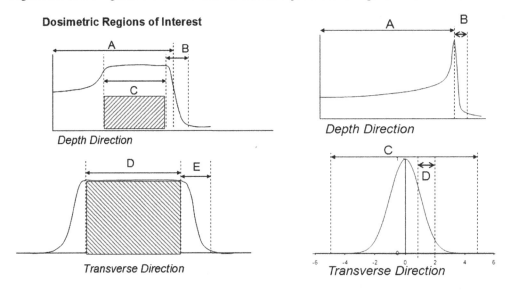

FIGURE 12.66   Scattering dosimetry.   **FIGURE 12.67**   Scanning dosimetry.

12.4.1   Longitudinal Spreading

The scattering modality is characterized by the SOBP and the scanning modality is characterized by the BC and/or BP. In practice the BC comes directly from the accelerator and the SOBP from equipment used after the accelerator, but starting with the parameters from the accelerator. Some of the potential errors can include the following:

- Use of the wrong hardware, such as the wrong ridge filter
- Use of the wrong RMW or track on that wheel
- Beam turning on and/or off on the wrong range modulator step
- The incorrect beam current modulation signal and/or the wrong beam current modulation accelerator output despite the correct signal
- Unexpected signals that could affect the beam timing relative to the phase of the RMW rotation.
- Incorrect positioning of a binary lamination or a wedge degrader
- Literally the wrong energy being extracted from the accelerator or degrader/ESS system

12.4.2   Transverse Spreading

The scattering modality is characterized by a spread out (sometimes uniform) transverse distribution, while the scanning modality is characterized by the pristine Gaussian distribution (if an error didn't modify the Gaussian distribution). Some of the potential errors can include:

Scattering
- A wrong scatterer inserted
- Any scatterer placed in the wrong position (that isn't flat)

- The beam trajectory misplaced relative to the scatterer positions
- The collimator is misaligned
- The wrong collimator is installed
- The compensator is misaligned
- The wrong compensator is installed
- The collimator or compensator is damaged.
- The beam source size is incorrect and the penumbra will be modified
- The beam distribution is incorrect
  - Some material in the beam path has changed
  - Some material (e.g. a cable) has fallen into the beam path
  - The IC is misplaced or damaged

Scanning

- The beam is in the wrong position
  - Some magnetic field(s) is incorrect (beamline or scanning system)
  - The IC is misplaced or damaged
- The beam dose is incorrect
  - The beam current is unstable, beyond the ability to detect and control it
  - The magnetic field ramp is incorrect or unstable
  - There is heightened electronics noise
- The scanning map is incorrect
- The scanning magnetic field settling time is too long and the beam is turned on before it settles
- The IC and electronics readout time is too long
- The scanning speed is too fast compared to the beam turn off time
- The vacuum window is ruptured
- The scanning beam size is incorrect
  - Some material in the beam path has changed
  - Some material (e.g. a cable or beer can) has fallen into the beam path
- The beam is not symmetric and the dose distribution will be incorrect
- The beam energy is incorrect
- An incorrect range shifter or ridge filter is used
- The gas or gas pressure in any of the chambers is incorrect

This is a partial list of the errors that are possible considering the components that have been introduced in this chapter. Remember that the contributions of the errors summarized in previous chapters must be combined with these.

12.5   EXERCISES

1. Following on from Equation 12.2, derive the values of the weights w_1, w_2 and w_3 in terms of the BC amplitudes b_1, b_2 and b_3.
2. How long will it take to deposit a modulated dose across a lateral dimension split in thirds. The dose in the first third = 1, second third = 3, third third = 1? Do this for constant speed and again for constant dose rate. Assume the maximum dose rate is that used in the middle third of the constant speed case. Compare the relative times to the example in the text.
3. Approximate the increase in penumbra for a beam with a 1 cm sigma moving at 20 m/s and a beam turn off time of 500 μs. Assume a linear ramp. (Computer not needed.)
4. How much charge is deposited by the beam moving at 20 m/s with a current of 10 nA in a 1 cm wide and deep (end of range) volume (just a BC)? The beam energy is 160 MeV.

5. For a pulsed beam delivery, what is the final accuracy if a pulsed beam is delivered with 4 pulses? Use a 10% error per pulse. A total dose of 2 is desired.

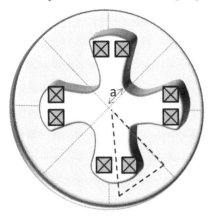

FIGURE 12.68   Exercise 8 hint.

6. How long would a treatment take with a pulsed beam that has average current of 0.5 nA if the pulse rate were 60 Hz, with 8,000 spots? It is desired to deposit 40 Gp total. What is the maximum current allowable? Consider only the beam current contribution to the treatment time. The target is 1 liter of water.

7. A scanning magnet for protons is designed with a maximum field of 3 kG? Assume a 5 cm full gap, a pole length of 0.3 m, pole width of 0.2 m and a power supply with 100 A.

8. What is the inductance of this scanning magnet?

9. What extra power supply voltage is needed to ramp this magnet 1/3 of its range in 10 ms?

10. Derive an expression similar to Equation 12.14, for a quadrupole. Use Figure 12.68 if some help is needed.

13 Accelerator Systems

Particle therapy technology serves to create a beam which can achieve the clinical goals. This is the simple axiom that should be adhered to. The device that increases the charged particle energy (at relativistic speed) to the desired value of range penetration is called the accelerator. Different technology can result in different beam properties. Understanding this can greatly help in determining how to design the system to deliver a useful and safe treatment.

Part of the reason for the size of the particle accelerator and beamline systems results from the Lorentz force law. If it is necessary to change the direction of a particle or focus a beam, it must be bent. The radius of curvature of the bend in a magnetic field is repeated here.

$$B(\text{kG})\rho(\text{m}) = 33.356\,P(\text{GeV/c})/Z \qquad (13.1)$$

In this equation, a particle with a charge of Z and momentum P (in GeV/c) is bent in a magnetic field B (in kilogauss) through a radius of curvature of ρ (in meters). Particles of therapeutic energies must have a range of about 30 cm in water. For protons, this is close to a proton kinetic energy of 230 MeV, and for carbon ions, this is close to a kinetic energy of 430 MeV per nucleon. As discussed in Chapter 8, particles with therapeutic energies travel at relativistic speeds. If the bending radius scales with the momentum, one can understand how the scale of ion therapy equipment (for protons and heavier) grows. A normal electromagnet can reach a field of about 16 kG before saturation. A 230 MeV proton in a 16 kG field has a bending radius of 1.45 m, and a carbon ion of 430 MeV/μ has a bending radius of 4.2 m. Superconducting technology provides the option to reach magnetic fields of several Tesla (1 T = 10 kG) with a commensurate reduction in bending radius.

13.1 ACCELERATOR TECHNOLOGY

Accelerators are devices which produce and shape an electric field in such a way as to accelerate the charged particle. From the Lorentz law, the electric field is key since the force from the field has to be in the direction of the particle's motion for it to gain energy. An electric field can be formed in different ways. For example:

- $E \sim dB/dt$; Maxwell's equations indicates that a changing magnetic field can produce an electric field. This was used for the betatron, but beam physics limitations restrict this technique to lower energies and light particles.
- $E \sim$ Applied voltage; An applied voltage can be either AC or DC; however, for clinical energies of hundreds of millions of volts obtaining these fields in DC involves serious engineering challenges. AC fields generated with microwave techniques are most common, thanks to the work done in the field of microwave radars.
- Finally, one can create an electric field, or basically reform the electric fields that exist in atoms owing to the existence of the charged particles in the atom with, for example the use of lasers.

Practically, one has two choices when accelerating a particle. One can send the particle through the accelerator once (single-pass schemes) or multiple times. A LINear ACcelerator (LINAC) shapes the electric field in such a way as to accelerate a beam in a linear path (just once), and thus,

DOI: 10.1201/9781003123880-13

the length is proportional to the strength of the electric field and the energy gain desired. The power required is also related to the length. Conventional linear accelerators do not produce sufficient electric field strength to enable the construction of a compact system, although nonconventional techniques have been investigated (such as a dielectric wall accelerator [DWA]). So far technology limits that option. Even more unconventional is the use of an incredibly powerful laser to reshape the electric field of an atom and obtain very high accelerating gradients. Technology and details currently limit this approach too. Of course the machine doesn't have to be strictly linear, but the overall length of the single-pass accelerating system is still determined by the acceleration gradient (MeV/m). The term radiofrequency (RF) comes from the fact that the frequencies used are in the radio range, like that of your local FM station or radars or microwave ovens.

One way to reduce the size of the machine and power required to accelerate charged particles is by efficient reuse of the electric field (multipass acceleration schemes). Therefore, circular machines such as a cyclotrons, synchrotrons or recirculating LINACs (a nostalgic favorite of the author) and related devices can be used. In general, due to the energy of the particles required for clinical use, these accelerators are larger than a conventional photon LINAC. However, modern superconducting cyclotrons and modern synchrotrons can actually fit in the same room as a modern photon LINAC.

The time during which the patient occupies the room depends upon the choice of the activities defined by the clinic, but usually includes in-room patient alignment which could take several minutes. In this case, the accelerator beam usage could be a small fraction of the required room time. Thus, for maximal use of the machine it could be more economical for it to feed up to several rooms if the number of patients being treated supports that choice.

There is also the opportunity to combine elements of different types of accelerators to achieve a balance among beam performance, cost and size. Effort has been applied in the areas represented in Figure 13.1. As an example, note that elements of LINACs and cyclotrons are combined in a cyclinac solution. Also, within each type of system there are variations on the theme. For example, the fixed-field alternating gradient (FFAG) accelerator can be implemented in a scaling or a nonscaling method, and there are pros and cons to both.

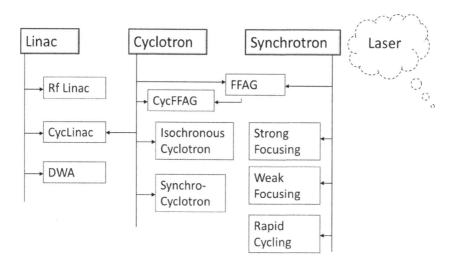

FIGURE 13.1 Types of accelerators.

For most of the history of charged particle therapy, and in current practice, the cyclotron and the synchrotron has been used. While other schemes continue to be investigated, only these two time-tested devices will be covered in this chapter.

13.2   TIME STRUCTURE OF ACCELERATOR BEAMS

The time structure of the beam is sometimes important in the consideration of the applicability of an accelerator. It is useful to define a few terms. The duty factor of the machine is the ratio of the time the beam is on T_b to the total period between beam extractions T_o as depicted in the upper portion of Figure 13.2. Duty Factor = T_b/T_o. During the time the beam is defined as 'on' (T_b shown as a value of one, or the gray hatched box), it could be on continuously as in a DC current, or continually due to other effects such as RF acceleration. In the latter case, the beam can only be on during the time period when the appropriate accelerating electric field is present as in the lower portion of Figure 13.2. The lower portion shows an expanded view of the first pulse of 'on' beam. This shows a sinusoidal electric accelerating field. The beam is represented by the gray filled box on top of the sine wave (dashed curve). In this particular configuration, the beam is only accelerated when the sine wave is at a maximum. Particles to the left and right of the maximum are accelerated less than those in the center.

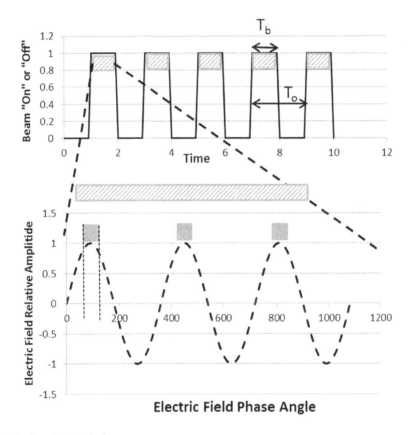

FIGURE 13.2   Accelerator timing.

There is generally a range of energies that can be accepted in the accelerator, which will define the energy spread in the beam when it is extracted. The phase location for the accelerated beam is different for different accelerators.

An example of the RF structure of a cyclotron is shown in Figure 13.3. A diamond detector, which is very fast, is used to detect a very low flux of protons. The oscilloscope timing is 20 ns/division. The cyclotron has an RF period of about 10 ns, which can be seen in the upper trace. The lower trace is for a proton flux that is 1/10th that of the upper trace. It is at the level where there are only 2, 1 or 0

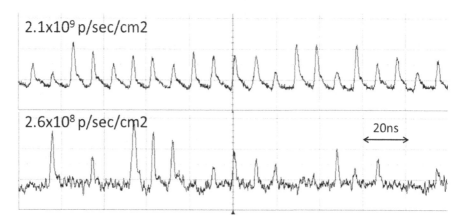

FIGURE 13.3   Pulse structure of cyclotron from a diamond detector.

protons within each RF cycle. Not all the RF 'buckets' are full, and the pulse amplitudes are indicative of the number of protons. This indicates the statistical nature of the production of protons at the source. Also note how there are discrete pulses, not a continuous trace, indicating the phase angle acceptance length of the cyclotron system.

13.3   CYCLOTRON-BASED BEAM PRODUCTION

The charged particle beam path in a simple cyclotron is shown in Figure 13.4. An electric field is created in the center gap (direction at a particular time, shown by the solid arrows), and on either side, a magnetic dipole is positioned (shown by the rotated **D**-shaped shapes). The beam is injected into the center of the cyclotron (shown by the filled circle) and accelerated by the electric field in the center toward the dipole. In this type of cyclotron, the RF structure is tucked under the magnet D shapes, so the RF structure has been called a 'Dee'. When the beam leaves the electric field gap region, it enters the magnetic field region and is bent 180 degrees and reenters the electric field region at the correct time to be accelerated in the opposite direction (the electric field is an AC field – there is a time dependence). The size and weight of a cyclotron is determined by the bending radius of the charged particles which depends on the particle mass, charge, energy and the strength of the cyclotron magnetic field.

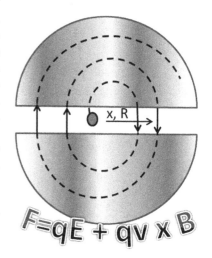

FIGURE 13.4   Cyclotron.

It is interesting to note that the technology to produce proton beams of sufficient energy for clinical therapy has advanced over the past decades. In fact, as shown in Figure 13.5, the weight of a cyclotron required for this has been reduced logarithmically. This has resulted from both efficient design improvements and superconductivity with magnetic field strengths from 3 to 7 T, or more. The use of a cyclotron for the acceleration of heavier particles for therapy had been thought to be too expensive but more recently a cyclotron that would weight about 600 tons, and only 6.3 m in diameter, has been designed by IBA that would accelerate ^{12}C ions. The final shape of the components of the cyclotron includes consideration of all of these and other issues

The energy required to maintain a DC field is higher than that required by an AC field. However, the AC field is time dependent and the field direction alternates as in Figure 13.2 (lower curve). The

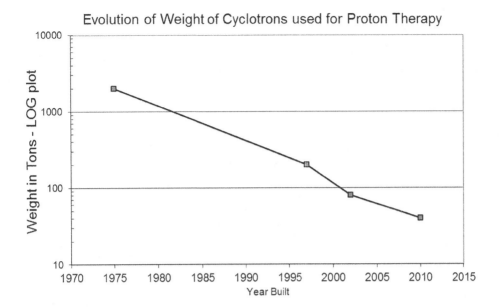

FIGURE 13.5   Reduction of medical cyclotron weight.

frequency of the electric field must be synchronized with the time it takes for the charged particle to follow the path shown in Figure 13.4 and then arrive in the gap region to get its energy boost.

Looking at the path of the particle in the magnetic field, the centripetal force keeping it in a circular arc is given by the magnetic field force so that

$$\frac{mv^2}{r} = qvB \tag{13.2}$$

where
m = mass of particle being accelerated (if you're asking if this is relativistic or not, good for you. But please wait)
v = speed of particle
r = radius of curvature of the orbit
q = charge of the particle
B = magnetic field

This can be rewritten as

$$m\frac{v}{r} = qB; \; \omega = \frac{v}{r} = 2\pi f \; \therefore \; \omega = \frac{qB}{m} \tag{13.3a,b,c}$$

where v/r = the angular frequency which is defined as $\omega = 2\pi f$ where f is the inverse time to move 360 degrees, or the frequency of the revolution of the particle (or some multiple of that). This gives the miraculous result of the cyclotron equation: $\omega = qB/m$. This relation indicates that if the mass and charge are constant then the frequency and magnetic field are a constant ratio and independent of the particle energy. This was the principle that led E. Lawrence, in 1929, to the invention of the cyclotron. With respect to the beam optics formalism, this condition is sometimes referred to as an isochronous cyclotron. It is not related to the path length $(z/\delta) = 0$, the usual isochronous condition, but equal travel time and this can happen even if the path lengths are not equal. It is more related to the time independence of the orbit, almost like $(\Delta t/\delta) = 0$ (if that were a thing). The time to travel the required path length is always synchronized with the RF period. The dispersion (x/δ), is not zero, since the radial position of the beam depends on its momentum. The radius R, from the center of the cyclotron is related to momentum and field by

$$R = \frac{P}{qB} \tag{13.4a}$$

$$dR = \frac{dP}{qB} \tag{13.4b}$$

$$dR = \frac{R}{P}dP \tag{13.4c}$$

$$\text{But } \Delta x = (x/\delta)\frac{dP}{P} \therefore (x/\delta) = R \tag{13.4d}$$

where x and R are used semi-interchangeably. The horizontal distance is x in Figure 13.4 and R, the radius of the particle orbit, will define where in x the beam enters the region between the magnetic poles.

Now, returning to the good question some of you asked while reading the previous paragraphs. If the energy increase is large enough, the effective mass increases due to relativistic effects and the angular frequency is no longer energy independent since m is viewed as energy dependent. Thus, to compensate for the mass increase, either the magnetic field and/or frequency cannot remain constant and still support the continuing acceleration of the particle. The particle will slip in phase with respect to a fixed-frequency electric field and no longer gain sufficient energy to be accelerated to the desired level.

If that is the case, the cyclotron can be built with the appropriate magnetic field pattern to increase the magnetic field with radius in synchrony with the mass increase and the orbits would be time synchronized. This will turn out to be easier said than done. Alternatively, the frequency of the electric field can be modulated in synchrony with the energy increase and the particles' orbit increase, thus forming a 'synchro-cyclotron'. In the latter case, only a short pulse of particles will be able to track the modulated RF as it is changing. Particles that, for example, lag behind will be subjected to a frequency that is not appropriate for them.

13.3.1   Phase Slippage

If the frequency of the RF is not well matched to the magnetic field, or the system is not quite isochronous enough, the particles may not complete their acceleration. Let f = the frequency of the RF and $1/f = T_{rev}$ which is the time for one revolution. When the particles go half way around, they face the other direction and the RF field has changed sign too so the timing is still good. Now change the RF frequency. The particle travels a revolution in T_{rev}; however, since the RF period, T_{Period}, has changed the new RF period does not match T_{Rev}. The fractional change in the period is related to the frequency change.

$$\frac{dT_{Period}}{T_{Period}} = \frac{-(1/f^2)df}{1/f} \rightarrow \frac{-df}{f} \tag{13.5}$$

Note that $(T_{rev} - T_{period}) \times 2\pi/T_{Period} = \Delta\phi_{RF}$ which is the change in the phase angle of the sinusoidal RF wave that the particle experiences the next time it gets to the gap with the RF field.

$$\therefore \Delta\varphi_{RF}/2\pi = -\Delta f/f \tag{13.6}$$

After the particle has circulated N turns, the total phase slip is given by

$$\Delta\varphi_{RF} = -2\pi N \Delta f/f \qquad (13.7)$$

When the particle slips a total of $\pi/2$, the acceleration stops, and if it didn't get out by then, say goodbye to that particle. Upon subsequent turns, it will start to decelerate since the phase is still slipping, for a while. Therefore, the maximum allowable error in the RF frequency is given by: $\Delta f/f \ll 1/(4 * N_{total})$, if this is to be prevented from happening before the beam is extracted which takes N_{total} turns.

13.3.2 CYCLOTRON FOCUSING EFFECTS

If it is attempted to maintain a constant electric field frequency, B must increase as γm_o.

One method of increasing the magnetic field with radius is represented in Figure 13.6 with a magnetic gap that decreases with larger radius. Since the boundary condition of a magnetic field is to leave the iron, normal to the boundary, this curves the magnetic field (black dashed arrows) resulting in components of the magnetic field in the radial direction (horizontal lighter dashed arrows B_r). Using the Lorentz force law, it is possible to explore the direction of the forces resulting in the following analysis. First, which way is the particle moving? If the main magnetic field (using B_z for now) is pointing down and it is known that the particle is contained within the cyclotron and, therefore, is experiencing a force toward the center, with the right hand rule, it is clear that the particle is moving into the paper on the right hand side and out of the paper on the left hand side. Unfortunately, from the Lorentz force law, it can also be seen that the radial component of the field B_r, on either side of the medium plan, effectively defocuses the ions that are not in the median plane. The force direction is up for particles moving into the paper on the right that are above the median plane and down if they are below the median plane. Without mitigation the particles would not be contained, vertically during acceleration. In the early days, the cyclotron did not work at higher (relativistic) energies, and it was necessary to build synchro-cyclotrons with electric fields in which the frequency changes with time, as the particle is accelerating. Now you know the rest of that story. But wait, there's more.

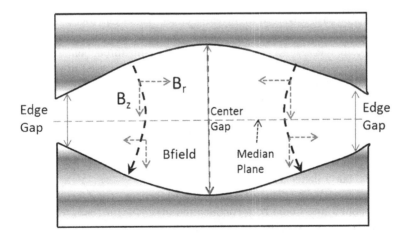

FIGURE 13.6 Cyclotron with compensated field.

The isochronous cyclotron allows for beam to be accelerated and extracted continuously. In fact, if there is sufficient stability, the beam current injected into to the cyclotron can be modulated, which will result in a modulated output only delayed by the transit time, which is on the order of tens of microseconds. The synchro-cyclotron implementation results in short pulses of low duty factor beam that can be accelerated and will track the modulated RF. Particles arriving too early or late will not be synchronized with the electric field and won't be accelerated. The output beam, from this system, is pulsed, with a repetition frequency given by the technology of the system used to vary the RF.

Fortunately, this is not the end of the story. Considering the three-dimensional (3D) shape of a cyclotron, it is desired to somehow increase the $\int B \cdot dl$ along the path of the particle orbit as it gains energy. Instead of just increasing the overall magnetic field strength around the whole circumference, this can be done by increasing the length through which the beam travels in a higher magnetic field as the radius increases. Consider the geometry of Figure 13.7(left). There are some 'hills' and some 'valleys' in the lower half of the cyclotron shown. The gap between the top and bottom halves is smaller for the hills and larger between the valleys. One such gap is shown between the poles in the side view of Figure 13.7(right). Therefore, the magnetic field in the hill region is higher. In addition to that, consider the path shown by the lines in Figure 13.7(left). The dashed path is longer over the hill so the overall integral is higher for higher energy particles. These wedges gave birth to the sector cyclotron concept. Given what's been discussed so far, these wedges could be more of an uncurved triangular shape. The shape of the wedge, or the increasing distance across the hill with energy, is given by the rate of mass increase of the particle as the energy increase takes the particle out to higher radius orbits. But is this enough?

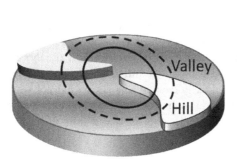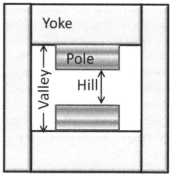

FIGURE 13.7 Cyclotron inner magnet structure (left) and side view of gaps (right).

To contain the beam, it is still necessary to maintain focusing throughout the acceleration cycle, so another mechanism is used. Fringing fields for focusing can be used in this cyclotron design. Fringing fields are created through the use of the hills and valleys since moving from the hill to the valley, or from a region of higher field to lower field is similar to the dipole fringe field discussed in Section 10.8. L.H. Thomas identified this azimuthal varying field (AVF) focusing solution in 1938. Figure 13.8 depicts this configuration. The upper part of the figure shows two triangular hills (H_A and H_B) on either side of a valley. The dashed line is a particle trajectory showing a radius of curvature of the particle in the hills, but a flatter trajectory (less curvy – well perhaps a bit exaggerated by being too straight) while in the valley. The dotted line would be a continuation of the curvature if there was no valley. Below that sketch is a graph of the magnetic field strength as seen by the particle moving along the dashed trajectory. There is a higher field in hills A and B (H_A and H_B) and a lower field in between. From Section 10.8, the fringing field is not enough. There needs to be an effective pole edge rotation to fully create a

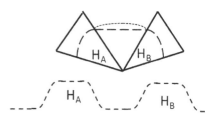

FIGURE 13.8 Fields along an orbit with fringe field.

controlled focusing effect. This results in the spiral sectors of higher energy cyclotrons, as seen in Figure 13.7(left) and why the hills are asymmetric triangles. In modern cyclotrons, the beam containment focusing is achieved through what is effectively a combined function magnet.

The fringe field of a dipole exerts a force perpendicular to the direction of bend due to the component of field perpendicular to the main field direction. The magnitude of this force is dependent upon the ratio of the hill and valley gaps and the detailed shape of the transition. The magnetic field as a function of the radius R and the azimuthal angle θ can be written as follows:

$$B_z(R,\theta) = B_0(R)\ \Phi(R,\theta);\ where\ \Phi(R,\theta) = 1 + f(R)g(\theta) \tag{13.8}$$

Φ is called the modulation function, and it describes the variation in the azimuthal magnetic field (the hill and valley effect). The flutter amplitude f (a function of the radius R [and therefore the energy]) describes the hill to valley ratio (and therefore the effect of the fringe field), and this could change as a function of radius (e.g. as the triangle width grows) or remain constant. The azimuthal field variation g is almost a step function whereby $g_{max} = 1$ and $g_{min} = 0$ and may actually change turn by turn. (The flutter amplitude is the amplitude modifier of the azimuthal step function.) It is essentially the lower curve in Figure 13.8. It deviates from a step function due to the finite extent of the fringe field falloff effect and the shape of the pole. Alternatively a sine function can be used as an approximation. Consider a system in which the cyclotron has N sectors. In that case

$$B_z(R,\theta) = B_0(R)\big[1 + f(R)\sin N\theta\big];\ or\ replace\ the\ sine\ with\ a\ stepfunction \tag{13.9}$$

Figure 13.8 can be redrawn a number of ways as in Figure 13.9. The lowest image is that of a side view looking at H_A, H_B and the valley in between. The field is stronger in the hill region with the smaller gap shown by the solid field lines. The weaker valley field is shown by the dashed lines. Some fringe field is shown by dashed gray lines. The strength of the field is plotted above in terms of the flutter amplitude with g as a step function (with no fringe field); thus, it's sharper than the lower

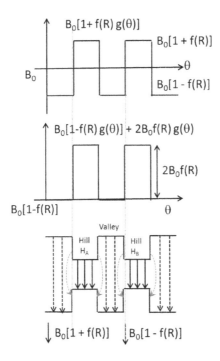

FIGURE 13.9   Flutter fields.

curve of Figure 13.8. The field amplitude is plotted two ways, the upper curve essentially taking the average field B_0 as the amplitude about which the field is plotted and the middle curve is the same plot but starting with the lower, valley field. Either way, the field is higher in the hills and lower in the valleys. The middle plot effectively renormalizes to a constant field $B_0(1 - f(R))$ with a field increase of $2B_0f(R)$ in the hills. Think of this as a dipole renormalized with field $2B_0f(R)$ inside the poles and 0 field outside the poles to get the effect of the fringe fields as discussed in Section 10.8.

Because of the step in field strength and a pole edge rotation, there can be focusing. Now consider the edge focusing effects of a sector focused cyclotron due to the fringe field. Assume y is vertical and z is in the nominal particle direction of motion. Going back to the coordinates used in Figure 10.23 of Section 10.8 (except s is replaced by z here), there is a force in the vertical (y) direction given by

$$\Delta p_y = \int F_y \, dt = \int dt q v_z B_x \qquad (13.10)$$

Figure 13.10 should help to fill in the rest of the story. Taking hill H_A from Figure 13.8, the beam is traveling out of the hill in direction z (solid gray line). The pole face (smaller dashed line) is rotated an angle β with respect to the direction perpendicular to z. The normal to the pole face (longer dashed arrow line) is given by script capital Z in Figure 13.10 and $(\mathbf{Z})$ in the text.

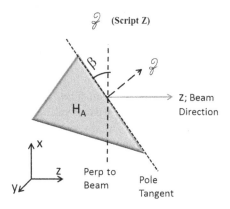

FIGURE 13.10   Cyclotron pole edge angle.

The main magnetic field B_y is the uniform field B_0 and the superimposed field of $2B_0f(R)$.

Because of the pole face rotation angle β, there is a component of magnetic field in the direction of the particle motion and in the x direction transverse to the particle motion shown on the left side Figure 13.11.

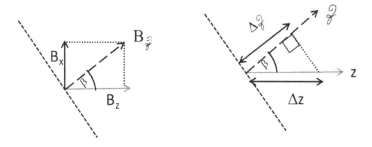

FIGURE 13.11   Pole edge focusing components.

With the right angle between B_x and B_z, these are related as follows:

$$B_x = B_z \sin \beta \qquad (13.11)$$

It is useful to convert the parameters in the vertical force (Equation 13.10) to be in the direction of the normal to the magnet pole edge, $\mathbf{Z}$. The time dt is related to the distance traveled by the speed of the particle in the z direction (v_z) as

$$v_z \, dt = dz \tag{13.12}$$

and the distance Δz is related to the distance along the $\mathbf{Z}$ direction, with the right angle shown on the right side of Figure 13.11 by

$$d\mathbf{Z} = dz \cos \beta \tag{13.13}$$

(Okay a little loose with dz and Δz.) The reason to do this is that the main magnetic field follows the pole face and the fringe fields extend along $\mathbf{Z}$. This will come in handy soon. Continuing from Equation 13.10 with the recent relationships results in

$$\Delta p_y = \int q \frac{d\mathbf{Z}}{\cos\beta} \, B_Z \sin\beta \; \left(\text{where } \mathbf{Z} = \textbf{\textit{ScriptZ}}\right) \tag{13.14}$$

This leaves

$$\tan\beta \int q B_Z \, d\mathbf{Z} \; \left(\text{where } \mathbf{Z} = \textbf{\textit{ScriptZ}}\right)$$

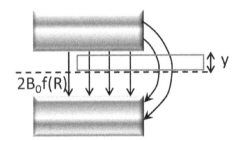

$2B_0 f(R)$

FIGURE 13.12 Geometry for Ampere's law.

to be evaluated. This is a form in which Ampere's law could be useful if the diagram in Figure 13.12 is used. It is also the reason that the coordinates were transformed in this direction. In that diagram, the integral is taken around the light gray solid rectangle.

$$\oint B_Z \cdot d\mathbf{Z} = 0 \tag{13.15}$$

This equals 0 since there are no charges inside. Then the usual trick is used. The long path on the midplane has no field component in the direction of the path, so for that part, the dot product is 0. The long path above the midplane is just $\int B_Z dZ$ (where $Z = \mathbf{Z}$ inside the equation, script Z wasn't available inside the equation editor). Leave this for a minute. This leaves the short vertical ends. The right hand short line would normally be 0 since there would be no field out there in a normal dipole, and it is allowed to draw the rectangle as long as was needed. But in this case, out there means in the valley, so there is a component of magnitude $[y][B_0[1 - f(R)]]$. This leaves one final vertical term equal in the hill to the field times that distance $= [B_0[1 + f(R)]][y]$. The sum of the three is still zero. Since the two y terms are opposite directions, they combine to be $[2B_0 f(R)][y]$ which, therefore, equals the magnitude of integral of $B_z d\mathbf{Z}$ the upper horizontal line, but opposite in sign.

Recall the definition of the focal length of a magnet. It is the distance from that magnet at which a ray crosses through 0 when it starts with a y offset (e.g. y, or $2y$ for that matter). This is related to the speed of the particle along the main direction of travel v_z and the speed of the transverse motion of the particle v_y so that with similar triangles, $f = y \, v_z/v_y$. With this force, the focal length (fl – using this abbreviation here since there are already too many fs in this section) is given by (using the favorite formula),

$$fl = \text{focal length} = \frac{\gamma m_0 v_z}{q 2 B_0 f(R)} \frac{1}{\tan\beta} = -\frac{R/2}{f(R) \tan\beta} \tag{13.16}$$

where R is the radius of the beam in the cyclotron. If $\beta = 0$, there is no focusing. This provides a useful relation for estimating the effects of focusing in the cyclotron, when combined with the matrix formalism for optics, remember that the matrix element for a thin lens is given by

$$\begin{pmatrix} 1 & 0 \\ -\dfrac{1}{fl} & 1 \end{pmatrix} \tag{13.17}$$

As the particle momentum increases, including the beam energy and the mass of the particle, the bending radius of the particles inside the cyclotron increases and the focusing strength required to contain the beam increases requiring the pole edge rotation to increase. Figure 13.13 shows one orbit in the middle of the cyclotron with the pole edge parameters and an orbit at the outer edge. If the pole edge rotation increases with radius to increase the focusing effect, the solid line showing the outline of the hill changes from a triangle in Figure 13.10 to a spiralized triangle. This is called spiral sector focusing. It is what was originally drawn in Figure 13.7(left).

When the energy is sufficient, the particle reaches the outer radius and is extracted from the cyclotron and directed to the treatment room. That is another story, not for this book.

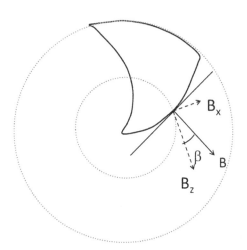

FIGURE 13.13   Spiral sector.

13.3.3   CYCLOTRON PARAMETERS

13.3.3.1   Energy

The final energy of the particle accelerated in a cyclotron is related to the number of turns and the amount of energy it gains during each turn. It's useful to note that the number of turns designed will affect the spacing between the turns. It is this spacing that is generally taken advantage of to extract the beam. The septum used to extract the beam has a finite thickness and will intercept more beam if the spacing is smaller.

In general, a cyclotron is capable of accelerating and extracting different particles and different energies if properly designed. However, the mechanism for changing the energy can require various modifications to the settings of the cyclotron and/or components. Regarding the particle type, the frequency equation derived earlier involves the ratio of charge to mass. Therefore, for example, using the same frequency electric field, a cyclotron may be able to accelerate ^{12}C and ^{4}He, with a charge to mass ratio of 1/2. But a proton has a charge to mass ratio of 1 and would require a different frequency but perhaps harmonics can play a role. In addition the issue of the magnetic field increase with mass as the energy increases should be considered. Practically speaking the time it takes to modify the cyclotron parameters for acceleration of one ion species to another, as done in nuclear physics laboratories, can sometimes range from minutes to hours. This is one of the reasons that the cyclotrons used for proton therapy (none that have been installed in hospital environments have been used for heavier particles yet) are fixed-energy accelerators. Cyclotrons for proton radiotherapy typically produce protons with kinetic energies between 230 and 250 MeV. To reduce the beam energy for the purposes of setting the appropriate range for treatment, the beam extracted from the accelerator is degraded in material taking advantage of the energy loss of charged particles that traverse matter. The degrader becomes a very important component for the preparation of the charged particle beam for therapy. The choice of final accelerator energy is quite strategic. It is necessary to achieve the maximum clinical range which essentially sets the maximum accelerated energy. It is also necessary to achieve the lowest clinical range. Since the energy must be degraded, the higher the fixed energy of the cyclotron, the more the energy must be degraded to achieve the lowest energy and this will affect the other beam properties including the beam current available. There are many trade-offs associated with this, some of which will be covered in subsequent sections.

13.3.3.2   Beam Phase Space Area

The beam accelerated in a cyclotron has a reasonably small phase space area. The extracted beam in a proton therapy cyclotron has a phase space area of about 5π-mm mrad which corresponds to a beam size that can be focused down to submillimeter with reasonable focusing strength. For most therapeutic ranges, this is not the beam that is transported to the patient. After the beam has penetrated the degrader, owing to beam scattering and range straggling, the beam phase space parameters are modified. A system is required to tailor the transverse and longitudinal phase space to match parameters that can be used for therapy, such as that described in Section 13.4.

13.3.3.3   Current

Proton therapy cyclotrons can accelerate microamps of beam current. The beam extracted from the cyclotron is anywhere from 25% to 90% of the accelerated beam. Most of these cyclotrons are designed to extract in the range of 300–500 nA accounting for beam loss in the degrader system. Again, for most of the therapeutic beam range, the maximum current that can be transported to the patient is much less, but is still very high relative to what is needed for conventional therapy. Cyclotrons can also be used for more novel therapies, such as FLASH which requires tens of Gy to be delivered in less than a second.

13.3.3.4   Time Dependence

In the case of an isochronous cyclotron, the RF can be left on, at fixed frequency indefinitely. During that time, the beam can be continuously accelerated. Lower energy and higher energy beam are in the cyclotron at the same time (at different places). The beam can be turned off by adjusting the ion source parameters or the RF settings. During the time the beam is called on, a portion of the beam within the maximum portion of the electric field wave, is accelerated as discussed in Section 13.2. Sometimes there are variations in the cyclotron RF, magnetic field and/or the temperature of the cyclotron (leading to small mechanical dimension changes) which contribute to changes in the synchronization between the beam and the RF electric field. This can lead to beam loss, but is relatively easy to adjust if detected. Not everyone agrees on which adjustment is best.

13.3.3.5   Operability

A cyclotron has relatively few parameters to adjust. As indicated in this chapter, owing to the cyclotron equation, the magnetic field and the RF are the main parameters. There are other adjustable parameters which serve to compensate for initial construction errors, which are generally held constant and are not considered adjustable. As mentioned in the previous paragraph the magnetic field and the frequency can be used as adjustments if cyclotron performance is drifting.

There are some other interesting cyclotron behaviors that may sometimes be important to address. One, in particular is the ability of a cyclotron to produce beam without an ion source injecting beam. While the cyclotron vacuum can be quite low, it is not zero, and the residual air in the cyclotron can be ionized by the high electric field, generating ions which can then be accelerated. This only happens, of course, with the RF on. Generally the RF is set to an amplitude below the threshold for this, when it is intended to halt the delivery of beam.

13.3.3.6 Size

As noted in Figure 13.5, there has been a marked reduction in weight, size and cost of cyclotrons used for clinical proton therapy in the last 50 years. Figure 13.14 shows some examples of therapy machines. Cyclotrons for nuclear physics at Indiana University and iThemba LABS in South Africa used a 2,000 ton separated sector cyclotron that accelerated protons with energies that could be used for therapeutic treatments (and eventually were used for that purpose). See the person standing in the upper middle section of the photo. The first special purpose system was built by IBA for MGH with a radius of 4 m, weighing in at 200 tons, already a factor of 10 reduction from the physics laboratory design (photo courtesy of Ion Beam Applications). Superconducting technology was used by Accel and then developed by Varian to construct an 80 ton cyclotron. (Photo courtesy of Varian Medical Systems, Inc. All rights reserved.) Finally in 2010, the Still River now called MeVIon cyclotron was constructed. This followed the precedent set by the Henry Blosser superconducting cyclotron for neutron therapy, by building a much smaller cyclotron with much higher magnetic field so that it could be mounted on a gantry. It weighs about 35 tons. (Photo courtesy of MeVIon Medical Systems.)

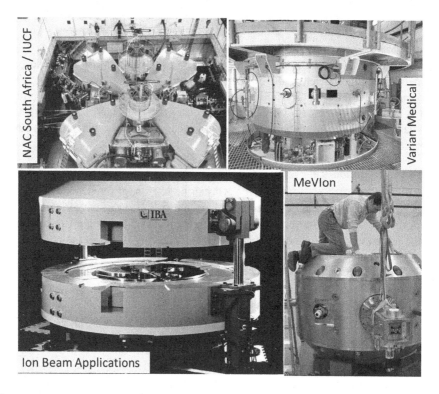

FIGURE 13.14   Real cyclotron examples.

13.4   DEGRADER AND ENERGY SELECTION SYSTEM

A particle therapy cyclotron is a fixed-energy cyclotron. The beam energy can be modified by a variable thickness degrader. If beam transport is used, this is coupled with an achromatic energy selection system (ESS). This degraded beam is then transported to the treatment rooms. There are a number of conditions necessary for this beam transport system (BTS). Implicit in this discussion is the fact that this equipment will operate in a hospital environment and it is desirable to simplify the operation and minimize the number of required settings.

13.4.1   Beam Conditions Resulting From a Degrader

A charged particle beam experiences interactions in matter as discussed in Chapter 9. The matter of the degrader has the effect of significantly increasing the beam emittance and energy spread compared to the cyclotron extracted beam properties. A BTS will only accept an emittance equal to its admittance. To minimize apertures, reduce cost and achieve a beam size suitable for therapy, it is desirable to minimize the beam size downstream of the degrader. This can be done using limiting collimators. But in addition, the emittance growth of the beam caused by the degrader can be optimized.

A degrader that is thick enough to reduce the beam energy sufficiently will cause significant beam scattering which increases the beam's angular divergence. Intuitively it might be thought, that will dominate the beam emittance and nothing can be done. The former is true, but the latter is not. The degraded beam emittance is a function of the beam conditions at the input of the degrader, which can play a role in the choice of the degrader geometry. The beam properties can be calculated using the beam transfer matrices. Various parameters can be observed, including the effects of the emittance, beam size, angular divergence as well as the location of the forward projected input beam waist and the effective backward projected beam waist of the degraded beam.

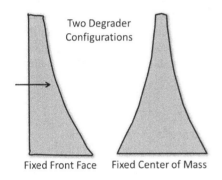

Two Degrader Configurations

Fixed Front Face   Fixed Center of Mass

FIGURE 13.15   Degrader configurations.

There are a variety of possible geometries for a degrader. The optimal choice will depend on the dependencies of the final beam emittance and the desired beam optics. Take, for example, the two geometries shown in Figure 13.15. It might be thought that the optimum situation is to focus the beam at a waist at the front of the degrader and to maintain a fixed-focus upstream of the degrader the front face should be flat. Alternatively it might be better to focus the beam to a location inside of the degrader and the fixed center of mass option would be better, so that the virtual source position might not change as a function of desired energy.

Figure 13.16 shows the degraded beam emittance as a function of degraded energy, starting with energy of

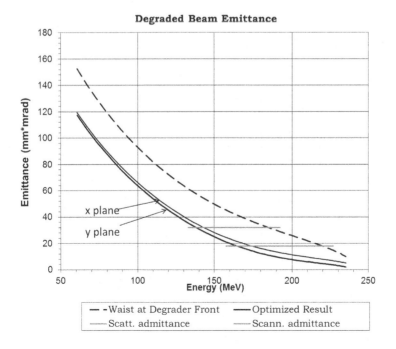

FIGURE 13.16   Optimizing emittance of degraded beam.

230 MeV. In this figure, there are three curves and two horizontal lines. The upper black dashed curve is the degraded emittance resulting from an input beam focused to a waist at the entrance to the degrader. The lower (solid black) curves are the degraded emittance of the beam (x and y planes) with an optimized input phase space. The horizontal lines are the scanning and scattering beamline admittances. Emittances larger than this will not fit through the beamline for those modalities.

Many calculations could be done to give some idea about the behavior (and they were indeed done), but these do not necessarily give the kind of insight that might be possible. To do that, it's helpful to return to the transfer matrices. It is repeated here, and will be repeated again, that most sorts of beamline design that are attempted without a look at the analytical conditions and constraints will likely not be optimized.

Consider the symmetric beam sigma matrix

$$\begin{pmatrix} \sigma_{11}{}^0 & \sigma_{12}{}^0 \\ \sigma_{12}{}^0 & \sigma_{22}{}^0 \end{pmatrix} \tag{13.18}$$

where $\sigma_{11}{}^0$ is the square of the beam size at the point of entrance of the degrader (at location 0). (The superscripts are used in this section to save space instead of (0).) To propagate the sigma matrix into the degrader material which causes multiple scattering, divide up the degrader into N layers of thin thickness L. Within each layer, the beam drifts a distance L and experiences multiple scattering with an rms angular growth of θ_m. If it's small enough the order doesn't matter. The new sigma matrix, after the first ($N = 1$) drift is

$$\begin{pmatrix} \sigma_{11}{}^0 - 2L\sigma_{12}{}^0 + L^2\sigma_{22}{}^0 & -\sigma_{12}{}^0 + L\sigma_{22}{}^0 \\ -\sigma_{12}{}^0 + L\sigma_{22}{}^0 & \sigma_{22}{}^0 \end{pmatrix} = \begin{pmatrix} 1 & L \\ 0 & 1 \end{pmatrix} \begin{pmatrix} \sigma_{11}{}^0 & \sigma_{12}{}^0 \\ \sigma_{12}{}^0 & \sigma_{22}{}^0 \end{pmatrix} \begin{pmatrix} 1 & 0 \\ L & 1 \end{pmatrix} \tag{13.19}$$

It can be noted that the beam divergence didn't change; in the absence of any focusing forces and that the diagonal terms are the same. Remember also that the emittance of the beam, in this formalism is given by the determinant of the sigma matrix, or $\varepsilon_0{}^2 = \sigma_{11}{}^1\sigma_{22}{}^1 - (\sigma_{12}{}^1)^2$, where the superscript refers to the evaluation at location 1 (unless it is an exponent – take care) and the subscript 0 on ε is at the start. At this point, if the drift length per layer is thin enough, the multiple scattering angle can be added in quadrature with the beam angles. Dropping the subscript m (for multiple scattering), the matrix after the first layer can be rewritten as

$$\begin{pmatrix} \sigma_{11}{}^0 - 2L\sigma_{12}{}^0 + L^2\sigma_{22}{}^0 & -\sigma_{12}{}^0 + L\sigma_{22}{}^0 \\ -\sigma_{12}{}^1 + L\sigma_{22}{}^1 & \sigma_{22}{}^0 + \theta^2 \end{pmatrix} \tag{13.20}$$

(Yes the theta is squared – that is not a superscript like on the sigmas.) Now it remains to continue this process for each layer. This is where the author needed to be and now asks the readers to be strong. For example after three layers the resulting matrix is

$$\begin{pmatrix} \sigma_{11}{}^0 - 6L\sigma_{12}{}^0 + 9L^2\sigma_{22}{}^0 + 5L^2\theta^2 & -\sigma_{12}{}^0 + 3L\sigma_{22}{}^0 + 3L\theta^2 \\ -\sigma_{12}{}^0 + 3L\sigma_{22}{}^0 + 3L\theta^2 & \sigma_{22}{}^0 + 2\theta^2 \end{pmatrix} \tag{13.21}$$

It keeps going, but for those that were strong of heart a rewarding realization manifests. This was very satisfying for the author. For the Nth layer

$$\sigma_{11}{}^N = \sigma_{11}{}^0 - 2NL\sigma_{12}{}^0 + (NL)^2 \sigma_{22}{}^0 + \left[\frac{1}{3}N^3 - \frac{1}{2}N^2 + \frac{1}{6}N \right] L^2\theta^2 \tag{13.22}$$

Similar relations for the remaining matrix elements can be derived. The main point here is that the expression is analytic in closed form. Strictly speaking, it is not necessary to use fancy codes for this calculation and perhaps some insights can be obtained. Focus back to the point in question – which is to investigate the behavior of the emittance of the beam as it goes through the degrader. Given the closed form of the sigma matrix, and the fact that the emittance of the beam is the determinant of that matrix, the emittance can be written after traversing N layers. However, this is calculated using the input sigma matrix parameters, the one written with the superscript 0. It is the objective, to minimize the emittance after passing through the degrader by optimizing the sigma matrix at the start of the degrader. For that the upstream beam magnetics might have to be adjusted appropriately. It is not correct to simply vary one of the beam sigma matrix parameters to obtain a reduced final emittance because the emittance of the incoming beam is a constraint and Louiville's theorem is satisfied until the beam enters the degrader. One useful approach is to start the incoming beam matrix earlier at an upstream distance L_0, from a waist somewhere inside the degrader so that the sigma matrix, at the start of the degrader (the new 0 matrix) is given by

$$\begin{pmatrix} \sigma_{11}{}^0 - 2L_0\sigma_{12}{}^0 + L_0{}^2\sigma_{22}{}^0 & -\sigma_{12}{}^0 + L_0\sigma_{22}{}^0 \\ -\sigma_{12}{}^1 + L_0\sigma_{22}{}^1 & \sigma_{22}{}^0 \end{pmatrix} \tag{13.23}$$

The length parameter L_0 can be a variable to adjust the phase space at the degrader entrance without changing the input beam emittance. This parameter can also be used to adjust where, in the degrader, a waist will occur. (Remember Section 10.12.) With this change, the final beam parameters, including the emittance, after the degrader can be written in terms of the new length parameter, which is what will be optimized while maintaining the appropriate sigma matrix conditions.

Consider the evolution of the beam size for different L_0 values without scattering in Figure 13.17. It is important to note that the final beam size may not be an indication of the relative value of the final emittance. The location of a waist, if any will affect the upstream and downstream beam sizes. The plot shows that as the waist is driven further and further downstream, that the beam size, at the end, does decrease, until it will increase again with larger L_0s. To have done this, with a constant

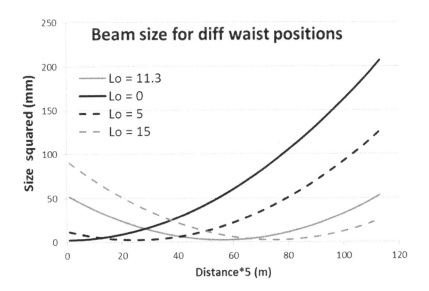

FIGURE 13.17 Beam size as a function of distance.

input emittance, the beam size at the entrance has to increase. Now look at the downstream emittance (after the degrader, including scattering) which is proportional to

$$\propto \left[1+(N-1)\left[\sigma_{11}^{0} - 2L_0\sigma_{12}^{0} + L_0^2\sigma_{22}^{0}\right] - (NL)(N-1)\left[\sigma_{12}^{0} - L_0\sigma_{22}^{0}\right] + \cdots \right] \quad (13.24)$$

Optimize this by differentiating with respect to L_0, and a remarkable result is obtained. This was really worth the effort (many pages of algebra) and most would probably have bet against this.

$$\frac{L_0}{NL} = \frac{1}{2} \quad (13.25)$$

Yes, that's it. It's really physics magic. To minimize the final emittance of the beam, after multiple scattering, it is necessary to tune the incident beam to have had a waist at a position halfway into the degrader, if the degrader didn't exist. Hindsight indicates that the multiple scattering in the material increases the rms angle by a specific amount, independent of the angles of the beam. It is helpful to minimize the fractional increase of the beam emittance in the degrader. The emittance growth is minimized around the waist position. When the beam size is decreasing, near a waist, it's basically fighting against the effect of the increasing divergence from the scattering. A waist too early reduces the distance inside that degrader that the beam is at a waist and a waist too late allows the beam to grow too much.

Some of these parameters are plotted in Figure 13.18. The second vertical line is the end of the degrader. The solid black line is the square of the beam size with the multiple scattering and the dashed black line is without scattering. The solid gray line is a measure of the emittance and note how it rises, flattens in the middle (when the beam is closer to a waist) and then rises again.

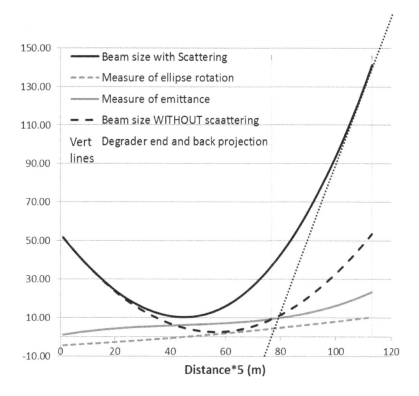

FIGURE 13.18 Beam parameters of the degrader optimization.

The dashed gray line is the degree of ellipse rotation. Now it's clear that the triangle degrader is better than the flat face so that it's not required to change the focus as a function of final energy since the waist has to be at the center of the degrader. The diagonal dotted line in the figure extending beyond the chart is a line to guide the eye to the back projected source position which is about two-thirds into the degrader (near the first vertical line). So the virtual source position does not occur in the exact center of the degrader.

Going back to brute force calculations, Figure 13.19 summarizes the back projected waist position results for both types of degraders with constant input beam conditions for all degraded energies. The waist position varies by more than 7 cm over the energy range for the case with the degrader, that is constructed with the front face fixed in space. This implies that the effective beam source will move as a function of energy, and the downstream beamline optics will have to accommodate this variation. In the case with the degrader that is constructed with the center of mass fixed in space, the variation in waist position is considerably reduced. In addition, the waist position is compared with a point two-thirds from the front of the degrader (gray dotted line). The higher energy portion of this graph is less accurate since the degrader is much thinner. These confirm the analytical analysis.

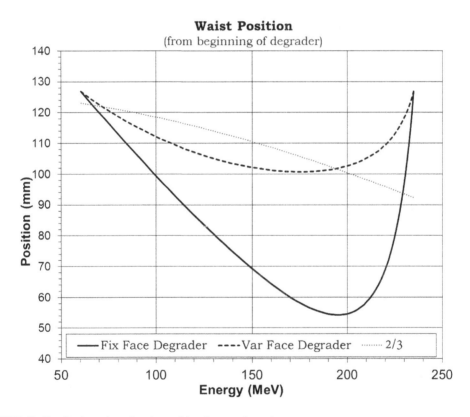

FIGURE 13.19   Back projected waist position for two degraders.

13.4.2   ENERGY SELECTION SYSTEM

In addition to the emittance growth through the degrader, the beam energy spread also grows because of range straggling. If the degrader, for example, reduces a 230 MeV proton beam to 100 MeV, it has lost 130 MeV. The range spread (Equation 9.42) will have increased by about 0.14 cm (owing to the 130 MeV lost) and relative to the 100 MeV left this is a range spread of 1.9%. Converting this to the parameter relevant to the beamline $\Delta P/P$, $\pm 0.55\%$ is obtained. Adding this to the energy spread in the beam already and the total may start being difficult to be transported through a cost efficient beamline given the transverse coupling of the momentum to the transverse position as a result of bending fields [e.g. (x/δ)]. Therefore, it is necessary to reduce that spread. The only convenient way to do that, in an environment with limited space and funding is to strip away the unwanted beam. (There are tricks with more complicated beamlines, such as an energy compression system which can sometimes reduce the spread, but this involves additional beamline elements, RF systems and controls.) To remove the unwanted energies, it will be necessary to physically separate them, or in other words, to create a transverse dependence on the beam momentum, or create a dispersion, as in a nonzero (x/δ). It is further desired that after these momentum gymnastics are complete to end with a neat dispersionless beam by the time it reaches the beam delivery system.

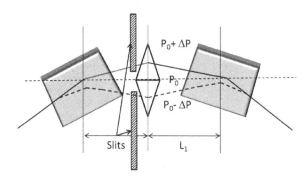

FIGURE 13.20   ESS achromat.

This can be achieved with the achromatic system shown in Figure 13.20 (touched on in Section 10.6.3.4), with a physical aperture, called slits. Consider an approximate form of the transfer matrices for this system. The bending dipole is approximated by a 'thin' wedge (unlike what is shown in the figure). Higher momentum rays bend less (indicated by $P_0 + \Delta P$) and lower energy rays are bent more, thus resulting in a physical separation with momentum dependence (x/δ). The dipole has zero length but does bend the beam through an angle θ. In this approximation, the transfer matrix element is given by

$$
\begin{pmatrix}
(x/x) & (x/\theta) & (x/\delta) \\
(\theta/x) & (\theta/\theta) & (\theta/\delta) \\
(\delta/x) & (\delta/\theta) & (\delta/d)
\end{pmatrix}
=
\begin{pmatrix}
1 & 0 & 0 \\
\dfrac{-\sin\theta}{\rho} & 1 & \sin\theta \\
0 & 0 & 1
\end{pmatrix}
\tag{13.26}
$$

Similarly a thin lens approximation, with a focal length $f/2$, will be used with the transfer matrix for half of the quadrupole element given by

$$
\begin{pmatrix}
(x/x) & (x/\theta) & (x/\delta) \\
(\theta/x) & (\theta/\theta) & (\theta/\delta) \\
(\delta/x) & (\delta/\theta) & (\delta/\delta)
\end{pmatrix}
=
\begin{pmatrix}
1 & 0 & 0 \\
-\dfrac{2}{f} & 1 & 0 \\
0 & 0 & 1
\end{pmatrix}
\tag{13.27}
$$

Near that quadrupole is a physical aperture called a slit since it is an aperture in only the x direction. The slit will stop rays that have a momentum difference from the central momentum of an amount dependent on the physical opening of the slit (see the solid and dashed lines in Figure

13.20). This amount is determined by (x/δ). The full transfer matrix through the system with drift lengths L_2 and L_1 is calculated by multiplying the matrices together in the correct order. Two thin lenses to create the one physical one are used for reasons that will become clearer later.

$$
\begin{pmatrix} 1 & 0 & 0 \\ -\sin\theta & 1 & \sin\theta \\ \rho & & \\ 0 & 0 & 1 \end{pmatrix}
\begin{pmatrix} 1 & L_2 & 0 \\ 0 & 1 & 0 \\ 0 & 0 & 1 \end{pmatrix}
\begin{pmatrix} 1 & 0 & 0 \\ -\frac{1}{f} & 1 & 0 \\ 0 & 0 & 1 \end{pmatrix}
\begin{pmatrix} 1 & 0 & 0 \\ -\frac{1}{f} & 1 & 0 \\ 0 & 0 & 1 \end{pmatrix}
\begin{pmatrix} 1 & L_1 & 0 \\ 0 & 1 & 0 \\ 0 & 0 & 1 \end{pmatrix}
\begin{pmatrix} 1 & 0 & 0 \\ -\sin\theta & 1 & \sin\theta \\ \rho & & \\ 0 & 0 & 1 \end{pmatrix}
\qquad (13.28)
$$

This works out to be (where $\sin\theta$ has been replaced by the letter s, to fit on the page)

$$
\begin{pmatrix}
\left[1-\frac{s}{\rho}(L_1+L_2)+\frac{sL_1L_2}{\rho f}-\frac{L_2}{f}\right] & \left[L_1+L_2-\frac{L_1L_2}{f}\right] \\
\left[\frac{-2s}{\rho}+\frac{s^2}{\rho^2}(L_1+L_s)-\frac{s}{\rho f}(L_1+L_s)-\frac{s^2L_1L_2}{\rho^2 f}\right] & \left[1-\frac{s}{\rho}(L_1+L_2)-\frac{L_1}{f}+\frac{sL_1L_2}{\rho f}\right] \\
0 & 0
\end{pmatrix}
$$

$$
\begin{pmatrix}
\left[s(L_1+L_2)-\frac{sL_1L_2}{f}\right] \\
\left[2s-\frac{s^2}{\rho^2}(L_1+L_2)-\frac{sL_1}{f}+\frac{s^2L_1L_2}{\rho^2 f}\right] \\
1
\end{pmatrix}
\qquad (13.29)
$$

The simplifications used allow this matrix to fit on one page in portrait orientation and do not detract from the points that can be made. It is the goal to, at least, have zero dispersion at the end of this system, and to have nonzero dispersion inside the system. Inside the system the unwanted energies will be removed and what's left will be combined at the end. For a zero dispersion, $(x/\delta)=0$. This requires

$$
\left[s(L_1+L_2)-\frac{sL_1L_2}{f}\right]=0 \qquad (13.30)
$$

The condition is met if

$$
\frac{1}{f}=\frac{L_1+L_2}{L_1L_2} \qquad (13.31)
$$

This seems like a reasonable solution, with the focal length of the lens of reasonable strength (a focal length on the order of the drift length or in the range of meters). However, is this enough, or is it necessary to become greedy? In this solution, (θ/δ) is nonzero. That means that if there is additional drift length after this system, the (x/δ) term will grow. Therefore, it is desired that, not only should the system to be dispersionless at the end, but the angular dispersion should also be zero. The combination of the two requirements creates an achromatic system. There will be no further transverse spread due to momentum downstream of this system, until the beam enters another dipole. To have $(\theta/\delta)=0$

$$
\left[2s-\frac{s^2}{\rho^2}(L_1+L_2)-\frac{sL_1}{f}+\frac{s^2L_1L_2}{\rho^2 f}\right]=0=\left[2s-\frac{s^2}{\rho}\left[L_1+L_2-\frac{L_1L_2}{f}\right]-\frac{sL_1}{f}\right] \qquad (13.32)
$$

From the condition for $(x/\delta) = 0$, the term inside the inner square brackets on the right is zero, which leaves

$$\frac{1}{f} = \frac{2}{L_1}$$ (13.33)

Which is another way of saying $L_1 = L_2$ and that basically the symmetry of the system gave rise to the canceling of (θ/δ). Symmetries in beam optics are very powerful things, and they can be used as a condition. This symmetry also has consequences for the optics in the middle of the system.

Looking at the middle of the system, $(x/\delta) = L_1 \sin\theta$. If $L_1 = 1$ m and $\theta = 30$ degrees, the dispersion is 0.5 m or 0.5 cm/%. If the beam momentum spread (not the same as the beam energy spread) is limited to ±1%, it would be necessary to install an aperture with a total width of 1 cm.

This is not the end of the story. The beam has a finite size and the energy spread mixed by the dispersion spreads it even further (see Section 10.12.2.1). In Figure 13.21 to the left, three small beams with no momentum spread but different momenta are shown in three different locations in a region of dispersion. (An actual beam would be a sum of these beams with the appropriate distribution.) Note how the dashed rectangle, indicative of the slit location, encompasses most of the solid line beam and a bit of the other energy beams. This is a form of momentum resolution. The purpose of the slit is to limit the beam momenta that go through to the remainder of the system. In this case, some of the energy from the outer beams will sneak through the slits, but not much. If the monochromatic beam size (the size of a single energy beam) is increased, as in Figure 13.21 to the right, then through the same dashed rectangle, much of the central beam is lost and a significant amount of beam from the other two energies passes through the aperture. To minimize unwanted beam loss and maximize the energy selection efficiency of this achromat, it is necessary to minimize the monochromatic beam size at the location of the aperture.

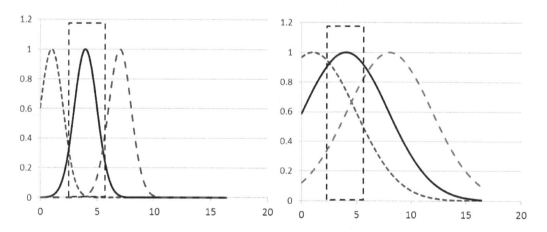

FIGURE 13.21   Monoenergetic beam size and momentum resolution.

Upon further examination of the achromat shown in Figure 13.20, the matrix elements (x/x), (x/θ) and (x/δ), (also known as R11, R12 and R16) are shown in Figure 13.22. Of note is that (x/θ) (black dashed curve) increases to a maximum in the middle of the system. As an aside, but an important one, also note that (x/θ) increases and decreases, while (x/x) (gray dashed curve) which starts out at 1, decreases, going through zero and becoming a negative number. In other words,

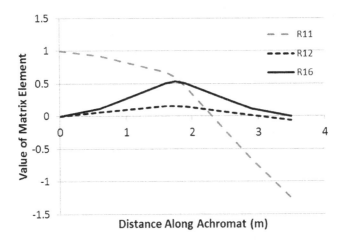

FIGURE 13.22 Achromat optics.

(x/θ) is a sine-like function and (x/x) is a cosine-like function. In this achromat, these functions have gone through a phase advance of about 180 degrees, if they were sine-like. Now back to the main point. The degrader, immediately preceding the achromat, caused the beam to undergo multiple scattering which resulted in a larger beam angular spread. The larger (x/θ) in the middle of the achromat will result in a larger beam size due to the larger θ. The beam envelope size squared (σ_{11} and σ_{22} - not the transfer matrix functions) are plotted in Figure 13.23. These show that the x plane beam size (the lower curve that does not end up at 0) is maximized at the center of the system. In this case, the beams of different energies would look like those shown in the right side of Figure 13.21, and there would be significant beam loss if it was attempted to isolate a smaller range of beam energies. This system is not optimized to be an energy selector even though it is achromatic. In the case where the initial beam angles are large, it would be better to minimize (x/θ) at the energy aperture.

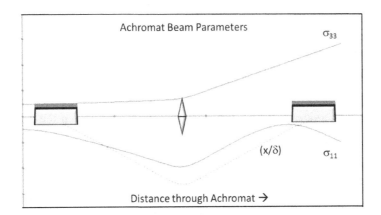

FIGURE 13.23 Beam size plots for achromat.

Up to this point, only the x (horizontal) plane has been discussed. Of course the y (vertical) plane must be considered. The vertical beam envelope is also shown in the upper section of Figure 13.23. This shows an increasing vertical beam size since the single quadrupole, which is used to focus the x plane to achieve achromaticity, will then defocus the vertical plane. This means more focusing will need to be added to the achromat to ensure that both the horizontal focusing is appropriate for the energy selection, but also that the vertical beam will be contained. Figure 13.24 shows an example of a system starting from the degrader and going through the ESS. There are two quadrupoles after the degrader to help adjust the beam at a collimator before the achromat to start cleaning up the larger emittance from the scattered beam. That actually helps the momentum resolution since it cuts off some of the angular divergence to which (x/θ) is sensitive. Then there are four quadrupoles in the ESS. The dashed line shows the dispersion, the lower curve is the horizontal beam size including the effect of a $\pm 1\%$ energy spread and the upper curve is the vertical beam.

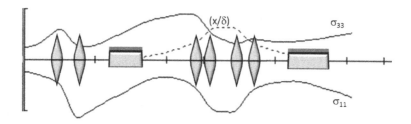

FIGURE 13.24 Full 2d achromat (non-optimized (x/θ)).

13.4.3 Degrader to Beamline Collimator Conditions

The energy collimation needed to restrict the transmitted energy spread is taken care of by the achromat part of the ESS. It is also important to collimate the monochromatic part of the beam to manage the growing beam size and localize the beam losses which helps to reduce the any losses in the remainder of the beamline. At the same time, it is helpful to produce a beam phase space which is independent of the beam energy so as to allow for a single downstream optical solution and create a beam which will be suitable for the beam delivery system. Collimation of the beam can be used to limit the emittance, so that it's consistent with the beamline admittance. In the case of scattering from the degrader, the beam angles grow and that or the effect of that is what needs to be limited. The beam angles can't be limited physically, unless it is allowed for the beam angles to propagate into physical beam size. This was touched on in Section 10.12.1.

There are two optical conditions helpful for locating a collimator which will limit the angular extent. One location is where the effective drift length is large and the other is where the phase advance (remember the sine-like and cosine-like functions) from the effective source is 90 degrees. Then the angles have evolved into position offsets. Given the limited space and strong focusing required for the large beam, the latter solution is usually best. Thus, a collimator at this location is constraining the angular acceptance of the system, while placing an additional aperture at the degrader constrains the beam size acceptance of the system. For most energies, due to the large effect of the degrader, this collimation will result in a sharp cutoff of the beam tail in one phase space plane, but not the other. Therefore, another collimator 90 degrees downstream may be required to prevent beam tails from scraping in the beamline. Usually it is required to keep beam loss after the degrader/ESS system to be less than 1%. The optics situation is more complicated for beam sizes smaller than the physical aperture at the degrader and at high energy.

To reduce the beam losses after the ESS, all the beam that will be lost should be lost before then. The losses can be contained and shielded to minimize the activation of the rest of the beam transport elements. Figure 13.25 is a graph of the ESS beam transmission percentage through the ESS collimators as a function of beam energy for a particular implementation. This is a simulation, in practice it is a little lower than this (sometimes a factor of 2). This system uses a graphite degrader. Some systems use beryllium, although that is harder to machine, but the scattering is less (see Figure 9.22 and exercises).

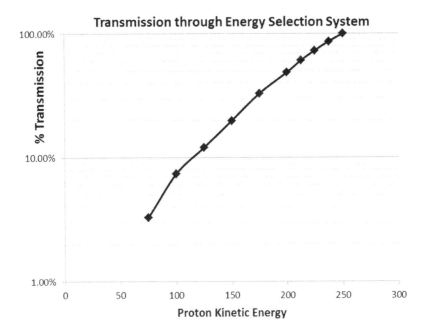

FIGURE 13.25   ESS transmission.

13.4.4   OPTICS TO DECOUPLE SOME BEAMLINE EFFECTS

The emittance to be transported through the beamline is different for the case of beam spreading via scattering or scanning. This is because the beam size near the end of the beamline is larger to achieve a smaller beam size at the target for the scanning case when compared to the scattering case. Therefore, to prevent beam loss at the end of the beamline, the beam collimation aperture is reduced so that a smaller emittance is transported and the losses near the ESS increase.

It is useful for the beam phase space at the entrance of the beam delivery system to be similar in x and y. The relevant conditions call for a solution which allows $\sigma_{11} = \sigma_{33}$ and $\sigma_{12} = \sigma_{34}$. This is actually not really necessary, but it simplifies many things, including treatment planning. It would also be good to ensure the same beam transport solution, independent of the emittance transported, or the setting of the slits and collimators as much as possible.

Using

$$\sigma_{12}(1) = R_{11}R_{21}\sigma_{11}(0) + (1 + 2R_{12}R_{21})\sigma_{12}(0) + R_{12}R_{22}\sigma_{22}(0) \tag{13.34}$$

$\sigma_{12}(1)$ can be made to only depend on $\sigma_{12}(0)$ (which is fixed), by making $R_{12} = R_{21} = 0$ (and similarly in the vertical plane). If these conditions are achieved, the emittance slits can be adjusted to keep the x and y beam sizes equal at the end of the beamline. Also since the effective beam source is at a waist, the beam will be at a waist at the end. This is operationally very interesting, since if there is beam loss or desire to obtain a smaller beam at isocenter, a good method of minimizing it will be to adjust the slits without having to retune the beamline.

The beam sizes squared divided by the emittance and the dispersion for the beam line are plotted in Figures 13.26 and 13.27 for the scattering and scanning case. The beam size can be determined from the transported emittances. In one example, the scanning case, accepts up to 18 mm mrad. In the scattering case, up to 32 mm mrad is accepted. The figures are divided into three sections. The first section includes those elements that comprise the ESS. The second includes those BTS elements that bend the beam into the treatment room. The third includes those elements that comprise the gantry. Using the conditions described earlier, it is possible to ensure that the beam line magnetic parameters (with the exception of the gantry) are unchanged in these two cases. This reduces the number of beamline tunes otherwise required. The beam sizes squared divided by the emittance are different, even in the part of the beamline with the same magnetics settings, since the emittance selected by the emittance definition slits after the degrader is smaller for the scanning case, while the beam size is, for most energies, determined by a collimator and is fixed. The maximum beam size for the scanning solution is larger than that for the scattering solution to obtain a smaller size at the target. The beam size is still adequately contained inside the vacuum system. The beam optics is such that the emittance can be adjusted with the slits and the beam phase space at the entrance to the gantry is the same in both planes, almost independent of the emittance selected.

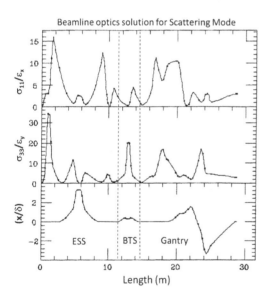

FIGURE 13.26 Beamline optics for scattered beam.

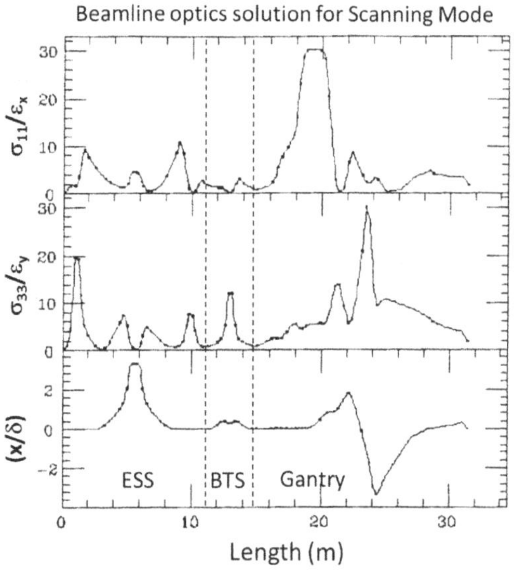

FIGURE 13.27 Beamline optics for scanning beam.

13.4.5 DEGRADER AND ESS SUMMARY

The use of a degrader can cause variations in the properties of the effective beam source if not properly designed. The requirements of scanning and scattering give rise to different beam phase space acceptances in the beamline. A useful input beam condition for the beam delivery system is that the phase space be identical in both planes. Despite these variations, it is advantageous to minimize the number of beamline tunes and minimize the beam losses for those tunes in a hospital environment.

A degrader with its center of gravity fixed in space together with the use of a fixed input beam tune for all degraded energies, results in minimizing the movement of the effective source location as a function of energy. The effective beam phase space downstream of the degrader can be adjusted almost independently of the input beam emittance. Minimization of the downstream emittance can result from a particular beam focusing condition upstream of the degrader. A particular optics solution has been discussed which allows the beamline magnetics parameters to remain unchanged when switching between scanning and scattering modes (except the gantry optics). The beamline optics shown results in a beam well contained inside the beam pipes and with the appropriate collimation results in low beam loss downstream of the degrader/ESS.

13.5   SYNCHROTRON-BASED SYSTEMS

The charged particle beam path in a synchrotron is shown in Figure 13.28. The synchrotron is a ring (or some closed shape) of magnets. The beam is injected from outside the synchrotron and then circulates around the ring repeatedly. The synchrotron has a device that creates an electric field at some frequency, which provides acceleration to the particles 'stored' in the ring in such a way as to keep the particles moving close to the same orbit (central trajectory), while the magnetic fields are changing to match the increasing beam energy. Thus, the beam is contained within the ring as its energy increases. When the beam reaches the desired energy, it is extracted. The time it takes to circulate one turn in the ring depends on the particle velocity and, therefore, it's energy; however, in the range of particle therapy and in the compact size synchrotron used it is usually between 50 ns to less than 1 μs. Therefore, if the beam were extracted in one turn, the pulse length would be less than 1 μs. In such an application, a 'rapid cycling' synchrotron has been proposed, but as yet has not been used for particle therapy. If it is desired to 'stretch' out the extracted current from the ring a variety of methods are possible. One such technique is resonance extraction and variants thereof. Many synchrotrons have been built and are operating for particle therapy. It represents over 30% of the installations.

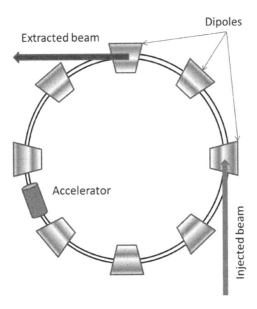

FIGURE 13.28   Synchrotron components.

13.5.1 Synchrotron Timing

There are a variety of ways to operate a synchrotron for particle therapy. Figure 13.29 indicates some of these. The top two traces show the energy of the beam in the synchrotron as a function of time. The third graph shows the charge of the beam that is stored in the ring at any given time. Finally, the fourth graph shows the beam current extracted from the synchrotron.

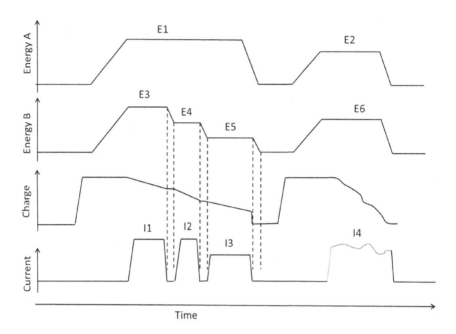

FIGURE 13.29 Synchrotron operation timing.

The synchrotron, at any given time, is set at a given beam energy. That is, the magnets in the synchrotron are set with the appropriate bending radius to enable a beam of a given energy to be circulating within the ring. The energy A trace describes the scenario in which the synchrotron is initially set to a minimum energy (namely that of the beam which is to be injected). A pulse of charge is injected in the ring at that minimum energy (as seen in the third graph). Next the magnets are ramped (while the accelerator is on). This takes some time owing to the inductance of the magnetic system and accelerator electric field strength. By the way, at the start, the charged particles are at a lower energy and moving slowly, giving rise to a certain current (moving charges) in the ring. As the particles are accelerated, they get faster and the internal current increases, although the total charge stays the same. So be careful asking about the current in a synchrotron.

The ramping of the magnets is done in such a way to be synchronized with the rate of acceleration the particles are receiving every turn. However, this is different than a cyclotron. The particles are actually located, relative to the RF at a synchronous phase. In a cyclotron if the beam deviates from near the top of the RF wave, it will not gain enough energy to make it to a subsequent RF wave top in time and will slip in phase. The synchrotron is not isochronous. This lack of isochronicity is called momentum compaction. Therefore, the particles are adjusted closer to the zero crossing of the RF wave such that, those that are too fast and, therefore, arrive early, experience a lower electric field and will arrive on the next turn at an appropriate phase to be accelerated more because their orbit time will have been a little longer. Thus, it has a restoring force which self-synchronizes the energy of the particles and it is thusly called a 'synchrotron'. This was introduced by Veksler in 1944 and MacMillan in 1945. The time for the energy ramp, as noted depends on the inductance and voltage available for the synchrotron magnets and the rate of acceleration. Once the desired energy level has been reached, other things can happen.

It is possible, at any given time to change the energy of the synchrotron. From the point of view of changing the magnetic fields, this is relatively straightforward. From the point of view of the beam, however, this is actually quite difficult. In the second graph (Energy B), the synchrotron changes energies three times during the time that Energy A kept a constant energy. It does take time to reduce the energy, again given the magnetics involved. Actually, in real life for reasons outside the scope here, it is probably easier to go up in energy instead of down. Going up can result in a beam emittance that is reduced (due to the increasing longitudinal energy). The period during which Energy A is set to $E1$ and Energy B is set to $E3$, $E4$ and $E5$ is called a period or a cycle. There is nothing, a-priori, limiting the time of this cycle other than the power capacity of the power supply and the cooling of the magnets that would be needed to maintain a field for a longer time. During this time for protons and heavier ions, the acceleration is turned off since they aren't losing much energy due to bremsstrahlung and can coast.

What gives meaning to the cycle is the extracted current. The amount of charge injected in the ring is finite. Once that charge is injected, at the start of the cycle, that's all the charge that will be available for the cycle (at least for the usual particle therapy synchrotron). That charge must be extracted which results in an extracted current, I, shown in the fourth graph (labeled Current) which reduces the charge in the ring according to $Q = \int I dt$. The rate at which charge is extracted from the ring will determine the beam current and the dose rate. In Figure 13.29, the charge extracted in the first cycle, $I1$, is not necessarily the same as $I2$ and $I3$. $I1$ is extracted for a longer time than $I2$ and the rate of extraction for $I3$ is lower than the others. At some point in time, the charge in the ring has been depleted (see the Charge graph). That depends upon how much charge was stuffed in, in the first place. Once that happens, the ring magnets must be ramped back down to their minimum value to allow another pulse of lower energy beam/charge to refill the ring. Again, the ramping down takes some time, and if it's ramped down from full energy, a lot of power is going somewhere, probably enough to power more than your house. This is the end of the cycle.

In the second cycle of Figure 13.29, the charge is extracted at a variable rate, after waiting a short time with the ring at full energy (which is possible), resulting in a varying extracted current which is also possible with a synchrotron.

13.5.2 EQUATION OF MOTION (GENERAL CASE)

As discussed in Chapter 10 (Equation 10.26), the equation describing the motion of a particle in a BTS with magnetic fields is

$$x'' + K(s)x + S(s)x^2 = \frac{1}{\rho(s)} \frac{\Delta P}{P_0} \tag{13.35}$$

where x is the deviation from the central trajectory of a particle in the transverse direction, and the derivative is with respect to the distance along the trajectory s. $K(s)$ is a function related to the quadrupole component of the magnetic fields, $S(s)$ is a function related to the sextupole component of the magnetic field; $\rho(s)$ is the bending radius (related to the magnetic field), and P is the momentum of the particle, while ΔP is the deviation of the momentum of the particle from the momentum that the BTS is set to transport, or the deviation from the nominal momentum P_0.

This equation of motion looks like the classical oscillator equation with a driving term. If a special case of this equation is considered whereby there are no multipole components (yet) of the magnetic fields above quadrupole, and the particle momentum does not deviate from the central BTS momentum, the following simplification is obtained:

$$x'' + K(s)x = 0 \tag{13.36}$$

A solution to this differential equation is of the form

$$x(s) = \sqrt{a\beta_M(s)}\cos\left(\Psi_M(s) + \delta\right) \tag{13.37}$$

From this, it can be shown (exercise) that

$$\Psi_M(s) = \int_0^s \frac{ds}{\beta_M(s)}, \text{ and} \tag{13.38}$$

$$\beta_M''' + 2\beta_M K' + 4\beta_M' K = 0 \tag{13.39}$$

Equation 13.38 can be regarded as the definition of Ψ and Equation 13.39 is the equation from which the so-called beta (β) function can be derived from the form of the beam transport quadrupole function.

13.5.2.1 Interpretation of the Parameters of the Equation of Motion

Having determined that Equation 13.37 is an acceptable form of solution for the equation of motion, it is useful to look at its functional form and interpret the physical meaning of the parameters.

A particle's transverse offset from a central axis at an azimuth s is $x(s)$. The particle's offset exhibits an oscillatory behavior with an azimuthal dependent phase and amplitude. δ is a constant offset to take care of the initial conditions dependent upon where one takes $s = 0$ ($x(0) = \sqrt{a\beta_M(0)}\cos\delta$). Graphically the solution can be represented by an oscillatory particle trajectory bounded by a maximum amplitude function as shown in Figure 13.30. The dashed lines are the boundaries $\sqrt{a\beta_M(s)}$ while the function $x(s)$ evolves within those boundaries (the solid gray line). The curve inside these boundaries represents an arbitrary trajectory of a particle with a particular set of initial conditions of offset and angle. No offset propagates on the black line of the s axis. The maximum offset of this particle at any given offset can never be larger than $\sqrt{a\beta_M(s)}$. However, at any given azimuth, the particle offset can be smaller than the maximum amplitude depending upon the phase advance of the trajectory $\Psi_M(s)$. If the trajectory offset crosses through zero at some azimuth s, it will also cross through zero at an azimuth that is a phase advance of a multiple of π from that location. Note that, from one zero crossing to another, the particle's x coordinate is zero, and this is consistent with $(x/\theta) = 0$ between those two azimuthal locations.

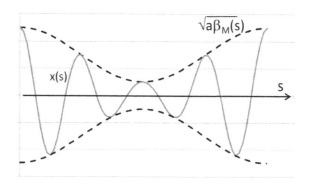

FIGURE 13.30 Bounded cosine trajectory.

It is very important to be reminded, that so far, only the trajectory of a particle in a magnetic field of the BTS has been discussed! The concept of a beam and the parameters describing the beam has not yet been mentioned here. Therefore, the β_M and Ψ_M functions are functions of the BTS only. These functions can be used to predict the trajectory of a particle once its initial conditions are known. The M in β_M stands for machine – as in a machine dependent function. Do not confuse these β_M functions with those β functions that are sometimes associated with beam size. That has not and will not been used in this book.

Since these equations deal with the trajectory of a particle, they can also be used to describe the trajectory of the centroid of a beam (but that's not the same as describing beam parameters). Measurements of the beam trajectory can then be used to determine the BTS parameters, or conversely, the BTS parameters can be adjusted to the appropriate optics by measuring the behavior of the centroid trajectory. Unless otherwise noted, the M will be dropped at this point and β will be assumed to mean β_M.

13.5.3 RELATIONSHIP BETWEEN THE TRANSFER MATRIX AND THE TRAJECTORY EQUATION

It remains now to relate the trajectory equation parameters to the more intuitive transfer matrix parameters. This derivation shows that between any two points in a BTS (1 and 2):

$$
\begin{pmatrix} x_2 \\ \theta_2 \end{pmatrix} = \begin{pmatrix} \sqrt{\dfrac{\beta_2}{\beta_1}}\left(\cos\Delta\Psi_{12}+\alpha_1\sin\Delta\Psi_{12}\right) & \sqrt{\beta_1\beta_2}\sin\Delta\Psi_{12} \\ -\dfrac{\left(1+\alpha_1\alpha_2\right)\sin\Delta\Psi_{12}+\left(\alpha_1-\alpha_2\right)\cos\Delta\Psi_{12}}{\sqrt{\beta_1\beta_2}} & \sqrt{\dfrac{\beta_1}{\beta_2}}\left(\cos\Delta\Psi_{12}-\alpha_2\sin\Delta\Psi_{12}\right) \end{pmatrix} \begin{pmatrix} x_1 \\ \theta_1 \end{pmatrix} \quad (13.40)
$$

where $\alpha = -(d\beta(s)/ds)/2$. This equation yields a wealth of information about the behavior of a BTS and the quantities that can be measured to ensure that the system is properly tuned. For example by inspection of the above matrix, it can be seen that

$$
(x/\theta)=\sqrt{\beta_1\beta_2}\sin\Delta\Psi_{12} \quad (13.41)
$$

(x/θ) (the effective drift) is oscillatory in the phase advance. It is now easy to see that if the BTS is tuned, so that between point 1 and 2, there is a multiple of 180 degrees of phase advance, it is equivalent to a zero effective drift length. By choosing the initial conditions appropriately, all the information about the BTS behavior can be measured. Conversely, knowing the BTS behavior, a particle's trajectory can be predicted.

With the transfer matrix in this form, and having a list of the betas, alphas and phase advances along the line, it is possible to determine the transfer matrix from any point (in the list) to any other point (in the list). Using the other form (the R matrix), it is only easy to determine the transfer matrix from the origin of the R matrix to another point downstream for which the R matrix happens to be known. Even if the R matrix is known at every point (relative to the beginning), it is not trivial to know it from any arbitrary point to another arbitrary point. Nor is it simple in that formalism to go in the reverse direction at an arbitrary position and infer potential sources of error upstream.

13.5.4 STABILITY OF A CLOSED MACHINE PARTICLE TRAJECTORY

A synchrotron is a beam transfer line. Only instead of transferring the beam from one room to another, the beam is being transferred from one place in the ring, back to the same place. Equation 13.40 can be rewritten in the form of the transfer matrix R, as:

$$
\bar{R}= \begin{bmatrix} \sqrt{\dfrac{\beta_2}{\beta_1}} & 0 \\ -\dfrac{\left(\alpha_1-\alpha_2\right)}{\sqrt{\beta_1\beta_1}} & \sqrt{\dfrac{\beta_1}{\beta_2}} \end{bmatrix} \cos(\Delta\Psi)+ \begin{bmatrix} \sqrt{\dfrac{\beta_2}{\beta_1}}\alpha_1 & \sqrt{\beta_1\beta_2} \\ -\dfrac{\left(1+\alpha_1\alpha_2\right)}{\sqrt{\beta_1\beta_2}} & -\sqrt{\dfrac{\beta_1}{\beta_2}}\alpha_2 \end{bmatrix} \sin(\Delta\Psi) \quad (13.42)
$$

Remember that M has been dropped, but these parameters are still functions of the transport system (of the ring) but not the beam. Also the 12 subscript on the phase advance has been dropped.

In a 'closed' machine, the beam is meant to travel through the same beamline repeatedly. Define an orbit to be a trajectory which is repetitive and stable. Two conditions are relevant for a machine in which the particles pass repeatedly.

1. There should exist within the machine, an orbit which is closed (x_c); e.g. the centroid of the beam should obey
    a. $x_c(s + L) = x_c(s)$, where L is the circumference of the orbit
        i. s is the coordinate representing a particular azimuth
        ii. x_c is the 0 offset orbit with respect to the 'closed orbit' (central trajectory)
2. A beam will traverse the same magnetic system repeatedly and the beam size should remain smaller than the apertures (or smaller than imposed by some other criteria).

Now, consider the requirement for stability of the trajectory of a particle. In a ring, a particle will repeatedly traverse the elements that give rise to the above transfer matrix until it is extracted. It must be the case that repeated transformation of the particles coordinates by this matrix result in a bounded solution. Going around the ring N times means that a particles trajectory after those N trips will be given by $\mathbf{R}^N$. N can be 1,000,000 or 10,000,000 or much more. In a ring, the magnet system keeps repeating and the machine functions at any given point keep repeating so $\beta_1 = \beta_2 = \beta$, etc. Therefore, Equation 13.42, around one complete turn, reduces to

$$\overline{R} = \begin{bmatrix} 1 & 0 \\ 0 & 1 \end{bmatrix} \cos(\Delta\Psi) + \begin{bmatrix} \alpha & \beta \\ -\dfrac{(1+\alpha^2)}{\beta} & -\alpha \end{bmatrix} \sin(\Delta\Psi) \qquad (13.43)$$

Simplifying how **R** is written, this becomes: (In text a thick single bar represents a matrix.)

$$\overline{R} = \overline{I}\cos(\Delta\Psi) - \overline{J}\sin(\Delta\Psi) \therefore \overline{R^N} = \overline{I}\cos(N\Delta\Psi) - \overline{J}\sin(N\Delta\Psi) \qquad (13.44)$$

A condition for trajectory stability is that $\overline{I}^2 = \overline{I}$ and $\overline{J}^2 = -\overline{I}$ or that after the Nth revolution, the particle's offset is still bounded. The form of the transformation matrix that will theoretically predict the beam behavior turn after turn is of particular importance in a machine with many turns of beam. The forces acting on the beam result from fields which satisfy Maxwell's equations. The transform matrix used here is an approximation, or a truncation to some order. If the beam is to circulate for some many millions (or more) turns, the model describing the beam behavior must be symplectic (and this gets back to the innuendos of Section 10.3.2 and the trajectory stability conditions just mentioned earlier). There are methods to create a symplectic matrix from a truncated matrix, or there are methods of tracking particles in a repetitive machine which take higher order effects into account and allow for accurate particle tracking. The theoretical predictions must satisfy these criteria. It could be that as a result of higher order effects, the beam phase space is very distorted, but the symplectic criteria guarantees that unless acted upon by an outside nonconservative force the area will remain constant. That is not covered in this book. Instead, some simpler cases are explored.

In a closed machine, if all is mostly well, a particle will continue to circulate for many turns. In each turn it will pass through the machine focusing elements, and if offset from ideal conditions, it will oscillate (with stability). This is the same effect as a trajectory offset in a single-pass well-focused beamline. The focusing strength can be quantitatively represented by the phase advance around one turn of the closed machine's magnetic lattice. From Equation 13.38, going around one turn

$$\Psi(s) = \oint \frac{ds}{\beta(s)} \equiv 2\pi Q \qquad (13.45a)$$

write $\Psi(s) = Q\varphi(s)$, with $\varphi(s)$ being defined from 0 to 2π $\qquad$ (13.45b)

Q is defined as the tune. This is the number of betatron oscillations (given the function called β) that a particle will undergo in one revolution of the ring. Q can also be written as $Q = N + v$, where v is the fractional part of Q and N is the integer part of Q. Therefore, unless the ring's optics is specifically designed for an integer only (multiples of 2π) tune, a particle will oscillate through N full periods and a fraction v more. Most of the time the tune will refer to only v and not Q.

Assume that the black solid line in Figure 13.31 is the unperturbed closed orbit $x_c(s)$. Look at the offset of a particle by following the dashed black line. This is sort of Figure 13.30 wrapped in a circle with the s axis as the black circle. This particle has oscillated three times around the circumference, so it has $Q = 3$. The dotted gray curve is similar, but with a larger amplitude. Therefore, that ring's tune is a multiple of an integer. If it were not integer, the particle wouldn't come back to the same place exactly. Noting the particles position at that place in the ring, for example if v were 0.1, it would take ten turns for the particle to come back to the same coordinates at that particular location. The particles offset at a location as a function of time can be Fourier analyzed, and the result would give the ring tune as a frequency (as well as the beam revolution frequency in the ring).

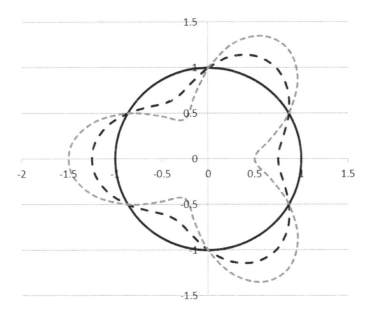

FIGURE 13.31   Betatron oscillations.

13.5.5   And We're Out of Here

It is even more interesting to make a Poincare map of the (x,θ) coordinates of a particle at various locations around the ring. While only direct measurements of position offset are possible it is possible from two position measurements to infer the angle of a particle at a previous location (if it is known what was in between). A Poincare map at a specific location in the closed machine is shown in Figure 13.32. One particle is represented by the circles filled with numbers and the number represents the turn number in which the particle had these (x,θ) coordinates. The particle's coordinates change each turn. Since in this example the tune is close to 0.3, then the particles coordinate offset changes two times before getting close to the original coordinates on the 3rd turn. As the number of turns increases, the locus of points describing the particle coordinates, the Poincare map, sweeps out an ellipse.

Since

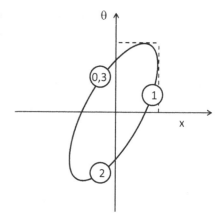

FIGURE 13.32   Poincare map.

$$x(s) = \sqrt{a\beta(s)} \cos\left(\Psi(s) + \varphi_0\right) \qquad\qquad (13.37\ \text{Repeated})$$

(except δ is now called ϕ_0 since that will be the initial condition), then

$$\theta(s) = \sqrt{a}\, \frac{\beta'(s)}{\sqrt{\beta(s)}} \cos\left(\Psi(s)\right) + \sqrt{a\beta(s)} \sin\left(\Psi(s) + \varphi_0\right)\Psi''(s) \qquad\qquad (13.46)$$

Now $x(s)$ and $\theta(s)$ are known and the Poincare map ellipse can be derived. It's perhaps no surprise that the parameters of the ellipse are related to the beta function of the circular machine at that point.

Recalling Equation 13.45a if β is changed (e.g. a focusing element is changed) at any point in the ring, the tune Q will change. It's instructive to reexamine the transfer matrix around one turn of the ring, and include another focusing element as a perturbation. This perturbation can be a thin element inserted into an existing quadrupole, or an additional air core magnet, for example with a focal length equivalent to $1/\delta$ (okay – reusing δ since it's not being used right now). Thus, the perturbation will represent additional quadrupole strength. A new transform matrix for one turn of the ring is then determined. Let $\mu = 2\pi Q$ instead of Ψ, since it's a turn around the ring. Starting with Equation 13.43:

$$\overline{R} = \begin{pmatrix} 1 & 0 \\ \delta & 1 \end{pmatrix} \begin{pmatrix} \cos\mu + \alpha\sin\mu & \beta\sin\mu \\ -\gamma\sin\mu & \cos\mu - \alpha\sin\mu \end{pmatrix} \qquad\qquad (13.47)$$

where $\gamma = (1 + \alpha^2)/\beta$. Evaluating the above matrix yields the following result:

$$\overline{R} = \begin{pmatrix} \cos\mu + \alpha\sin\mu & \beta\sin\mu \\ [\delta \cdot \cos\mu + \delta \cdot \alpha\,\sin\mu - \gamma\sin\mu] & [\delta \cdot \beta\sin\mu + \cos\mu - \alpha\sin\mu] \end{pmatrix} \qquad\qquad (13.48)$$

The tune of the ring can be determined from the trace of the matrix since

$$\text{For the original matrix } (13.47)\ \tfrac{1}{2}\ \text{Trace } R = \cos\mu \qquad\qquad (13.49a)$$

$$\text{For the perturbed matrix }\ \tfrac{1}{2}\ \text{Trace } R = \cos\mu_{\text{new}} = \cos\mu - \delta\beta\sin\mu/2 \qquad\qquad (13.49b)$$

If μ is only changed a small amount and $\mu_{new} = \mu_o + \Delta\mu$ then: $\Delta\mu = -\beta\delta/2$. Therefore, the beta function can be determined by measuring the change in tune cause by a perturbation δ.

Even more importantly, what if $\delta\beta$ is such that $|\cos\mu - \delta\beta (\sin\mu)/2|$ is greater than 1. The traces with and without perturbation are plotted in Figure 13.33. This figure shows the unperturbed trace (solid black – beautiful cosine), the perturbation (gray dashed) and the sum. In this case, there are two regions of unphysical $\cos\mu_{new}$, both near a tune of 180 degrees (half integer) (see hatched region). Remember here μ is the tune, the phase advance around the whole ring, not the phase at specific locations in the ring. So this is a question of what values of the tune are stable with the perturbation. Near 180 degrees, $\cos\mu$ is not real and the determinant of the matrix is not 1 and the trajectory becomes unstable. This means, that if an offset of the closed orbit is made, the particle's coordinates will not be restored or oscillatory and will not necessarily stay within the ring. In other words, it is ejected, but perhaps somewhere not planned for.

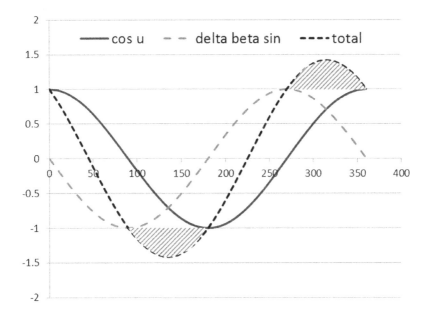

FIGURE 13.33   Trace value with and without perturbation.

Extracting a beam is not as simple as making all the particles in the beam unstable. Only a part of the beam needs to be extracted at a time to stretch this out over time. There are a number of ways of doing it. For example using an electrostatic field at a frequency that can resonate with the tune, some of the particles inside the ring can get 'shaken' and their offset from the central orbit would increase until they are extracted by a physical separator, like an electrostatic septum.

Another more classic method is resonant extraction involving the use of higher order multipoles. It's not the simplest, but it is classic. Imagine that there is a perturbation in the ring that kicks the beam a certain amount at a particular location. The perturbation could be the mini quadrupole that was installed for Equation 13.49b. If the tune is an integer, the next time the particle returns to the same location, it is traveling in the same trajectory as the previous turn. Therefore, the perturbation adds (a particle that passes through a quadrupole off axis is kicked) and the perturbation amplitude grows and grows with each successive turn. If the ring tune were half integer, then, for example the particle would be at a location in the ring with a positive offset. On the next turn the particle would be on the opposite side of phase space (not necessarily position but it could be position) and it would have a negative offset. If there were a perturbation of the same sign on both sides of the central orbit (e.g. a dipole), the effects after two turns could cancel. But what if the perturbation on the left side of the closed orbit were opposite that on the right side (e.g. the perturbed quadrupole or other odd function), after one turn, the perturbation, for this half integer tune would add to the perturbation. Therefore, the combination of the ring tune and the type of perturbation will combine with the first pertubation to increase the beam offset. Extracting it a little at a time and controlling it during that time, is the real trick.

What is a perturbation? It could be unwanted, which is not good because it wasn't part of the design. Or it could be part of the design. Figure 13.34 is a plot of the magnetic field of a quadrupole (linear, solid black line), sextupole (quadratic, gray dashed line) and octupole (cubic, black dashed) as a function of the distance from its center. Note how, with the right settings, the linear field is stronger than the higher order fields, until a particle is offset beyond a certain amount (see the open circles for those locations). In the case of this figure, if the particle is at a coordinate greater than 3.5 or less than −3.5 the octupole overcomes the quadrupole. It can be imagined that this can create a stable and unstable region inside the ring depending upon which element is stronger at a given location.

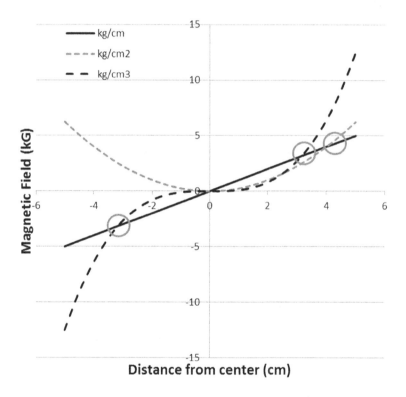

FIGURE 13.34   Multipole perturbation strength.

Consider the effect of an octupole on the coordinates of a particle in an approximate calculation.

$$\begin{pmatrix} x_1 \\ \theta_1 \end{pmatrix} = \overline{\boldsymbol{R}}_0 \begin{pmatrix} x_0 \\ \theta_0 \end{pmatrix} + \overline{\boldsymbol{R}}_0 \begin{pmatrix} 0 \\ kx_0{}^3 \end{pmatrix} \qquad (13.50)$$

where the last term is an angular 'kick' coming from the field of an octupole with strength characterized by k. (This is for the octupole here, not the quadrupole.) The transfer matrix $\boldsymbol{R}$ is the ring transfer matrix. The coordinates x_1 and θ_1 are the coordinates of a particle's after one turn of the ring, including the octupole kick. After two turns, that particles coordinates are given by

$$\begin{pmatrix} x_2 \\ \theta_2 \end{pmatrix} = \overline{R}_0 \begin{pmatrix} x_1 \\ \theta_1 \end{pmatrix} + \overline{R}_0 \begin{pmatrix} 0 \\ kx_1^{\,3} \end{pmatrix} \tag{13.51a}$$

$$= \overline{R}_0 \left[\overline{R}_0 \begin{pmatrix} x_0 \\ \theta_0 \end{pmatrix} + \overline{R}_0 \begin{pmatrix} 0 \\ kx_0^{\,3} \end{pmatrix} \right] + \overline{R}_0 \begin{pmatrix} 0 \\ kx_1^{\,3} \end{pmatrix} \tag{13.51b}$$

$$= \overline{R}^2{}_0 \begin{pmatrix} x_0 \\ \theta_0 \end{pmatrix} + \overline{R}^2{}_0 \begin{pmatrix} 0 \\ kx_0^{\,3} \end{pmatrix} + \overline{R}_0 \begin{pmatrix} 0 \\ kx_1^{\,3} \end{pmatrix} \tag{13.51c}$$

Consider the case where $\mu = 2\pi/2 + 2\pi\varepsilon$, where ε is a small value. Therefore, the tune is near a half integer plus a little bit. The transfer matrix can then be evaluated near that half integer tune.

$$\overline{R} = \begin{pmatrix} \cos\mu + \alpha\sin\mu & \beta\sin\mu \\ -\gamma\sin\mu & \cos\mu - \alpha\sin\mu \end{pmatrix} \cong \begin{pmatrix} 1 & \beta 2\pi\varepsilon \\ \dfrac{2\pi.\varepsilon}{\beta} & 1 \end{pmatrix} \tag{13.52}$$

Neglecting terms of ε^2 and higher the coordinates of the particle after two turns are given by

$$x_2 = x_0 + 2\beta(2\pi\varepsilon)\theta_0 \tag{13.53a}$$

$$\theta_2 = \theta_0 - 2\frac{2\pi\varepsilon}{\beta}x_0 + 2kx_0^{\,3} \tag{13.53b}$$

These equations are very interesting. Plotting the change in angle after two turns $\theta_2 - \theta_0$, as a function of its x position shows this interesting behavior in Figure 13.35.

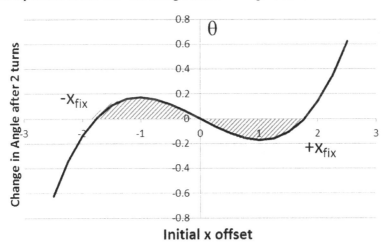

FIGURE 13.35   Half integer resonant coordinates.

A particle starting at $x_0 = 0$ will, of course, get no angular kick. But also, a particle starting at the two points $x_{\text{fix}} = \pm\sqrt{2\pi\varepsilon/k\beta}$ has no net change in angle. Therefore, it is called a fixed point. At this point, although the tune of the ring is ε away from a half integer resonance, a particle starting at this point feels like the ring is treating it as if it were exactly a half integer tune. Looking at this curve in more detail reveals other interesting features. A particle between $\pm x_{\text{fix}}$ is basically in the hatched region. A particle starting in this region on the plus side, after two turns has a net negative change in angle, or a restorative force toward the central orbit; similarly for the hatched region on the left side. So particles in between the fixed points are stable. A particle to the right of $+x_{\text{fix}}$ starting off at a positive offset position has a positive angular kick, thus reinforcing its attempt to leave the central orbit and wants out of there. This is an unstable region. Controlling the location of the stable and unstable region (e.g. β and k) is an important part of ring extraction design. The unstable particle migrates further and further away from the center until it passes into a component (an electrostatic septum and/or a magnetic septum) which will help it out. Depending upon the ring tune, and the particle trajectory, the particle orbit can be stable or unstable. Combinations of tunes, perturbative quadrupoles, sextupoles and octupoles will create the appropriate stable and unstable regions.

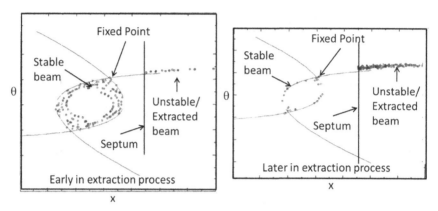

FIGURE 13.36 Extraction process and separatricies.

An example of this type of extraction is shown in Figure 13.36. This is not for a particle therapy synchrotron, but for an electron stretcher/storage/synchrotron ring, one that is near and dear to my heart. The graph on the left shows the phase space at a particular location in the ring. Points representing particles and intersecting parabolas representing the phase space boundaries between stable (inside) and unstable (outside) regions are included. In this figure, extraction has begun since there are already some particles to the right of the septum. The parabolas are oriented so that the unstable particles grow in amplitude in the x part of phase space, not θ, so they can physically reach the extraction septum. The right figure is different because the strength of the exciting quadrupole has increased, so the stable region has shrunk and some particles which used to be in the stable region are now outside – more particles can be seen outside the septum. Since the septum is helping to extract the particles, the (x,θ) phase space will have a sharp edge on that side (temporarily).

The extraction time dependence of the number of particles extracted from the ring is shown in Figure 13.37. Extracting the beam from a synchrotron is a semi-stochastic process. There are a variety of methods of extracting the beam. One can imagine a pail of water in which water is stored and in which a spigot is inserted into the bottom with a valve to control when, and at what rate, the water is extracted. The water would be expected to come out smoothly, but if it were filled with air bubbles and shaken, the result would be different. Figure 13.37 shows an example of the time dependence of a form of resonant extraction. Getting the output to look even this uniform was no simple trick (if I must say so myself). Current medical systems are even much smoother. This example was for high duty factor, not medical use.

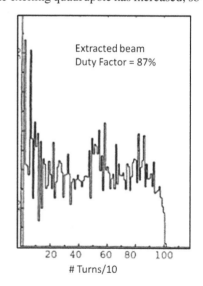

FIGURE 13.37 Extraction time structure.

It might be worth a word or two to note that some parts of these resonance methods can be, and sometimes are, used in a cyclotron to improve the extraction efficiency.

13.5.6 SYNCHROTRON BEAM PARAMETERS

13.5.6.1 Energy

In general, a synchrotron is capable of accelerating and extracting different particles and different energies if properly designed. Given that the accelerating voltage, frequency and magnetic fields of the guiding magnets can all be dynamically varied, most ion species can be accelerated to a range of desired energies. Therefore, no degrader and ESS is needed. For protons, synchrotrons typically accelerate beams from 70 MeV (or a little lower) to up to 330 MeV. Larger systems can accelerate most ion species with, for example ^{12}C at 440 MeV/nucleon. Some synchrotron extracted beams can have an energy spread so low (e.g. 0.01%) that the beamline and/or gantry design can be rethought and may not need to be achromatic.

13.5.6.2 Beam Phase Space

The beam extracted from a synchrotron has a reasonably small phase space area. Since no degrader is needed the beam extracted retains the beam phase space of the synchrotron extraction mechanism, until it passes through a window someplace. The extracted beam phase space is very dependent upon the system design and extraction technique. Some proton synchrotrons can have a phase space area of as small as 0.1π mm mrad which corresponds to a beam size that can be focused and maintained in a beamline with smaller apertures and magnet gaps using reasonable focusing strength. This beam will be transported to the patient and modified, or not based upon the beam delivery method chosen.

13.5.6.3 Synchrotron Charges and Currents

During the time the particles are stored and accelerated, they must all live nicely with each other. However, they are all charged up, and they repel each other. This effect is called the space charge force and is represented in Figure 13.38. The distribution of the particles inside the dashed circle is Gaussian. The black filled circle is one of the particles. The black line is the force that the black particle feels depending upon its position. If that particle were in the center it would not feel any forces since they all cancel. The force grows as it is offset from the center, but when it gets far enough away, the force from the further particles is reduced and the solid curve drops off.

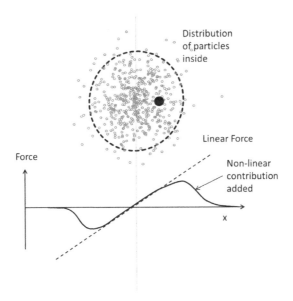

FIGURE 13.38   Space charge force.

This repulsion is shown closer up in Figure 13.39(left) and its magnitude, as a function of the speed of the particles is shown in the lower (solid black horizontal line) in Figure 13.39(right). As the particles move together in the ring, they are a moving charges or a current and since parallel currents attract as shown in the right side of Figure 13.39(left), this attractive force partially cancels the repulsive force. The black dashed line in Figure 13.39(right) shows that the attractive force depends on the magnitude of the current. Current is charge per unit time, which is related to the speed in the ring. As the particles move faster, the deleterious repulsive space charge effects are reduced as shown by the dotted curve (total force) in Figure 13.39(right). Thus, the worst-case situation occurs during low-energy injection and the number of charged particles that can be stored in the ring depends upon the injection energy of the particles.

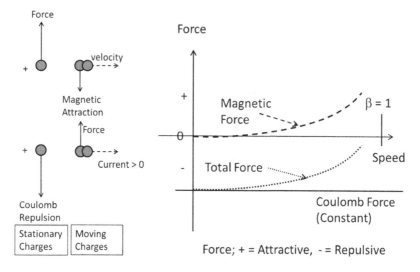

FIGURE 13.39   Attraction and repulsion (left) and speed dependence on forces (right).

Figure 13.40 is a graph of the number of protons that theoretically can be stored in various medical synchrotrons (the limitation arising from the space charge forces only). Recall from Chapter 11 that 1 Gy/min in that portion of a liter ⇒ 40 gigaprotons/min ⇒ <1.3 Gp/acceleration cycle (assuming

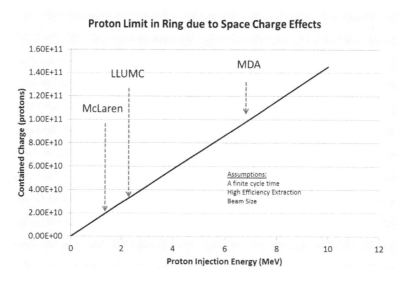

FIGURE 13.40   Synchrotron space charge limits (theoretical).

a 2-second cycle). Thus, a connection between a prescription and an accelerator constraint is obtained. At low energies, the speed is low, and the force is mainly Coulomb or repulsive, which can be viewed as a defocusing force. But from Figure 13.38, this force is position dependent. The effective focusing that a beam can 'feel' comes not only from the ring magnetics, but also due to the forces that the beam experiences from the particles in the beam. This can cause perturbations resulting in unstable motion. In this case, the repulsive force depends upon the particle position, and this results in a spread of focusing conditions particle by particle, or a spread of particle tunes in the beam. So again, the ring is treating individual particles differently. If this tune spread reaches from a stable operating point into a resonance region, this will cause some of the particles to be lost. This is considered one of the limitations of the stored current. This intensity limitation is called the Laslett tune shift effect. When the attractive and repulsive forces are not balanced it modifies the effective focusing effect of the magnets in the ring and effectively changes the tune according to Equations 13.54.

$$\Delta v_x = \frac{r_p NR}{\pi \beta^2 \gamma^3} \frac{1}{a_x (a_x + a_y)} \frac{1}{v_x} \ (\text{relativistic form}) \tag{13.54a}$$

$$\Delta v_x = \frac{r_p NR m_p}{2\pi E_k} \frac{1}{a_x (a_x + a_y)} \frac{1}{v_x} \ (\text{non} - \text{relativistic form}) \tag{13.54b}$$

where:
$r_p = 1.5 \times 10^{-18}$ m (classical proton radius)
$R = C/2\pi =$ the effective ring radius (m)
$\beta = v/c$
$\gamma = 1/(1 - \beta^2)^{1/2}$
$a_x =$ the x beam radius ($\sim \sigma_x$) (m)
$a_y =$ the y beam radius ($\sim \sigma_y$) (m)
$v_x =$ the ring tune in the x (extraction plane) direction
$N =$ no. of protons in the ring
$m_p =$ mass of proton

The importance of these effects (limitation of charge in the ring) may depend upon the clinical situation. A few examples follow.

- The beam-spreading modality plays a significant role. If scattering is used, the beam current incident on the beam delivery system may need to be on the order of nanoamps, whereas for scanning, only tenths of a nanoamp may be required.
- It may be desired to reduce the number of fractions required to deliver the total dose, in which case the dose per fraction would be increased, thus increasing the desired dose rate (so that the treatment time per fraction does not increase).
- Considerations of target motion inside the patient may affect the time constraints on the beam delivery and, therefore, the beam current requirements.
- The instrumentation and the beam analysis time will affect the dose rate that can be safely applied.

13.5.6.4 Timing
There is a possibility of a strong interaction between the time distribution of the beam extracted from the accelerator and the beam delivery method. Beam delivery for particle therapy has now evolved into beam scanning as this enables the most conformal dose distribution possible. The interaction between the synchrotron beam timing and the scanning beam delivery is useful to consider.

Beam Delivery and Accelerator Timing

There are essentially two styles of scanning beam delivery as discussed in Chapter 12. One is called dose-driven and the other is called time-driven. The first essentially integrates the dose at a given location before moving on to the next location. The second assumes that the beam current has the desired stability and uses time as the variable to identify the dose deposited at a given location. If, for example, the beam is moved continuously, the amount of dose deposited along the beam path in any given time interval is determined by the beam current, the scanning speed or both. If these are not precisely correct, the dose deposited will not be precisely correct.

In dose-driven beam delivery, one must stop the beam when the desired dose has been reached. There is, however, a time required for the beam instrumentation to measure and analyze this dose, and a time required for the beam to be turned off. Thus, there is a time lag between the time the system decides to turn off the beam and the time the beam is actually turned off. Therefore, either one can anticipate that the beam to be delivered in this time frame will be known and begin the turn-off process earlier, or one can lower the beam current to ensure that the dose that will be delivered in this time frame will not be significant. In addition, in both time-driven and dose-driven continuous scanning, the beam turn-off time will occur while the beam is moving, and one must account for the locations that receive a dose during the turn-off time.

The earlier considerations highlight the importance of knowing the beam extraction stability and the turn-off time, as discussed in Section 13.5.5. The smoothness and controllability of the extracted beam can determine which method of beam scanning can be applied. The smoothness can also determine the anticipated time required to turn off the beam. If the extracted beam current is unstable, the highest possible beam current should be taken into account.

Organ Motion Timing

In some cases, the target moves. It is required to reduce the dose delivered to healthy tissue, and this poses some challenges when the beam delivery has to be done in a time-dependent way. If the dose were simultaneously delivered to the entire 3D volume of the target (as is almost the case with scattered-beam delivery), the location of the moving target during the beam delivery is the only consideration. One could, for example, deliver dose to the entire volume within which the target was moving; then healthy tissue would be irradiated, but the target would receive the desired dose in the shortest possible time. If the beam is gated on only when the target is in the beam path, the macroscopic timing capability of the accelerator and the time frame of the motion have to be taken into consideration in determining the length of the treatment. Figure 13.41 shows an example of

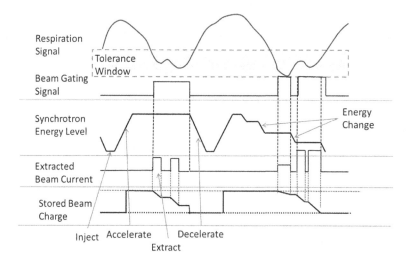

FIGURE 13.41 Synchrotron timing.

this situation for a synchrotron. A large improvement in the efficiency of a synchrotron for beam treatment was achieved with the development of a variable-cycle synchrotron at NIRS and Hitachi. These accelerators can synchronize both the injection and extraction with the target motion.

The upper trace shows the respiration signal coming from the patient, and the trace just below indicates the time that beam delivery to the patient is allowed. This is when the respiration phase is within the tolerance window (gray shaded dashed outline). The next lower trace indicates the energy level of the synchrotron. There is a time for ramping up the energy, a flattop during which extraction is possible and a time for ramping down the energy to enable new low energy beam to be injected before the cycle repeats itself. Thus far, there are two constraints, the breath phase and the synchrotron flat top time period. In addition, the second cycle of the synchrotron shows four energy levels. During the period of energy change, no beam is extracted, so that is a third constraint. The next trace (4th) shows the extracted beam current with different pulse widths and intensities. Finally the lower trace indicates the charge remaining in the synchrotron which is reduced during each extraction interval. Owing to the breathing constraint, both the first and second cycle waited some time, while beam was available, for the respiration signal to be okay. This increases the overall time for the irradiation. Sometimes, depending on the timing, the patient will have to wait for the synchrotron to refill and accelerate. One of the advances of the last few years is using the respiration signal to feed forward and give a signal in anticipation of when beam might be needed, so the cycle will start early and be ready when needed.

Some types of beam cycles for synchrotrons include rapid cycling with fast, one-turn extraction or very short pulses of a periodic (e.g. 30 Hz) beam, where each pulse can be at a different energy. Alternatively, one can use slow extraction, pulling out particles from the accelerator as needed, at the beam current needed until they are used up as has been addressed in this chapter. If there are, for example, 10^{10} particles in the synchrotron, it may take seconds to use up those particles at the normal currents identified earlier. If more are needed, one has to wait for the time it takes to inject and accelerate another bunch. Note that a breathing cycle can take about 3 seconds.

13.5.6.5 Cost

Reduced capital cost of particle therapy systems may contribute to its becoming more widely used. This includes a reduction of both the building and the equipment costs associated with a facility. Today a variety of synchrotrons are being used in medical treatment (see examples in Figure 13.43).

Proton synchrotrons have already been reduced in size by ProTom and Hitachi (to name two), with diameters on the order of 5 m. One facility for proton therapy has already been constructed in an existing conventional radiotherapy clinic at Massachusetts General Hospital as shown in Figure 13.42

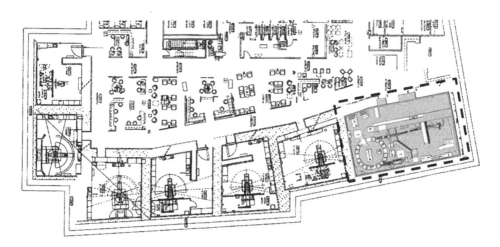

FIGURE 13.42   Proton therapy alongside of photon therapy.

(see the lower right room surrounded by the dashed black line, next to the nearby linacs). No new building was built for the machine; the components were brought in via elevator. This is a first of its kind milestone and is arguably a holy grail of particle therapy (at least the author believes so). Synchrotrons for heavier ions (e.g. carbon) are much larger; however, smaller systems utilizing superconducting technologies are always under investigation.

13.5.7 EXISTING SYNCHROTRON SYSTEMS

The Hitachi synchrotron shown in Figure 13.43 (courtesy of Hitachi) is representative of the size of current synchrotrons used in medical facilities for proton therapy. Concepts for smaller devices have been advanced, such as the one developed at the Lebedev Institute and licensed by ProTom International and installed at McLaren and MGH. (Photo in Figure 13.43 is courtesy of McLaren Proton Therapy Center.)

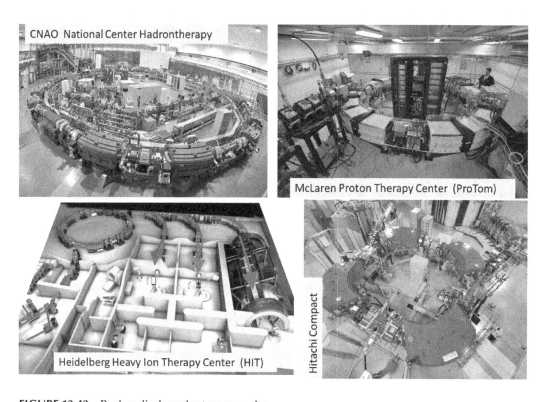

FIGURE 13.43   Real medical synchrotron examples.

The synchrotrons used for heavy ion treatments are larger, as shown in Figure 13.43 and have a high level of functionality and flexibility. The first heavy ion medical synchrotron was built at the HIMAC facility in Japan which began operation in 1994. In that facility, the maximum beam energy is 800 MeV/μ. This gives a range of 30 cm in water for Si ions. The diameter of the synchrotron is about 40 m. However, since that time designs have evolved to reduce the size of a carbon ion accelerator leading to the design of the Heidelberg Ion Therapy Center shown in Figure 13.43 (courtesy of the Heidelberg Heavy Ion Therapy Center) with the first of its kind, heavy ion gantry and CNAO (photo courtesy of the National Center of Oncological Hadrontherapy in Italy). This synchrotron in this facility has a diameter of 20 m and can accelerate carbon ions to 400+ MeV/μ. Smaller ones also exist now.

13.5.8 The Future of Synchrotrons

Synchrotrons have the advantage of possible energy variability without a degrader and a small beam phase space. Work continues with the aims of obtaining adequate intensity storage while minimizing the cost of higher energy injection and smoothly controllable extraction, with a flexible accelerator cycle that can change the beam energy extracted during a single cycle. Many of these have been implemented but thus far, all of these capabilities have not been used in any one synchrotron for treatment. There may be increasing demand for faster, high dose rate, treatment without compromising accuracy. For heavier particles, the use of superconducting technology can reduce the size, but the cost and the rapidity of change of the magnet excitation will be affected.

Some of the considerations for future and currently ongoing synchrotron development include aspects of cost and beam parameters. From the cost perspective, the size and cost of proton synchrotrons have been reduced to fit in conventional linac rooms, but the size and cost of heavier ion machines is still large and high. Perhaps it will continue to be consistent with a laboratory environment, or perhaps work with superconducting technology will enable a significant size reduction. Cost is also related to the injection components and the energy of the injection, which will have an effect on the stored charge, as well as the elements that can be accelerated. Perhaps, there will be a breakthrough in the determination of which ion is most useful beyond protons and perhaps it won't be as heavy as carbon, but that remains to be seen. Some say He and Li could be useful.

From the perspective of beam parameters, there are several considerations. Higher energy protons, such as the 330 MeV capability of the ProTom synchrotron allow for proton radiography and tomography. Charged particle imaging can become an important technique which enables direct determination of the target stopping power. However, that would require the system availability for both imaging and treatment and those logistics would benefit from some study. In the heavier particle regime, work has started in the consideration of using helium as an imaging particle while in the process of treating with carbon. Thus, multi-energy, multiparticle switching, or simultaneous use, could become useful. This could lead to the need for quick energy changes and answering the question about whether the use of superconductivity will negatively affect that. In all cases, improvements in extracted current control and irradiation time are always welcomed.

13.6 ACCELERATORS

There are many types of accelerators and each of these produce beam with different properties. Clever scientists have found methods to effectively tailor the various beams for use in medical particle beam delivery. Some of the parameters are inherent in the accelerator physics and engineering, and some require beam modifications external to the accelerator. Identification of the constancy of a parameter or dynamic behavior of the system can help to determine the beam delivery method and can provide an insight into the level of complexity of the operation of the system. The time structure of the beam is an important quantity, and some parameters associated with the pulse structure of the various accelerators are included in Figure 13.44.

This figure has eight traces with the horizontal axis being time. The upper trace for each machine is the energy level. The synchrotron cycle that has been discussed is indicated by the separation of $E1$ and $E2$ in the upper trace. The cyclotron extracted energy is constant and uses a degrader to adjust the energy. Within a pulse, the third synchrotron trace can be similar to the second cyclotron trace when multi-energy extraction is used. However, there is still a gap when the charge of the synchrotron runs out. The dose rates, for both the cyclotron and synchrotron, are adjustable and variable. It is a challenge for either of them to maintain sufficient control for continuous scanning, but some cyclotron-based systems have achieved that. The pulsed synchrocyclotron has different challenges which have been overcome with today's faster, solid state RF controls and multi-pulse delivery is used.

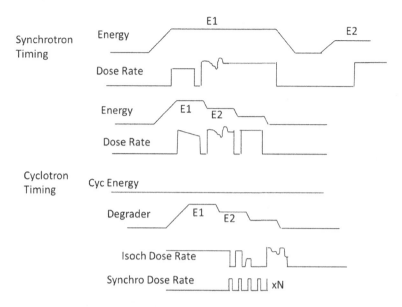

FIGURE 13.44 Accelerator timing comparison.

It used to be said that a cyclotron had fewer knobs than a synchrotron and was therefore simpler, but nowadays both these systems enjoy automated operation. About one-third of all particle therapy systems are synchrotron based. The future development of accelerators must be directed toward meeting the demands of optimal, safe delivery of particle therapy at a cost that is competitive with conventional radiotherapy systems. It is no longer adequate to identify an accelerator and then ask how it can be used for particle therapy; this has been done. The focus now has to optimize the delivery of particle therapy, including the beam parameters, timing, size, and cost. Thus far, there is no single standout technology that can accomplish all this, but the technology to achieve operation with the desired parameters does exist. Costs and size are being reduced. Still the field is ripe for new insights and developments. It might be useful to note that the accelerator is actually not the most costly part of a particle therapy system and may not need the most work. That might be the gantry which will be discussed in the next chapter.

13.7   WHAT COULD GO WRONG – 6?

The number of components that have been introduced in this chapter certainly dwarfs that of those discussed in Chapter 12. However, the beam parameters at the target are not all equally sensitive to the parameters produced by the accelerators. For example, in the case of beam scattering, the beam profile that is extracted from the accelerator is almost irrelevant. However, in the case of scanning it is quite relevant. As with the beam delivery modality, the errors will depend upon the components used. Most previous 'what could go wrong' sections referenced the input beam parameters from the accelerator, so now it's time to consider these. The cyclotron and synchrotron will be considered separately.

Cyclotron

The cyclotron has relatively few parameters owing to the cyclotron equation, but there are some, including corrector magnets, the RF system and the magnetic field.
- The RF could be wrong (frequency or amplitude) so beam current may degrade and the dose rate may be reduced causing a longer treatment time or increased to an unexpected high rate.

- The magnetic field could be wrong, so the wrong energy may be extracted if the cyclotron equation is still satisfied. (e.g. the cyclotron could get hot).
- With the wrong magnetic field or RF, the beam extraction could become unstable
- The ion source could be unstable and the beam current will be too unstable relative to the timing set in the rest of the system.
- The degrader could be in the wrong position; thus, the degraded beam energy is wrong and some might get through the ESS.
- The cyclotron electrostatic septum may be unstable, shorting or otherwise compromised, thus affecting the stability and trajectory of the extracted beam.
- The cyclotron gradient corrector could shift affecting the trajectory and focusing.
- The energy defining slits could be in the wrong position, thus the beam current is not what is expected and the energy spectrum within the beam is incorrect; therefore, the distal dose falloff is affected for shallow fields.

Synchrotron

It is sometimes said that the synchrotron has more parameters of adjustment than a cyclotron, but that can depend on how it is configured. It has an RF system, like the cyclotron, but with much lower power. It has multiple magnets, but they may all be wired in series, so it is like one bigger magnet, like a cyclotron. It also has correction coils to ensure an appropriate beam orbit, like the cyclotron, but perhaps there are more such correctors. It has a septum to extract the beam, like the cyclotron. Where the comparison diverges is that the extraction system from a synchrotron is not passive and there are some components that depend upon the particular extraction implementation. Also there is an injector accelerator with its own set of controls.

- The extraction mechanism could be compromised and the beam current stability could be outside of tolerance.
- The beam energy that is extracted may be incorrect.
- The injector accelerator could be unstable or may be the wrong energy and affect the stored charge.
- The wrong energy may be set for extraction.

As can be seen, there are a number of potential consequences of accelerator errors. However, the beam current and energy may be the most consequential. Therefore, instrumentation with redundant backup are used to ensure that the settings are correct.

13.8 EXERCISES

1. How long does it take for a proton beam to be accelerated in a 235 MeV cyclotron with an electric field of 200 kV per turn? Assume a magnetic field of 2 T.
2. A cyclotron is accelerating a proton beam up to 235 MeV (with 200 kV/turn) with a magnetic field of 2 T. A beam is 0.7071 mm wide (Gaussian σ). The electrostatic septum is 0.1 mm thick. What is the turn spacing at the 235 MeV extraction radius? If the distance of the peak of the beam from the septum just before extraction equals the distance on the other side (during extraction), and if there are no losses other than what is intercepted by the septum (if it touches it its gone), what is the extraction efficiency? (Why was 0.7071 mm chosen?)
3. Extra Credit: Derive Equation 13.22.
4. In the thin wedge dipole achromat (with L_1 = 1 m as in Figure 13.20), what is the energy width portion of the proton beam size at 130 MeV (obtained by straggling of a proton beam through a carbon degrader having started at 230 MeV), at the center of the achromatic (where the slits would normally be placed)? The dipole bend angle is 30 degrees.

5. Show that Equation 13.37 is an acceptable form of the solution of Equation 13.37.
6. Derive Equation 13.38.
7. Derive Equation 13.40.
8. What is the phase advance necessary between point 1 and 2 to measure the ratio of the beta function between those points and how could this be done? ($\alpha_1 = 0$).
9. Calculate the space charge limit for two synchrotrons. The injection energy is 6 MeV for one and 2 MeV for the other. The beam size is 1 cm × 1 cm. The tune is near 0.96 and is within 0.04 of that resonance. The radius of the ring is 2.5 m.
10. What is the extraction energy sensitivity to the magnetic field of a cyclotron ($\Delta E_k/E_k$ vs. $\Delta B/B$? (Extraction occurs at a fixed radius.)

14 Gantries

14.1 INTRODUCTION

It was in the early 1960s that the medical electron linac was mounted on a rotatable structure. It was desired, at that time, to treat a patient while lying down (e.g. in a prone or supine orientation). That is not to say that most practitioners don't desire the same today, but it is to say that things have changed since the 1960s. This will be discussed further.

The development of advanced three-dimensional (3D) treatment planning software allowed for the conceptualization of treatments using beam-to-patient angles that maximized the dose delivered to the target volume and minimized the dose received by nearby critical structures. The equipment used for generating and directing the charged particle beam for therapy is larger and heavier than electron and photon systems. Creating a structure that produces the desired beam angles, involves either mounting the accelerator on a gantry structure or starting with a fixed accelerator and transporting the beam to the desired orientation. In the case, where the beam is extracted in the horizontal plane, the purpose of the gantry is to bend that beam into the vertical plane and enable the beam angle, relative to the target, to be rotated in that plane.

A combination of a medical gantry and patient positioning system is required to direct a particle therapy beam to a supine patient from a variety of angles. The requirements of beam direction and size for precision beam treatment result in stringent criteria for the magnetics and mechanical structure of the gantry. Varieties of ion beam gantries have been designed, constructed, and are being planned.

It is a matter of some interest to reduce the size and cost of particle therapy systems. Methods can include finding ways to deliver high-quality treatments in a system without a gantry, or to develop a gantry compact enough to fit into a space naturally found in a hospital environment. There can be many variations of the generic gantry that bends a beam out of the beamline plane and produces a net bend of 90 degrees. The desired functionality of a gantry is dependent upon its clinical use and this will influence the gantry design. It is one of the most expensive parts of a therapy system. Costs would be reduced if it weren't necessary. (Hold on to that thought.)

14.2 ACCELERATOR ON A GANTRY

Normally, the path of a beam from an accelerator is in the horizontal plane – parallel to the floor. There are notable exceptions like when an accelerator is mounted on a gantry. This has been done for X-ray therapy with an electron accelerator mounted on a gantry since the 1950s. However, the development of a hospital-based particle accelerator, with sufficient energy useful for particle therapy, took much longer.

14.2.1 CYCLOTRON ON A GANTRY

In the early 1970s, Henry Blosser began to file patents for the design of a superconducting cyclotron. While the concept for such an accelerator with sufficient energy for proton therapy was an intention, the first implementation was for a fast neutron therapy machine built by Blosser and operated at Harper Hospital in Detroit. The original concept as depicted in the patent is shown in Figure 14.1, but the actual implementation was mechanically different as shown in Figure 14.2. The system was operated for many years and treated many patients.

DOI: 10.1201/9781003123880-14

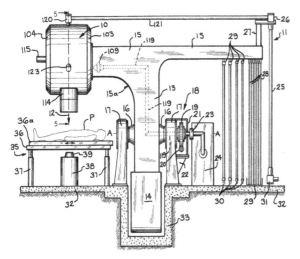

FIGURE 14.2 Cyclotron on a gantry actual.

FIGURE 14.1 Cyclotron on a gantry patent.

In modern times, the MeVIon system, with a magnetic field of 8.5 T accelerates protons up to 250 MeV. This system is a cyclotron on a gantry that rotates 180 degrees and is a compact single treatment room system. It is used for proton therapy.

14.2.2 SYNCHROTRON ON A GANTRY

The concept of an accelerator mounted on a rotating gantry is not limited to cyclotrons. In the early 1990s, the author, and in the 2000s, Niek Schreuder, independently developed a concept of mounting a synchrotron on a gantry with a full 360 degrees of rotation. Such a single-room system would have no need of a degrader and could be very narrow, depending upon the type of injector. The left side of Figure 14.3 shows the concept developed and it could fit into a treatment room as depicted in the right side of that figure.

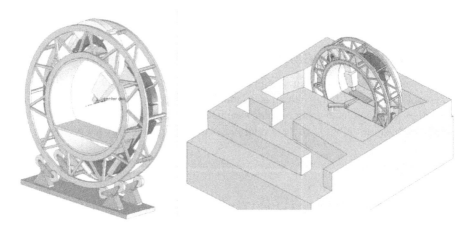

FIGURE 14.3 Synchrotron on a gantry.

It was learned after this that the idea had been broached in the early 1980s, but it's hard to find additional information. More recently, Danfysik has also identified this concept and filed for patents. As of the time of this writing, this has not yet been built.

14.3 GANTRY GEOMETRY

Most gantries in operation accept a beam from an external source, not internal to the gantry. The main purpose of this device is to direct the beam, from the horizontal plane to a vertical plane. Two extreme straightforward geometries can be envisioned as indicated in Figure 14.4. The beam can start in the same plane as the patient and the beam is directed upwards until it is high enough to accept the equipment necessary for treatment, after which it is bent downwards. Upwards and downwards are relative, since the system on which this is mounted can rotate and upwards can become leftwards or rightwards. Alternatively, the beam can come from another plane parallel to the patient, far enough away to include the necessary treatment equipment and simply be bent down towards the patient. (See the dashed lines in Figure 14.4.) There are possible geometries in between these two extremes, some of which have actually been built. There are also alternative geometries that accept the same beam input and result in the same beam output, but look different in between.

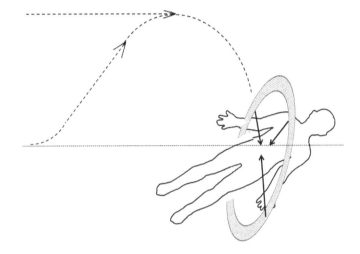

FIGURE 14.4 Concept of how gantry changes beam orientation.

A generic layout for a gantry that can be used for beam delivery is shown Figure 14.5. The number of bends shown is semi-arbitrary. If Bend 1 and 2 do not exist, then the beam is starting

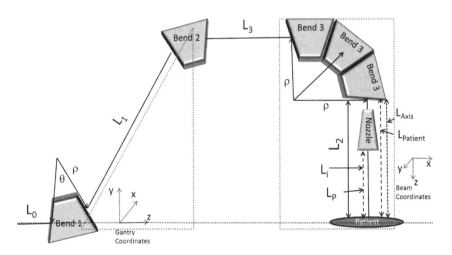

FIGURE 14.5 Generic gantry component layout.

from the plane above and the gantry can rotate about line L_3, and if they do exist, then the beam is coming from a plane nearer or the same plane as the patient and rotates about line L_0. Bend 2 can be combined with Bend 3 ($L_3 = 0$). Additional variations exist in which some of the lengths are not in the same plane as the rest of the gantry (e.g. corkscrew). This will be discussed a little more below. The parameters used in Figure 14.5 are defined in Table 14.1.

TABLE 14.1

Gantry Parameter Definitions

Parameter	Description
L_{axis}	Distance from the last dipole to the axis of rotation (not necessarily the same as the patient position)
$L_{patient}$	Distance from the last dipole to the patient
L_P	Distance after the nozzle for patient-related equipment
L_I	Distance after the nozzle for instrumentation, if any
ρ	The radius of curvature of the bending magnet
θ	Dipole bend angle

The choice of most of these parameters will have both physical and clinical effects. It is useful to characterize how these parameters will affect the size of the gantry and how a particular desired operation of the gantry will impact the choice of these parameters.

A purely geometrical approach helps to understand how some of these parameters contribute to the scale of a gantry. The bending radius of the gantry dipoles depends upon the type of dipole. A room temperature dipole can support up to about 1.6 T (or 1.8 T if the designer is good). Any fields higher than that require superconducting technology. The distance L_2 is critical to patient space and instrumentation and sometimes for beam spreading, so this is evaluated in the range from 1 to 3 m.

For the sake of internal consistency, in this chapter, the gantry coordinates of Figure 14.5 are used. In this book, these coordinates do not rotate with the gantry. The direction along the initial beam trajectory that enters the gantry is z. When the gantry is up (in the vertical plane, 0 degrees), the gantry is in the yz plane. When the gantry is sideways (90 degrees, in the horizontal plane), the gantry is in the xz plane. However, it is helpful to consider the beams-eye-view coordinate system (called beam coordinates in that figure). Those coordinates rotate with the gantry. In particular, imagine that the gantry is on its side (horizontal plane, gantry xz plane) and the beam is coming out of the paper. Then from the beam's point of view, z is coming out of the paper, x goes to beam left (this is in the bend plane direction, beam xz) and y goes up on the paper. The trick comes in following this as a function of gantry angle. If the dipole bend plane is considered, then the beam x direction is always in the bend plane and this can help. For example, when the gantry is in the vertical plane (0 degrees), then y is in and out of the paper and x is still left and right on the paper.

In terms of the stated parameters, the relevant gantry lengths are

$$\text{Total } Z \text{ Length} = 2\rho\sin\theta + L_1\cos\theta + L_3 + \rho \qquad (14.1a)$$

$$\text{Total } y \text{ length} = 2\rho(1 - \cos\theta) + L_1\sin\theta = L_2 + \rho \qquad (14.1b)$$

The volume assumes a rectangular geometry. The patient volume assumes a cylindrical geometry with access to the patient inside the last gantry dipole.

Figure 14.6 is a plot of the total z length and total y height of the gantry (when it is vertical) as a function of L_2 with $L_3 = 0$, for a bend angle, θ, of 45 degrees. The area plotted includes the triangle (not rectangle) up to Bend 2 and the rectangle beyond that (see the light dotted shapes in Figure 14.5). The scale of a rectangular-shaped room is from 2×4 ($L_2 = 0$) to 5×7 ($L_2 = 3$) m. The area taken up by the gantry is up to 20 m². This sets the scale for the space around the patient; it contributes more than a factor of 5 increase in volume, but is invaluable for therapy.

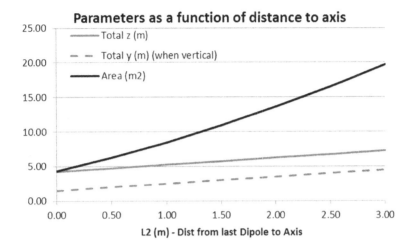

FIGURE 14.6   Lengths and area as a function of L_2.

Figure 14.7 shows the full 360 degrees rectangular volume of the gantry room (left scale) and the volume of 360 degrees cylinder around the patient ($2\pi L_2^2 \rho$) (right scale), which includes beam spreading and instrumentation. This sets both the scale of the cost of the room (more will be noted about concrete in Section 14.7.2) and the convenience of working with the patient.

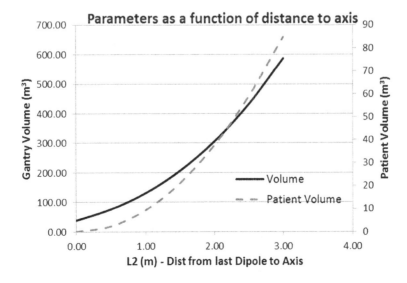

FIGURE 14.7   Volumes as a function of L_2.

Figure 14.8 takes a different route. Instead of being patient centric (L_2), these parameters are explored as a function of the dipole bending radius with $L_3 = 0$. Most of the values below the right edge of the curve would require superconducting magnets for clinical protons. All will require superconducting magnets for heavier ions. According to Figure 14.8, the rate of change in y is slower than in z. However, z does change because L_1 changes with the radius.

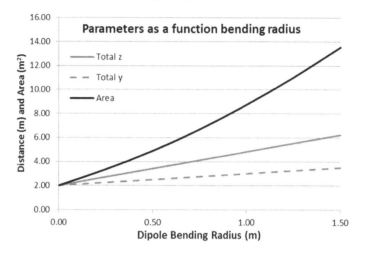

FIGURE 14.8   Lengths and area as a function of bending radius.

Figure 14.9 also includes the magnetic field strength required for 235 MeV protons. The volume of the room is reduced by a factor of 2 if the field is doubled. The ProNova gantry with superconducting magnets takes advantage of this benefit.

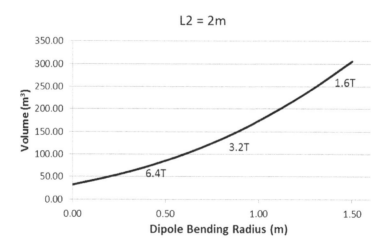

FIGURE 14.9   Volume vs. bending radius.

14.3.1   ISSUES AFFECTING THE GANTRY DESIGN PARAMETERS

The delivery of any beam will be characterized by the relative orientation of the beam and the patient target. While, in general, either the patient or the beam orientation can be manipulated, for the most part, nonplanar beams are delivered by rotating gantries that direct the beam orientation perpendicular to a patient in a prone or supine, horizontal, position. The choice of patient treatment related parameters is most important. This includes the space for patient treatment, allowable skin dose (related to source-to-axis distance [SAD]) and beam modifying treatment equipment. It is necessary to take all these issues into account.

14.3.1.1 Physical Implications of Clinical Issues

A number of parameters will determine the features and layout of the gantry without an integrated accelerator. These include:

Patient Movement During Gantry Rotation

If $L_{patient}$ is different than L_{axis}, then the patient will have to move with the gantry as it rotates, if it is desired to keep the beam pointing at the patient. Depending upon the distances, this could position the patient above the reach of the clinical staff or could require additional structures to be constructed (e.g. Riesenrad gantry). Note that for some angles, the patient positioning device could be adjusted to make $L_{patient}$ greater than L_{axis}, if, for example, a larger SAD was desired.

Planned Patient Movement During Dose Administration

Another consideration is the method by which the desired field size and shape is achieved. For example, to obtain a two dimensional field, it is possible to move the patient rather than the beam, in either or both directions. This can affect the necessary length L_{axis}.

Patient/Patient Positioner Workspace

L_p plus the length of the target in the patient is both the distance needed for clearance between the nozzle and the patient and the distance needed to place the patient in a desired orientation. This has clinical ramifications for access to the patient and treatment protocols. Typically, it is desired (but maybe not required) to treat the patient from all orientations. The combination of gantry movement and patient positioner movement is flexible. For example, if a gantry has 180 degrees of rotation, then there are likely more 180 degrees rotations of the patient positioner and depending upon how the robotic links are engineered, the space required can be different for different designs.

Beam Spreading Effective Source-to-Patient Distance

There are a variety of terms used in identifying the distance between the effective source of divergence of the beam and the target. The SAD is the distance between the effective source of the divergence of the beam and the gantry axis of rotation. The source-to-isocenter distance (SID) is the distance between the effective source of the divergence of the beam and the isocenter axis where isocenter can be the target center, or the gantry axis, depending on who is using the term. Finally the source-to-patient distance (SPD) may be defined as the distance between the effective source of the divergence of the beam and the patient, since the patient may extend beyond the SAD or SID. This parameter may determine $L_{patient}$ for certain types of beam spreading schemes due to the effect on the skin dose. Note that $L_{patient}$ does not have to be fixed. Primarily these terms define the ratio of skin dose to dose at depth since the beam is generally spreading out. (See Figure 7.20.) Reducing the surface dose to an acceptable value can determine the SAD/SID/SPD required. Alternatively, the magnetic optics of the gantry or beam transport can be designed to create a more parallel trajectory of the beam with effectively infinite SAD while the physical distance is small. The beam must be uniformly spread out to the desired treatment cross section. With the spreading devices after the last dipole, the effective SAD will be smaller than the distance from the last magnet to isocenter. The spreading can be done before the last magnet; however, that will increase the last gantry magnet size significantly. This distance must be large enough to reduce to an acceptable level, the difference between the skin dose and the target dose, and to provide the desired field size while not losing too much energy in the spreading.

Field Size

The desired field size could have an effect upon physical distances depending upon how the field size is achieved. In the double scattering scenario, the field size is dependent upon how much scattering is implemented and the SPD. In the scanning scenario, the angular bend of the beam is dependent upon the distances from the scanning magnets and/or any optics between the magnets and the patient. If $L_{patient}$ is not fixed, then the patient positioning system can be adjusted further from the spreading system to accommodate a larger field size. This latter point is quite powerful but has not been used in conventional therapy yet.

Patient-Related Subjective Space

Some questions can help to highlight important design choices.

1. How much planned patient movement is acceptable as a function of gantry angles?
2. What is the minimum volume needed to work with the patient?
3. What space requirements are there for the beam modifying system?
   a. After the last bend
   b. Before the last bend
   c. Before and after the last bend
4. What minimum SPD is allowable?

These parameters create among the largest challenges to making a gantry compact.

The available 'open' space within the gantry is determined by a variety of issues, including the clinical requirements for patient movement, treatment angle accessibility, therapist accessibility to patient and beam delivery equipment. Typically, the distances are about 2 m in length and 4 m in width if 180 degrees-patient rotation is needed.

The above patient-related considerations could lead to a parallel scanned beam after the final dipole (an SAD that is infinite or close to that), by scanning before last bend (to minimize components after last bend). This leaves a completely open and accessible floor space within the radius of the last dipole (cantilevered dipole, or gantry floor on support rings). This also leads to minimizing the overall radius. Also, this can only be done with scanning. An advantage of scanning is that there may be less need for a collimator or a compensator. However, the available space between the exit of Bend 3 and the isocenter is very limited and often compromises have to be made. Boundary conditions are determined by clinical requirements and physical requirements. For clinical reasons, at least 40 cm radius is needed for free and clear access to the patient in treatment position, available space around the isocenter and other devices for, e.g., control of patient position.

Additional physical space requirements include:
   a. 30 cm: scan magnet for y beam direction
   b. 20 cm: dose monitors and beam position monitors
   c. 25 cm: supplementary energy modulation (optional)
   d. 5 cm: alignment control (lasers etc.)
   e. 15 cm: field limiting collimator (optionally to obtain sharp penumbras)

This necessitates a total space of 50–65 cm for instrumentation in the nozzle. If a scan magnet is included for scanning in the x direction, another 30–40 cm will be necessary for mechanical and electrotechnical reasons. In addition to the 40 cm free space, this yields a total space allocation of $L_p = 90\text{--}115$ cm, or $L_p = 120\text{--}140$ cm if an x-scan magnet was used. For use of an x-scan magnet, one also has to consider the SID, since the source for x-scanning is located at its center. In this scenario, the minimum SID will be between 100 and 120 cm. A larger SID, which is more advantageous, will result in an increased gantry radius. Additional space is required if two scanning magnets are used. Thus, some clinical and mechanical considerations prevent the realization of a compact gantry below this SID.

14.3.1.2 Physical Implications of Magnetics

Magnet Contribution to the Overall Physical Radius

For the most part, the choice of dipole bending radius, ρ, is determined by the use of super-conducting magnets or not. If the choice is room temperature, the maximum field of about 16 kG is generally used giving rise to a radius of about 1.45 m for a 235 MeV proton beam or about 4.2 m for a 430 MeV/nucleon ^{12}C beam. Figure 14.10 is a plot of the fraction of the length of a gantry that comes from the bending radius portion. Above a bending radius of 1 m or so, over half the length of the gantry comes from this factor (actually 18 kG is possible, but you have to know what you're doing). The bending radius is a major contributor to gantry length. Clearly if there were no other disadvantage or expense, this is a significant parameter to optimize. Also, the importance of this parameter on heavy-ion gantries becomes clearer, but it's not until the radius is below 1.4 m that real gains can be realized.

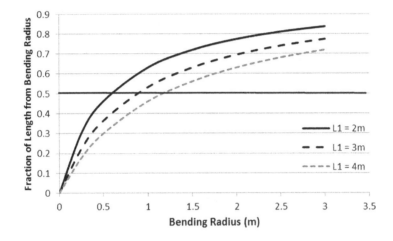

FIGURE 14.10   Gantry fractional length as a function of bending radius.

Beam Spreading Method

The beam spreading method contributes to the determination of the location and the optical configuration of the gantry magnetics. Scattering restricts the location of the beam delivery components to after the last gantry dipole since a dipole would compromise the uniform beam. The spreading technique also dictates the beam position tolerances and the beam focusing conditions. This will in turn create physical magnetics constraints. For example, it might be important to ensure that the beam size and position are independent of the gantry angle upon entering the beam spreading components.

If a beam scanning magnet in the beam y direction is inserted in the gantry section before Bend 3 (or combined Bend 2 and 3), the y aperture (the magnet gap) of Bend 3 may be rather large. The gap could be varied along the length, but at the exit of Bend 3 its aperture could be about 2/3 of the desired field size in the beam y direction. One could consider such a magnet design and such systems have been built. These are very large and slower magnets (recall Section 12.3.10.2). The power consumption of such a magnet could be on the order of 200 kW, so that superconducting coils may become interesting. If not compensated with a secondary achromatic system (achromatic from the scanning dipole to the isocenter – not the same as the full achromatic gantry), spreading in the beam energy in the beam x direction could result in an increase of the size of the scanned beam, dependent on the deflection angle. This may contribute to the beam size at the field edges, but usually other dispersion effects (Bend 3 may give 2–4 cm/%) are more important.

Magnetic Field Switching Speed

For some applications, it is required to change the beam energy transported by the gantry while treating a patient, e.g. for depth modulation. The magnets on the gantry must be capable of this. A larger gap dipole has a larger inductance and its magnetic field is harder to change rapidly.

Gantry Structure and Magnetics Related

1. *3D geometry:* What is the cost tradeoff between the building and gantry for the case of bending all in one plane, or taking some detours? The total length of the system is given by Equation 14.1a. For an in-plane gantry, the first bend is in the neighborhood of 45 degrees (exploration of this parameter has been done and it always seems to come out close to this answer). This angle is a balance between getting the distance needed surrounding the patient in a short overall distance without unduly increasing the cost of the magnets and power supplies. For a system in which the large bend is not in the same plane as the first bend (i.e. corkscrew, Section 14.6.2), then the total system length is $2\rho\sin\theta + L_1\cos\theta$. Several geometrical shapes have been proposed from the in-plane gantry, to the corkscrew, to the supertwist. Other geometrical schemes are possible, such as keeping the bends in plane but splitting the last magnet (Bend 3) further to allow more optical conditions, but this will probably increase the overall size a bit. The real advantage of the out-of-plane structures is the longitudinal length necessary for installation. This must be balanced with the additional cost of extra bending dipoles and more complicated beam optics. These costs are site-dependent quantities.
2. *Superconductivity:* Superconductivity is a fairly mature technology, although there are challenges when subjecting cryostats to motion and turning them upside down. The physical space required for the superconducting hardware, compared to the physical space required for a normal bending magnet, is more, and operationally it is different. The Toshiba gantry installed at NIRS recently has been a successful example of this.
3. *Gantry rotation angle:* Is the gantry required to rotate 360 degrees, or is 180 degrees or is another range of angles sufficient? Reducing the angular rotation to 180 degrees could lead to saving almost half the cost of the building space, but it leads to additional patient moves between fields. Several installations now rotate 180 degrees, starting with Paul Scherer Institute (PSI) gantry 2. So, the experiment has begun.
4. *Distribution of bends:* The length of the section of the gantry 'rising' out of plane is related to the gantry radius, the choice of the bending angles of magnets Bend 1 and Bend 2 and the desired optics in between. In some designs, Bend 2 and Bend 3 are combined. This would reduce the total length. In some designs, however, the space between Bend 2 and Bend 3 is used for scanning and other optical manipulations.
5. *Mechanical properties/isocenter requirements:* This category includes mechanical deflections and reproducibility, as well as non-reproducible positioning errors, particularly of beam sensing devices that can result in beam missteering.

14.3.1.3 Desirable Features

Some desirable features have been identified, which couple to the above issue-related questions.

1. The gantry should enable pencil beam scanning (spot or continuous scanning).
2. It would be nice if the patient table is positioned on the rotation axis of the gantry and is not required to move with gantry angle.
3. It could be desirable that no components move physically (magnets or patient table) during a patient field.
4. The gantry optics is such that tuning is deterministic and diagnostics are straightforward.
5. Some desirable clinical specifications include:
   a. Field size capability of 30×30 or 40×30, or 20×20 or 60×20 cm^2. Many ideas have been posed.
   b. Free space between the nozzle and patient should be at least 0.4 m.

Note that the above list is not universal, but it is useful to have this list to help define the gantry design priorities. The final gantry solution may be institution dependent since there are still so many subjective directions.

14.3.1.4 Compact Gantry Parameters

The limit of compactness of the overall system is driven by the previously described parameters. There are a variety of concepts evoked when the term compact is used. It's useful to distinguish between those parameters that affect the physical compactness of the gantry, and those clinical functions desired in a gantry that would also be nice to build in a compact fashion but may impose challenges. These conditions can be subjective, or site-dependent, and affect the gantry size.

A possible definition of 'compact' based upon what's been done to date can include:

1. A total gantry radius less than 3 m.
2. The gantry is 'less expensive' than a 'more expensive' gantry (those that have already been built).

Thus far, there has been no compact gantry that achieves all the capabilities outlined in previous sections, although this is a subjective statement.

14.4 MAGNETIC SPREADING GEOMETRY CONSIDERATIONS

In the case of particle beam scanning, with an unmodified beam, the potential gantry geometries expand. Scanning dipoles can be put either before or after the last gantry dipole(s). Additional focusing elements are possible to include. Inserting the scanning dipoles before the last gantry dipole allows the space between the last gantry dipole and the target to be minimized, at a cost of a large aperture in the gantry dipole to accept the diverging scanned beam. There are a variety of optical solutions that can control the angle of the scanned beam at the target. A particular SAD requirement does not necessarily flow down to a particular physical space, depending upon how the beam spreading is done and what optics configuration is designed. In particular, an infinite SAD can be achieved with point-to-parallel optics from the scanning position to the patient, but an in-depth analysis of the numbers, especially with scanning beams, may indicate the relative advantages and disadvantages of such a beam.

Some of the possible geometries are sketched in Figure 14.11, with identifying numbers 1–7. The gray curved elements are gantry dipoles, the even darker boxes are combined function scanning dipoles and the lighter boxes are single direction scanning dipoles. The ovals after the dipoles are focusing elements.

1. The scanning dipoles (x and y) can be positioned after the last dipole. (1, 2, 6, 7)
    In this case, the beam optics may be simpler, and the width of Bend 3 (and possibly Bend 2) may be smaller. However, obviously the scanning system takes up space in L_2.
2. The scanning dipoles can be positioned before the last dipole (Bend 3) (4).

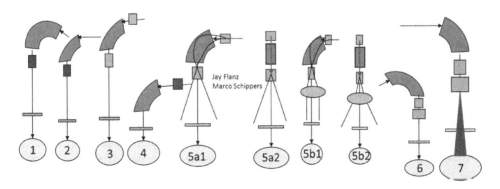

FIGURE 14.11 Some gantry geometries.

3. The scanning dipoles can be positioned before and after the last dipole (3, 5).

In this case, point-to-point optics is used between scanners, with the appropriate magnification and the appropriate SAD from the last scanner to the patient (Proposed by the author and Marco Schippers and studied at KVI).

Going through the figure one by one, the geometries are described as follows:

1. Beam enters a dipole with greater than 90 degrees of bend and exits that dipole into a combined function scanning magnet.
2. Beam enters a dipole with 90 degrees of bend and exits that dipole into a combined function scanning magnet.
3. Beam enters a scanning magnet that scans the beam in the bend plane of the gantry dipole which must accommodate the additional beam angles. The resulting extremes of the scanned beam trajectory, in that plane determines the gap in the subsequent orthogonal scanning dipole. Optics in the dipole can help adjust that gap size and adjust the scan angle of the bend plane.
4. This is a gantry with a short radius. The beam enters a combined function scanning dipole. Since both planes are scanned before the gantry dipole, the gantry dipole gap has to be larger and the magnet bigger. Optics in or near the gantry dipole can be used to create the desired scanned beam trajectory (e.g. large effective SAD or otherwise).
5. In this system, the scanning dipoles are split to before and after the bend.
   a. a1. A scanned beam (in the bend plane of the gantry dipole) is imaged to the second scanning dipole so that its gap does not have to be larger.
   b. a2. This is the same as a1, but looking at it in the orthogonal direction.
   c. b1. This is similar to a1 and a2, but with additional optics after the second scanning dipole to create a larger effective SAD.
   d. b2. The orthogonal direction of b1
5. This is similar to 1, but with two scanning dipoles after the gantry dipole.
6. This is similar to 2 but with two scanning dipoles after the gantry dipole.

An option worthwhile to study in more detail may be splitting of Bend 3 at 60 degrees and the insertion of the y-scan magnet in the gap. The second part of Bend 3 should have a large gap, so that a parallel displacement in two directions is possible, but that gap increase won't be as large. The Proteus 1 gantry does something similar. However, a small increase of the gantry radius is likely.

14.5   BEAM OPTICS CONSIDERATIONS

There are a number of beam optics conditions that are important, or helpful in the design of a therapy gantry. Some of these have been previously reviewed. The conditions can be summarized as follows:

1. Achieve the appropriate beam phase space parameters after the gantry consistent with the beam delivery technique used.
2. Achieve the appropriate SAD/SID.
3. Achieve an achromatic gantry to minimize the spatial dependence of beam energy.
4. Minimize the gantry sensitivity to deviations in input conditions. Therefore, optics conditions such as point-to-point optics or parallel-to-point optics are useful depending upon expected deviations in beam trajectory at the gantry rotation coupling point.
5. Allow for beam trajectory corrections within or near the gantry depending upon the information from the ionization chambers in the nozzle.
6. Minimize the gantry angle dependence on the beam output parameters..

14.5.1   Beam Phase Space Conditions at the Gantry Coupling

At the coupling point, the optics should become independent of the preceding beam transport system (e.g. Section 13.4.4 about the energy selection system [ESS]). At the coupling point, it's helpful if the horizontal and vertical phase spaces are equal so that the beam transport in the gantry and final beam size does not depend on the gantry angle. This implies that the coupling point should also be free of momentum dispersion, so that there is no correlation between position and/or angle with particle momentum.

If these conditions are not met by the upstream optics, matching could be done in the beamline section before the coupling point to accept the beam from the upstream beam transport system and transform it into something suitable for the gantry. This gantry section also allows the installation of emittance limiting apertures, although the consequences of neutron production should be considered. Assuming as above, that the beam phase space is properly prepared before the gantry, it will be the case that the gantry design and operability will be simpler.

14.5.2   Minimizing the Gantry Dispersion Function

14.5.2.1   Suppression of Dispersion

It is assumed that at the target, there should be very little dependence of range on position and so the dispersion should be minimized. Generally, making the system achromatic is also desirable for deeper fields (nonzero angular dispersion could increase the spread in ranges, although range straggling may win). Achromaticity is also helpful for other beam optics reasons, including the sensitivity to magnet manufacturing errors, power supply stability and the beam size in the gantry. This works if the magnets are wired in series. In addition, tuning errors, if they have any affect at all, might only cause scraping and reduce intensity, but not change the beam trajectory.

Somewhere in the gantry focusing is needed to control of the dispersion produced by Bend 1. Where it is located will affect some of the gantry physical parameters. This section can also be used to limit the spread of the beam energy since it is likely to have the highest dispersion. For example, an ESS can be included in the gantry optics. In that case, a slit system can be used and Bend 1 is then acting as an analyzing magnet. The resolving power is determined by slit apertures, magnification factor of the imaging and bending angle of Bend 1. So here a choice can be made between:

a. *Maximum beam transport efficiency to the target.* The energy spread will cause large beam sizes at places where the dispersion introduced by the bending magnets is large. So either the initial beam energy spread should be small (typically less than 0.5% rms) or in the total system the dispersion should be limited.

b. *Reduced beam transmission to the isocenter of an energy analyzed beam.* The advantage is a relatively small beam diameter, more or less independent of dispersion. The use of a second collimator behind Bend 2 can also be considered. This collimator should not cut the beam, but rather intercept the particles that were scattered at the slit before.

It's useful to mention another possibility. The energy spread in the beam in the gantry might be so small (e.g. coming from a type of synchrotron) that it may not be necessary to require achromaticity.

14.5.2.2   Momentum Bandwidth

One of the important contributions to the time that it takes to change the beam energy in a multi-room system, is the time it takes for the magnetic fields to change in the beamline. One of the key sections of the beamline is the magnetics on the gantry. While, in general, finding ways to reduce the time in the entire beamline can be helpful, more thought has been given to the gantry magnetics, partly because with each change of energy coupled to the range of gantry beam angles that are possible, the number of sets of magnetics parameters grows. One way to address this issue is to increase the range of energies that can be transported through the gantry without having to change the magnetic fields. This is sometimes called the (momentum) bandwidth of the gantry beamline or the energy acceptance of the gantry.

If the momentum of the beam centroid is being considered, then it's helpful to consider the momentum-dependent matrix elements in the system. From Section 13.4.2, the dispersion in the center of an achromat is given by $(x/\delta) = L_1 \sin\theta$. All systems under consideration are achromats of one kind or another. The achromat in that section had the condition

$$\frac{1}{f} = \frac{2}{L_{\text{1achro}}} \tag{14.2}$$

Although it's not exactly the same in a gantry, it's useful for understanding. In order to have a large bandwidth, it is necessary that the beam be transported in the evacuated tubes inside the magnetics without hitting the walls. The fact that a dipole is like a prism for light has already been described; which means that a range of energies are bent different amounts and without mitigation they will spread apart and hit a wall. For that to be contained, the dispersion inside the system, (x/δ), has to be minimized. For this to be minimized, simplistically, L_{1achro} (in Equation 14.2) and/or $\sin\theta$ have to be small. The value of θ is determined by the physical geometry so there is not much that can be done with that. However L_{1achro} could be a free parameter. Stepping back a little, the transport matrix in Section 13.4.2 used a thin prism approximation, with no length in the dipole. The dipole length will act to increase the effective length of the system and so it's not only L_{1achro} that must be minimized, but also the dipole length. The only way to make the dipole shorter is to increase the magnetic field, which becomes code for superconductivity. This enables the possibility to make very short achromats, which reduces the dispersion inside and increases the bandwidth. What does this have to do with a gantry? The other L_1 (gantry) in Figure 14.5 is constrained by physical geometry. So, one possibility is to replace the Bend 1 in Figure 14.5 with a small achromat with high-field dipoles and very short L_{1achro}. The ProNova SC360 superconducting gantry does this with a momentum acceptance of about ±3%, compared with a conventional gantry momentum acceptance of ±1%.

Range changes of about a few mm translate to about a 0.6% change in beamline momentum (Equation 9.38). The magnetic field of a gantry with a momentum acceptance of 3% would not have to change until after several energy layers. If fewer changes are needed, the disadvantage of the slower superconducting magnets may be mute. The momentum difference from 230 to 70 MeV protons is about 50%, so that sets the scale (although a percent is not the same as an absolute range number). Note that, it is also conceivable that the magnetic field of the gantry can change during a treatment, in principle, with no effect on the beam, if the bandwidth is sufficiently large. In such a case, while the magnetic fields do not require a change for every layer, they can be changed slowly, during a treatment, in anticipation of the next required energy change. This is important to consider when designing the energy acceptance of the gantry.

The above has assumed that all the nonzero dispersion is inside the small achromats. This does not have to be the case. If (x/δ) is allowed to grow after a dipole, a strong quadrupole, very close to that dipole, can change the sign of (θ/δ). This would have to be followed by another strong quadrupole close to the previous one so as to not allow (x/δ) to grow in the other direction which would make the beam size grow beyond the limits of the physical apertures. Of course, each time the strong quadrupole is containing the dispersion, it is defocusing the beam in the vertical plane. Therefore, alternating gradient focusing is required. Designs based on the fixed field alternating gradient (FFAG) principles have been proposed. One is based on superconducting technology, from the Daresbury Laboratory. Another very interesting proposal by

Dejan Trbojevic uses strong permanent magnets, thus, no power supplies, no superconductivity, no magnetic settings and no changes with energy. The result is an extremely light (a few tons) gantry. This is also applicable to heavier ion gantries. A carbon-ion gantry with a momentum acceptance of ±20% has been designed. The engineering tolerances are a challenge, but this system has yet to be developed.

14.5.2.3 Beam Profile for Beam Delivery

There are a variety of factors related to the beam profile at isocenter that are important.

For a double scattering system, the beam size must be appropriate on the scatterers. For a scanning/wobbling system, the beam size must be appropriate for the scanning modality. As the beam size is made smaller, and the penumbra of the scanned beam is therefore smaller; however, the dosimetry necessary to ensure the safe beam delivery becomes more complicated due to the smaller size of interest and the faster response necessary.

For some types of scanning systems, such as spot scanning, the symmetry of the beam combined with the overlap amount of each spot is important in obtaining a smooth transverse dose distribution. Aberrations of the gantry optics can contribute to the symmetry, or lack thereof, of the beam profile and this should be considered in gantry design.

14.6   GANTRY EXAMPLES

Gantries have been produced by industry or industry/laboratory collaborations. This was done during the earlier years (1980s) in the case of neutron gantries by Scanditronix and The Cyclotron Corporation (TCC). The earliest proton gantries were built by Science Application International Corporation for the Loma Linda University Medical Center (LLUMC), a team of Ion Beam Application (IBA) and General Atomics (GA) for Massachusetts General Hospital (MGH), and the compact gantry built at the PSI. Most recently, heavier ion gantries are now possible to construct, starting with the carbon gantry at Heidelberg and the superconducting gantry built by Toshiba at NIRS in Japan.

Consideration of gantry geometry possibilities leads to several paths. It is interesting to note that practically all these paths have been, or are being, pursued. A consensus about a preferred scheme has not yet developed, other than finding one which results in the least expensive overall system and yet meets all the clinical requirements.

Many geometries have already been discussed, including accelerators mounted on a gantry, and location of the gantry and patient positioner relative to the isocenter. This non-exhaustive list includes how the beam will be shaped (spread) to match the beam spreading system, the target and the SAD. The use of scattering vs. scanning affects the gantry geometry. In particular, the location of the scattering or scanning devices may be an important factor, if it is desirable to have a large effective SID (and this can be debated). This can be achieved by a large physical separation between the beam spreading system and the target after the last gantry dipole, or by incorporating these systems within the gantry optics and using close to point to parallel focusing from the spreaders to the gantry output (as at PSI). This latter technique tends to require large apertures and therefore heavy dipole(s).

14.6.1   HIGH-LEVEL GANTRY REQUIREMENTS

Now with over a hundred facilities worldwide, there are many gantries in operation. The parameters of the first few historical gantries are summarized in Table 14.2. Most of the other gantries are similar, with a few exceptions.

TABLE 14.2

Early Gantry Parameters

Gantry Parameter	Seattle	LLUMC	PSI	NPTC
Date	1984	1992	1995	1998
Incident particle	p	p	p	p
Energy (MeV)	50.5	250	270	235
Treatment particle	n	p	p	p
Energy (MeV)	21.3	270	85–270	70–235
SAD (m)	1.5	2.75	∞	2.25
Weight (t)	~43	~96	~120	~100
Rotation (degrees)	380	370	370	370
FieldSize (cm)	33×29	40 diam	$20 \times \leftrightarrow$	30×40
Isocentricity (diam mm)	4	0.7×1.6	2	2
Phase space (π-mm·mr)	n/a	2	40	30/18
Designer	Scand.	SAIC	PSI	GA
Beam delivery	Sn	W = Wobble/Sn/Sc	$\leftrightarrow$ = table move	Sc/Sn
Sn = Scan, Sc = Scatter				

$\leftrightarrow$ = *mechanical motion.*

14.6.2 THE CORKSCREW GANTRY

The corkscrew gantry geometry was introduced by Andy Koehler and Harald Enge. The goal was to reduce the overall size of the room housing the structure. The major bending occurs in the plane of rotation so that these magnets do not sweep out as large a volume of space along *z* as the gantry is rotated. This is shown in Figure 14.12(left). The 3D aspect of the beam trajectory led to the name '*corkscrew* gantry'. The result is that the building to house the gantry can be shorter longitudinally, but not so much in the other dimensions. This results in a somewhat reduced volume available for the patient positioner motion and for access to the patient within the gantry open space. It also required 360 degrees of magnetic bending (compared to 180 degrees for the in-plane gantry). This was the first gantry built for proton therapy at the LLUMC.

The corkscrew gantry starts with one pair of 45 degrees dipoles with four quadrupoles in between and another pair of 135 degrees dipoles with another quadruplet. The first pair bends

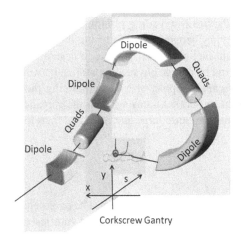

FIGURE 14.12   Corkscrew gantry layout (left) and photo of corkscrew gantry (right)

the beam out of the plane of the beamline. The result is a vertical beam when the angle is set to that shown in Figure 14.12(left). The second pair accepts the vertical beam and bends the beam 270 degrees in the orthogonal plane from vertical to horizontal toward the isocenter from the side. Figure 14.12(right) shows a photo of a newer incarnation of the corkscrew gantry built by Sumitomo Heavy Industries (SHI) (photo courtesy of SHI). The innocuous man standing shows an indication of the scale of the system.

This system uses two orthogonal bends, and the whole system is required to be achromatic, so each subset must be achromatic. Figure 14.13(top) is a plot of R_{16} and R_{26}. Since the two sets of bends are in different planes, each plane should have its own achromatic settings. In order to achieve overall achromaticity, four parameters are required to be adjusted to zero (two in each plane). There are eight variables (quadrupoles) in the system, thus leaving four more parameters in addition to the dipole pole edge rotations per plane. Each of the achromats are symmetric, so only two knobs each are available for the achromatic solutions (like the ESS earlier discussed), thus two knobs, in each plane, can be used for additional optics settings (maintaining symmetry) (some of the knobs are overlapping). This is adequate to ensure a reasonable beam size to be transmitted but does not leave much flexibility for subsequent optics changes. Figure 14.13(bottom) shows the results of a TRANSPORT calculation of R_{11} and R_{12}. R_{12} is more or less controlled by R_{16} but does not end up small. Therefore, the design is not concerned about large initial angles adding to the overall beam

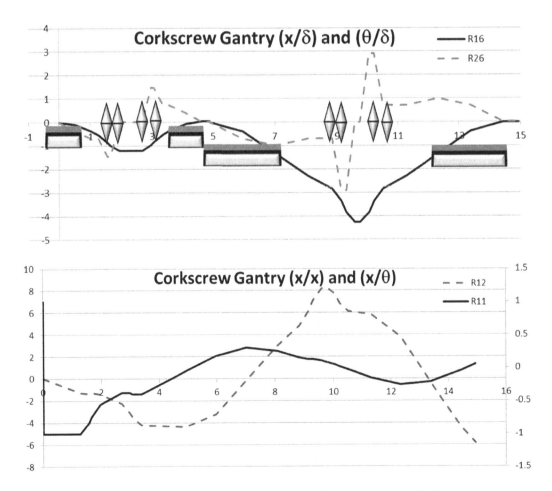

FIGURE 14.13 Dispersion functions in corkscrew gantry (top) and geometric optics for corkscrew gantry (bottom).

size, since the accelerator for this system is a synchrotron with small emittance. This optics tune produces a waist at 50 cm from the exit of the last dipole. The input is presumed to be a symmetric beam with up to 24π mm-mrad phase space area and $\pm0.5\%$ $\Delta P/P$.

The gantry is primarily supported by large rings. The bearing is supported from the floor and wall by support struts. The gantry is a cone-shaped structure made up of a 7-ft circular ring at one end and a 16-ft ring at the larger end. The plates are connected by struts. The assembly is fabricated in sections small enough to transport into the gantry room. The magnets are aligned to 0.2 mm individually and 0.4 mm gantry overall. Measurements of beam pointing accuracy to isocenter result in an iso shape of less than 1.6 mm diameter.

A variation of the corkscrew gantry is the *Supertwist* gantry suggested by Francis Farley. This starts with the corkscrew physical layout concept and departs from the two orthogonal bend solution by stretching out the corkscrew in such a way that the total gantry length is longer and each magnet twists the beam in a trajectory through this path. It then actually looks more like a corkscrew than the corkscrew gantry. This leaves more room for the patient. Much of the focusing is done with pole edge rotations and the system is achromatic.

14.6.3   IN-PLANE GANTRY

An in-plane gantry bends the beam in only one plane. With the gantry at 0 degrees, the beam is bent up over and then down. This gantry is longer than the corkscrew gantry and the tradeoff between the expense of a larger room compared to the expense of additional magnetic elements and power supplies is evaluated based partially on the local constraints of a facility.

A conventional in-plane gantry was first built for MGH/Northeast Proton Therapy Center (NPTC) by IBA and GA. The elements that rotate with the gantry begin with four quadrupoles in the plane of the beam switchyard. Their purpose is to match the beam from the beamline to the gantry optics with an emittance of up to 32 mm-mrad. Sometimes, this matching section is omitted in modern systems, but it provides a useful 'knob' to correct for asymmetric beam shapes at the entrance of the gantry.

The beam is deflected through 45 degrees (Bend 1), focused by five quadrupoles before it is bent through 135 degrees (Bend 2 + Bend 3) and directed toward the isocenter. The distance from the output of the 135 degrees dipole to the isocenter is 3.0 m. As shown in Figure 14.14(left), the rotating structure utilizes a configuration of rings, trusses and shell elements to support the magnets in a space frame. Perhaps you can find the person in this photo to get a sense of the size of the device. The structure is engineered like a radio telescope and while subcomponents may not be extremely stiff, the overall design creates a very accurate gantry pointing angle. The front ring is axially constrained. It's actually quite amazing how much a large structure like this will 'corkscrew' (even

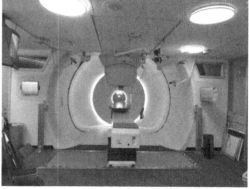

FIGURE 14.14   In-plane gantry photo (left) and in-plane gantry Tx room (right).

when it's not a corkscrew gantry) axially if not restrained. The truss elements are removable for ease of transport, assembly and possible repair. (A strut was actually removed and put back and it worked.) The structure is supported on both rings using a 'whiffle-tree' assembly. Figure 14.14(right) is a view of the treatment room showing the space available for patients and related equipment.

In order to achieve an achromatic tune, given that the two dipoles are bending in opposite directions (unlike previous achromats discussed in this book) and without a symmetric system, it is necessary to reverse the sign of the dispersion within the gantry. At least, two of the quadrupoles are required to do this as shown in the plot of R_{16} in Figure 14.15(top).

The gantry quadrupoles can be adjusted to produce a waist of 12 mm at the target (in air). This beam size is produced with air in the space between the last gantry dipole and the target, so there is some considerable multiple scattering at play. For scattering and wobbling, the quadrupoles can be tuned to produce an 8 mm radius waist at the center of the range modulator which is about 20 cm from the last dipole. For scanning, with a vacuum chamber beam sizes from 3 (at high energy) to 7 mm (at the lower energies) are possible. A beam size of 4 mm has been achieved with a helium chamber with He in only about 1/3 of the length.

As shown in Figure 14.15(bottom), there is strong focusing making both R_{11} and R_{33} smallish at the end, while R_{12} and R_{34} are maintained flat and parallel. The shape of R_{11} is quite correlated with

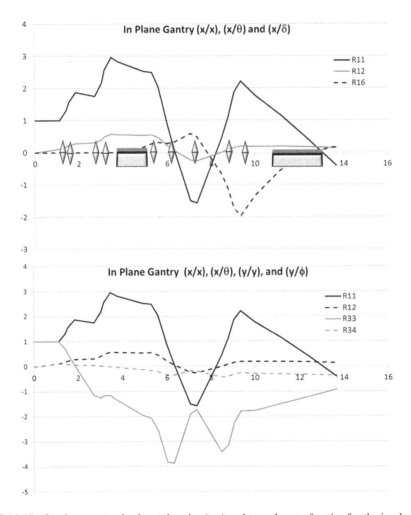

FIGURE 14.15 In-plane gantry horizontal optics (top) and two planes of optics for the in-plane gantry (bottom).

R_{16} since the achromaticity gymnastics is dominating. Still there is enough flexibility to adjust the gantry focusing for different beam delivery modalities with reasonable clinically useful beam sizes.

There is sufficient space in the gantry to include beam position and profile monitors that are capable of determining and correcting the beam trajectory angle and position to the required tolerance within the beam modification elements of the nozzle. This tolerance is basically submillimeter precision at the location of the scatterers for scattering, or at the isocenter for scanning. The magnets accept a momentum spread of ±0.5% $\Delta P/P$.

14.6.4 COMPACT GANTRIES

Various parameters have been explored that would enable a gantry to become more compact. Probably, the most practical explorations have taken place at PSI. The PSI gantry 1 design, the brainchild of Eros Pedroni, results from the special requirements of the beam at PSI and a general intention to design a gantry for potential users with inadequate space for a conventional in-plane gantry. In particular, it is desired to reduce the radius in the bending plane. As usual, in the optics design of a gantry, the special conditions of the accelerator and beamline that will be used must be taken into account. The optical design was performed by Harald Enge, yet another proton gantry in which he was involved. A large beam phase space results from degrading the beam energy significantly (from 590 MeV to between 270 and 85 MeV) just before the gantry. (This gantry was built before PSI included a stand-alone 250 MeV Comet cyclotron.) The gantry in Figure 14.16(left) is a 'compact' style that spans a diameter of 4.5 m and is designed to accommodate the spot scanning technique. The end of the last dipole is close to the gantry axis of rotation. The patient table is mounted directly on the front wheel of the gantry and moves with the gantry to keep the distance between the isocenter and gantry axis constant (L_{axis} and $L_{patient}$ are not equal).

The beam is focused to achieve unity magnification and achromaticity. The apertures can accommodate up to ±1% $\Delta P/P$. There are seven quadrupoles in this system to achieve the desired conditions. The scanning magnet is located before the last 90-degree dipole that is the limiting aperture of the system. Another condition requires that the scanning magnet to output optics satisfy point to parallel focusing. A near-infinite effective SAD is achieved. The distance between the exit of the last dipole and the isocenter is minimized, using pole edge rotations on the last dipole. This was the first gantry installed dedicated to scanned beams.

With experience from gantry 1, the PSI team went on to innovate. The eccentric gantry position was decided to be inconvenient and it was desired to increase the space around the patient. Advanced fast scanning had become important with small and precise beam parameters. Thus, gantry 2 was born as shown in Figure 14.16(right). It was the first gantry built to rotate only

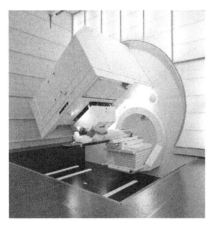 

FIGURE 14.16   Compact – PSI gantry 1 photo (left) and compact – PSI gantry 2 photo (right).

180 degrees. As noted earlier, this reduces the room volume and set the stage for an evaluation of that feature. A tour de force of optics is incorporated in this design for upstream parallel scanning and control of a small beam size with fast energy changes.

14.7 ADDITIONAL GANTRY CONSIDERATIONS

14.7.1 The Cost of Beam Size

Assuming that it is desired to have a specific beam size at the target and that the final magnetic element is at a certain distance from the target as in Figure 14.17, a larger emittance beam will have to be much larger than a smaller emittance beam at the location of that last magnetic element. The principles covered in Section 10.12 to transport the beam (this time going backwards) from the target to a point 3 m upstream are used to calculate the beam size and the results are shown in Figure 14.18. There are four curves. The solid curves are for a beam with emittance $\varepsilon = 18$ mm-mrad and the dashed lines are for a beam with emittance 5 mm-mrad. In current use today, the former would be the emittance from a cyclotron and degrader system into the treatment room, and the latter would be from a synchrotron. Of course, the cyclotron beam emittance can be further reduced at the expense of further beam loss. Each of these beams, in Figure 14.18, 'starts' (if you've seen the movie Tenet) at the isocenter and each of them is configured with two options, one with a beam size of 3 mm at the target position (in vacuum) and the other with a beam size of 9 mm. Therefore, the plots start at the left side with either 9 or 3 mm. The results show, that for the 18 mm-mrad emittance beam, the 3 mm beam size grows to a beam size at the end of the gantry of 18 mm or has to be 18 mm at the gantry in order to achieve 3 mm at the target. The beam with the same emittance but starting with a larger size wasn't squeezed as hard and does not grow as much, so is smaller at the end of the gantry. Both of the lower emittance beams do not grow large since their emittance is smaller. The required aperture of the dipole vacuum chamber could be six times the numbers presented (e.g. 1.8 cm → 10 cm).

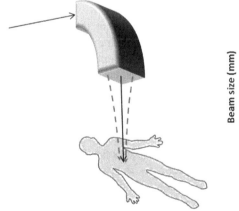

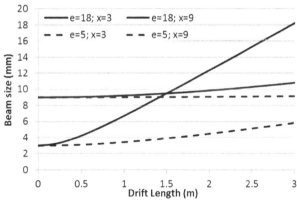

FIGURE 14.18   Effect of target beam size on gantry.

FIGURE 14.17   Gantry to target beam.

If the last magnetic element is on a gantry, then the size and weight of the gantry will be affected by this beam size constraint. Figure 14.19 shows the power and weight requirements for a gantry dipole located 3 m away from the patient as a function of the beam size desired. The upper solid curve is for a beam with an emittance of 25 mm mrad and the lower curve is for a beam with 5 mm-mrad.

There is a considerable cascade of consequences for the smaller beam size. With larger emittances, the beam size at the last gantry dipole will be much larger and so must be the gantry dipole

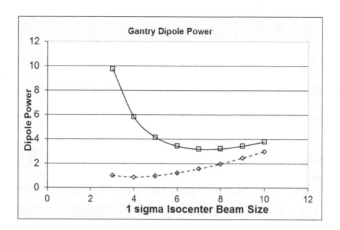

FIGURE 14.19   Power and weight of gantry dipole vs. beam size.

gap. This larger gap will affect the magnet weight since the magnet weight ~ gap^2. The gap affects the magnet current required that will in turn affect the power of the power supply required since power ~ current^2 ~ gap^2. In addition, if it is necessary to change the magnetic field quickly in that dipole (which is be in the case of scanning – although there are a number of possible alternatives), then the gap will have affected the magnet inductance and therefore the voltage required for the larger current changes since $V(t) = Ldi/dt$

For larger emittance beams, the power and weight requirements increase a factor of 3 when reducing the beam size from 6 to 3 mm. An even smaller optical beam size would be necessary when considering scatter from material in the beam path. (Instruments, windows, gas, range shifter/ ridge filter). Thus, the ramifications of larger versus smaller emittance beams should be appreciated (independently of how they are achieved).

14.7.2   Room Size Matters

There are a number of cost drivers for a gantry. These include the physical structure, the gantry magnetics and power supplies and components that are mounted or associated with the gantry as well as the cost of the room in which it is housed. The room size as well as the radiation in the room from the gantry beam helps to define the shielding requirements and the cost of the concrete (or other shielding material) and associated construction costs. One example was the choice of the gantry for the NPTC. It was originally assumed that the corkscrew gantry would be chosen, given the relationship between MGH and Harvard. However, from purely cost considerations, it was recognized that the magnetic bend angle was 360 degrees, or twice that of a conventional gantry, which must also take into consideration the cost of the power supplies and associated controls as well as optics flexibility. This was balanced against shortening the gantry room by about 1/3. Cost estimates showed that it was more economical to select the conventional, in-plane gantry, in this particular situation. If the magnet and power supply costs are reduced, a different decision could be reached, if the space is available.

Much work goes into gantry configurations that are deemed to be more compact. One of the complications is that the term has not been well defined. A compact gantry can be smaller than a non-compact gantry and generally has a radius less than 3 m, thus implying that it is good to have a radius less than 3 m. A compact gantry may minimize the volume of space needed to work around the patient. A compact gantry may necessitate more ancillary systems, such as superconductivity equipment in order to achieve a more compact structure. This comes back again to whether the added expense of auxiliary equipment, such as more magnets, smaller patient enclosures and superconductivity outweighs the costs, for example, of the building structure. Of course, if the space is absolutely not available, nothing can be done. But it's perhaps instructive to look at one aspect of a perhaps nonintuitive aspect of the building structure size tradeoff. Consider the gantry room structure depicted in Figure 14.20(left). There are, a least, two competing factors to consider, as highlighted in Figure 14.20(right). There is the size of the gantry room, and the resultant thickness, or volume of concrete needed to enable, for example, the roof to hold itself up. On the other hand, there is the distance between sources of radiation in the gantry including the patient, and the walls that influence the radiation that reaches the walls (the $1/r^2$ effect). This will determine the thickness needed to shield the gantry room from the external environment. Going back to Equation 7.36, μ was defined as ρ/λ. For concrete, $\rho = 2.302$ g/cm^3 and λ is between 71 and 50 g/cm^2 for concrete at 90 degrees from the source direction. Well, to make numbers easier take $\lambda = 50$ g/cm^2. Therefore, $\rho/\lambda = 2.3/50$. It might have been noticed that the 1/10 value thickness is $2.302/\mu$ and that the density of concrete is 2.3 coincidentally. So the 1/10 value thickness is 0.5 m.

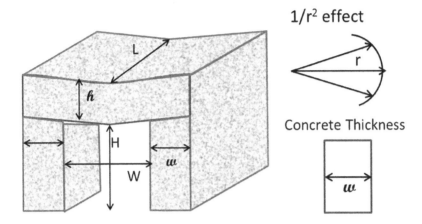

FIGURE 14.20 Concrete gantry housing (left) and tradeoffs (right).

If $W/2$ in Figure 14.20(left) = 5 m, then the secondary source has been reduced by a factor of 25 ($1/r^2$) by the time it reaches the walls. If it is desired to reduce the secondary radiation by another factor of 10,000, then 4 1/10 value layers are needed in wall thickness w or 2 m (4/0.5). So a total reduction factor of 250,000 is achieved. The question to ask is what is the relative value of the $1/r^2$ distance to the shielding wall thickness? Other than the space itself and the land it's on, the $1/r^2$ space adds cost to the roof above it. The wall shielding thickness, however, will be reduced if r is larger. Consider first just the effect on the shielding wall thickness. The results are shown in Figure 14.21. To achieve the same reduction in secondary radiation, an increase in the room radius of 4 m results in a wall thickness requirement reduction of 30%.

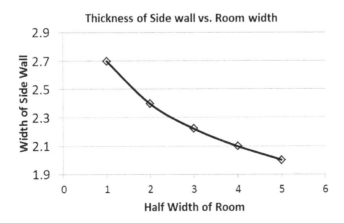

FIGURE 14.21 Wall thickness vs. room width.

The concrete on the roof must be designed to be able to hold itself up. The deflection of the roof, δ, under a load (probably its own load) is given by

$$\delta = \frac{5W^3}{384EI} \times (\text{weight of slab}); \text{ weight} = \rho_F WhL \qquad (14.3)$$

where
    E = Young's modulus of the material (For concrete $E \sim 20.7$ GPa.– it depends on the type of concrete)
    I = moment of Inertia = $(Lh^3)/12$
    ρ_F = Weight density. (The F indicates a force, not a mass) (For one particular type of concrete = 23,000 Nt/m³)
    h = thickness of roof
    W = width of open room

Therefore,

$$\delta = \frac{5(\rho_F WLh)W^3}{384E(Lh^3/12)} = \left[\frac{5\rho_F}{32E}\right]\frac{W^4}{h^2} \qquad (14.4)$$

If two configurations are compared, labeled by 1 and 2

$$\frac{h_1}{h_2} = \left(\frac{W_1}{W_2}\right)^2 \qquad (14.5)$$

The thickness of the ceiling slab then scales with the room width squared. This is not by any means a course on the design of a concrete roof, but just as a back-of-the-envelope exercise ensuring a deflection below 0.5 mm seems like a place to start (it could be required to be lower). (The correct condition is for the stress not to exceed the elastic limit, but this is a back-of-the-envelope book.) From the curves plotted in Figure 14.22, it appears reasonable to start with a 2 m thickness ceiling when the width of the room is 10 m. That's also not too far away from what might be required for shielding.

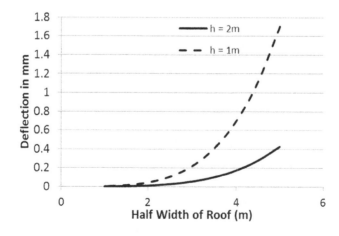

FIGURE 14.22 Ceiling deflection vs. room width.

Putting this all together to calculate the volume of concrete required as a function of room width, the chart in Figure 14.23 is derived. The thickness of the walls is determined by the radiation shielding requirements. The thickness (h) of the ceiling is determined by the span with a minimum of 1 m. The length and height of the room are fixed (to 10 m each). What is explored here is the relative increase in volume as a function of the width of the room. (Sometimes the goal of a compact gantry is to reduce the width of a room.) The volume increases almost a factor of 2 for a factor of 5 increase in room width. There is a break in the curve at the width where the ceiling slab thickness is constrained to be 1 m thick, even though deflection considerations can perhaps allow it to be thinner. It can be decided how to include these factors when considering a particular gantry design with respect to the benefit of its compactness.

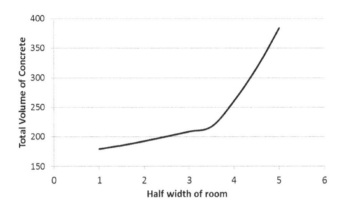

FIGURE 14.23 Concrete volume vs. room width.

Reducing the width of the room increases the wall thickness. Reducing the width of the room a factor of 5 only results in a factor of 2 change in the volume of concrete. Unless concrete is very expensive, the cost of building the room may not change very quickly as a function of the size of the room.

14.7.3 THERE IS NO ISOCENTER

The direction that the beam travels after it exits the gantry, toward the patient, is a result of several factors. Which set of factors play the primary role can depend upon the beam delivery modality used. For example, the patient collimator that is required in the scattering mode will fix the beam location with respect to the physical location of that collimator. Depending upon how close to the patient the collimator is placed, the trajectory angle of the beam can also play a role. In addition, the location of the second scatterer relative to the beam will be important in obtaining the desired transverse dose distribution (as shown in Section 12.2.2). In this case, the physical location of the second scatterer and the beam trajectory are concerns, and motion of the magnetic elements can create misalignments such that the beam trajectory may be modified. This, of course, also plays a major role in beam scanning. Thus, a combination of the beam trajectory and the location of physical beam elements are necessary to deliver the appropriate beam to patient. These physical elements are located on the gantry structure; therefore, the behavior of the structure as a function of angle may be important to the success of the design.

There are a variety of mechanical design concepts that can help ensure the correct location of the important components. One such example is a radio telescope style space frame design (shown in Figure 14.24(left)) wherein, while internal components of the structure might flex and bend, the overall system as a whole maintains the location of, for example, the patient collimator at the same physical location. Another example of a design would be a solid structure build to be very stiff and resistant to flexing (an example is shown in Figure 14.24(right)). In all cases, there will be some flexing and engineering analyses are done to estimate the effects and obtain an appropriate design. The answer to the question of what is important about a gantry structure has been assumed. That assumed answer is a word called 'isocenter'. All things are about that word and all things are labeled that word. The centroid location of the patient target is called isocenter, and sometimes there are multiple isocenters in the patient. The patient positioning device can be positioned at an isocenter and rotated about isocenter, but this is not necessarily the same isocenter as the isocenter in the patient. If there is a gantry, it can rotate about its own isocenter and direct the beam toward an isocenter. All of these isocenters sometimes overlap at a physical, mystical spot in the treatment room that can also be called isocenter. Perhaps they are all the same if the patient target is positioned at a specific location defined by the center of rotation of the gantry and the patient positioner moves the patient to that location about which it can rotate.

FIGURE 14.24   An IBA gantry (left) and a Hitachi gantry (right).

The patient positioner is mostly robotic and can rotate about any physically achievable geometry. While the gantry has a rotation axis, that is not necessarily indicative of the trajectory of the beam. In a perfect world, these would all be aligned, but this is not a perfect world, so it would be helpful to better understand the reality of these terms. Well, it would be helpful to better articulate the questions. It might be possible to create a more optimal, cost-effective design of these components by focusing on the actual situation and not the word. It can be useful to explore this as a physicist without a bias stemming from the word isocenter. At the highest level, a good question is how to make sure the beam trajectory is directed to the patient target and that the target prescription dose is achieved? It doesn't, at a high level, matter where in the room the patient is positioned, as long as the beam gets to the right place in the target. Therefore, from a practical perspective, all that matters is that it is possible to achieve the desired positions simultaneously and reproducibly. How the beam trajectory is adjusted to arrive properly at the patient, or how the patient is manipulated to be in the correct place to receive the beam is a matter of implementation.

Much time and effort is devoted to a clinical specification that the mechanical isocenter of a gantry should be within a certain tolerance. The requirement does not say anything (typically) about the beam, since the assumption is that the mechanical structure is the key factor. In the case of scattered beams, where the beam profile is constrained by a collimator on the gantry nozzle, there is some rational behind this requirement, since a fixed collimator is difficult to adjust. However, the patient positioner could be adjusted to compensate for reproducible deviations from a perfect line toward a room isocenter. Therefore, even in a scattered beam case, a perfect trajectory is not necessary, if it is repeatable. At MGH, we teamed with the Massachusetts Institute of Technology (MIT) to invent what is called the patient positioner correction algorithm (PPCA). For scanned beams, other possibilities are opened. If there are no physical devices constraining the beam delivery (unless a collimator is used), the beam can be adjusted to any physically achievable trajectory to meet the patient targeting requirements. This statement is only filtered by the fact that there is generally a downstream ionization chamber which is used, in addition to dosimetry, to measure the beam trajectory positions. This is a physical device, which, if moved, will give erroneous data relative to the beam position. Still gantry isocenter specifications, other than reproducibility are ambiguous.

Going back to the physics back-of-the-envelope perspective; a gantry structure can be divided into two components, a ring and a nozzle and their motions can be considered. The ring (and structure connected to the ring) will deflect under its own weight and as it rotates, will continue to deflect (it is not so much a rigid deflection – it is flowing, the shape is almost the same independent of angle, with the exception of when a part of the gantry has additional weight, like a nozzle or its counterweight). The nozzle itself, which is connected only at one end to the ring, will deflect relative to the connected end. First, ignore the nozzle and imagine a very light rod (which does not deflect) of constant length; then the following situation exists. Assuming that the originally designed structure was a circular ring, with a rod whose length is the radius of that circle, Figure 14.25 is the result of the imagination. Looking at the left side of the figure, since the gantry is squished, when the gantry is at 270 degrees, the rod does not quite reach the axis of rotation (it is along the major axis of what is now an ellipse). However, when the gantry is a 0 degrees, the rod extends beyond the axis of rotation (it is on the minor axis of the ellipse). When measuring the gantry mechanical structure against the specifications, the coordinates of the end of this rod are typically measured, or the outline of gray-shaded region in the center. So, what is being measured is the degree to which the gantry squished. Now departing from the cardinal angles, the shorter dashed line shows the rod at an angle between 90 and 360 degrees. Since the ring is not a circle, but closer to an ellipse, the rod does not point to the center, but to a location at some distance from the center.

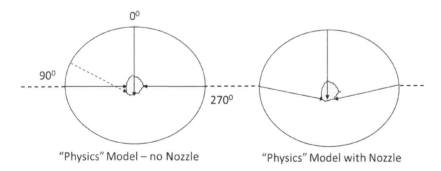

"Physics" Model – no Nozzle   "Physics" Model with Nozzle

FIGURE 14.25   Gantry ring deflection with massless rod (left), rod with mass (right).

The equation of an ellipse is

$$\frac{x^2}{a^2} + \frac{y^2}{b^2} = 1 \tag{14.6}$$

The slope along this ellipse at a given point on the ellipse (x_e, y_e) is

$$\text{Slope} = m = \frac{dy}{dx} = \pm \frac{b}{a} \frac{1}{\sqrt{(a^2 - x_e^2)}} x_e \tag{14.7}$$

A line perpendicular to this point has a slope $= -1/m$. Taking into account that x_e and y_e are related by $y_e = \tan\theta x_e$, the equation of the line (or of the rod) pointing perpendicular from the curve of the ellipse toward the inside of the ellipse is given by

$$y = \left[\pm \frac{a}{b}\sqrt{(a^2 - x_e^2)} \frac{1}{x_e} \right] x + \left[\tan\theta x_e - \left(\pm \frac{a}{b}\sqrt{(a^2 - x_e^2)} \right) \right] \qquad (14.8)$$

Knowing the equation of the line and the length of the rod allows the coordinates of the end of the rod to be evaluated. A specific example is one in which the unperturbed gantry radius is 2.5 m. Assume a deflection of the gantry such that the horizontal size is increased by 0.05% and the vertical size is reduced by 0.05%. This amounts to a deflection of 1.25 mm. The rod measurement would result in the plot of Figure 14.26. This curve may look simple, but it is actually a bit complex. For example, the lines in the lower right quadrant result from the rod measurement when the gantry is between 270 and 0 degrees, not below 270 degrees.

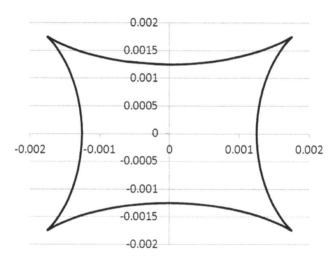

FIGURE 14.26   Coordinates of end of massless rod.

It's instructive to ask the question, does this figure represent something that is relevant? It is, after all, what is measured. Take, for example, the case when the gantry is at 270 degrees. Is it really the case, that in such a situation it would be necessary to move the patient 1.25 mm to the right? Does a 0.05% increase in path length of the beam really matter when that is a $1/r^2$ decrease of 0.1%? That's unlikely in most reasonable situations. Therefore, this curve may not be reflective of a clinical reality and may not be the best specification to rely on to set up a patient. Instead consider the minimum distance from a nominal center (0,0) to a line that is along the rod (it could be an extended line). That would indicate the minimum distance of the (sometimes extended) beam to the axis of the gantry. Figure 14.27 illustrates this concept. This figure has several lines (put on glasses if you need them). Two, coming from the upper right and middle right (lines 1 and 2), are lines perpendicular to the ellipse. These could represent the rod at two different gantry angles and the end of the rod is indicated by where it ends. Measurements from the axis center (shaded circle) to the end of the rods would give the solid black line in Figure 14.26. A solid black line in Figure 14.27 goes from the center to the end of line 1. Alternatively, the black dashed line from the center to line 1 is perpendicular to line 1 and shows the minimum distance to line 1, but still along the rod/beam path. For line 2 (solid gray), the line had to be extended (gray dashed extension) to find the minimum distance from the center along the path of the beam (extension of the rod) and the gray dashed line from the center

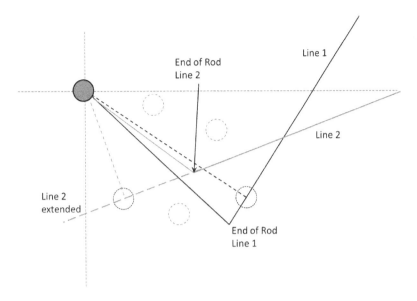

FIGURE 14.27   Minimum distance lines.

to this extended dashed line is the minimum distance. The open circles, in the figure on line 1 and line 2, show where the perpendicular lines of minimum distance intersect. Call this a new measurement. Other open circles (longer dashed) show where some other lines would intersect. Overall, in comparison with the end of rod measurement, the minimum distance measurements would result in the graph of Figure 14.28 (black dashed curve).

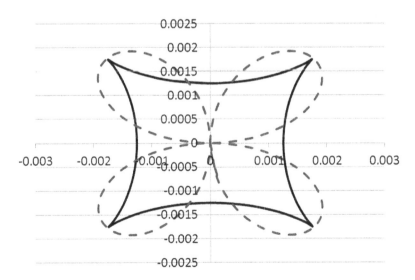

FIGURE 14.28   Coordinates of end of massless rod and minimum distances.

While the overall extent of the two curves is similar, their characteristics are not. In particular, the clover leaf pattern of the dashed (minimum distance curve) shows that as the cardinal angles are approached the effective error is reduced. While the clover leaf seems to bulge out of the solid pattern, it is an illusion since the shape is quite different as a function of gantry angle and most distances are smaller than the standard rod measurement distance when minimized.

The situation becomes a little more involved when the rod acquires Higgs bosons. The rod will deflect under its own mass as in the right side of Figure 14.25.

- When the rod is pointed down from above, the deflection of the ring makes it so that the end of the rod is below the original circle center.
- When the rod is pointing up from below, the deflection of the ring effectively pushes the rod above the nominal circle center.
- When the rod is point to the left, while the gantry is at 270 degrees, then the end of the nozzle (the part closest to the center) will sag downwards.
- When the rod is point to the right, while the gantry is at 90 degrees, then the end of the nozzle (the part closest to the center) will sag downwards.

Continuing the physics modeling, the rod can be considered as having an angular variation as a function of the absolute value of the cosine of the gantry angle. The behavior of the end of the rod, added to the end of rod measurement (not minimum distance) shown in Figure 14.26 results in the curve of Figure 14.29. This is an interesting curve (plotted, upside down so as to compare with a measurement plot that was available). When the gantry is at 90° or 270°, the nozzle is deflected downwards and stays downwards as the gantry rotates toward 180°, longer than the curve in Figure 14.26. When the lower right-hand side of the curve starts to go from the lower right edge to the lower center, this is delayed and results in the loop. When the curve is in the opposite quadrant, the result instead of the loop is to stretch the figure out thus the sharp edges are softened. A minimum distance analysis can be done for this data as well and the plot is pretty. A measurement of the Burr proton therapy center (BPTC) gantry results in the curve in Figure 14.30. The shape is very similar to that of the physics model although it is a bit flatter and rotated a bit. Additional components such as counterweights and heavy cables and trays combine with other items to deviate from the ideal physics model, but the result is otherwise encouraging from the point of view of understanding what is happening.

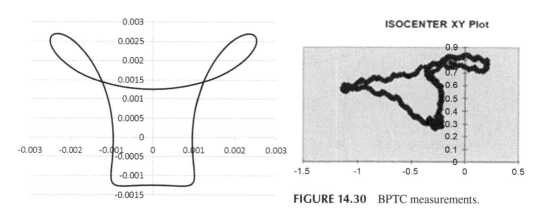

FIGURE 14.29   Coordinates at end of heavy rod.

FIGURE 14.30   BPTC measurements.

From a practical point of view, the extremes of these deviations from the ideal center of circle location may be considered the degree to which the gantry points away from what is termed the isocenter. However, from a truly practical point of view, the distance along the beam path is not as important as the transverse distance. At a deeper level, what matters most is the reproducibility of this pattern, so that one could predict the location of the components (discussed earlier) on the gantry. Perhaps the best correction to target the target is to follow the minimum distance curve. Well, perhaps more than perhaps.

In the ideal scanning mode, the beam is magnetically scanned and there may be no apertures and therefore no physical components that are connected to the gantry after the magnetic elements on the gantry (except the ionization chambers). Therefore, the nozzle sag factor is removed and what is most important is the alignment of the magnetic elements on the gantry so that the beam trajectory can be predictable and the desired beam path quickly obtained. One of the most time-consuming parts of the technical commissioning of a gantry scanning system is to determine the appropriate magnetics settings, including corrections to misalignments as a function of gantry rotation. Stability of the magnetic element alignment can help greatly in reducing this burden. It is noted that the use of collimators and compensators with scanning is not uncommon, and this analysis would need to be revisited in that context.

14.8   GANTRY – LESS OR MORE?

While the gantries that have been built or conceived are in fact wonderful ideas and innovations, it is the case that a small, inexpensive gantry, which contains all the features that have been thought to be desirable, has not yet been built. (This is indeed a subjective statement.)

There are gains to be made in space with superconducting magnets, especially for heavier ion systems (such as the Toshiba carbon-ion gantry in Japan). On the other hand, the physical implications of the patient space, instrumentation, beam pointing accuracy and clinically related constraints pretty much set the scale of the proton gantry. Building costs play a role in the overall system determination and new ideas in this are being developed. This will be site dependent and it remains to be seen if it will be a strong cost driver.

In the search for a full featured, compact gantry, or a compact gantry with reduced features, a list of parameters, questions and constraints has been outlined in this chapter. It can be used as shopping list for a gantry consumer who would like to know the implications of certain choices.

An alternative to this search is to consider that modern particle therapy may not need a gantry. Figure 11.9 is a chart of the number of fields treated as a function of gantry angle for one example institution (others are similar). While all angles are used, there are clear preferences. There are many possible reasons for this. But this plot spans a time period when scattered beams were used. In the current decade, while scanning beams have become the state of the art, the situation may have changed for the better. The increased conformality of such beams enables irradiation of complex geometries with considerably fewer fields. Recent works, such as those by Susu Yan, have shown that with appropriate imaging and positioning, most cases can be treated equally well with a fixed beamline or a gantry. Perhaps one way to make particle therapy available to more patients is to consider beam scanning with a gantry-less system.

14.9   WHAT COULD GO WRONG – 7?

The gantry adds one specific additional component and that is the structure. Alignment of the magnetics on the structure can be affected by the motions of the gantry, but magnet misalignment has been addressed previously. Perhaps the only really new issue is that of the pointing accuracy of hardware mounted to the gantry and reproducibility of the beam trajectory on the gantry (not necessarily the gantry itself), but this is highly unlikely to go into error. If it is not reproducible, then an issue exists. Of course, the likelihood of any of the errors previously listed was not discussed. This factor is addressed in Chapter 15. Therefore, there is nothing much new by the addition of the gantry other than the one big overriding factor. If the rotation mechanism isn't working, then treatments are likely to have halted and a nonworking machine may be associated with one of the highest classes of error. Of course, any of the previous subsystems discussed could fall into this category if they are not working. This is not a hypothetical situation. It took about three weeks for us to replace the large dipole on an in-plane gantry. One consideration may be the number of treatment rooms in a facility. A gantry affects only the delivery in one room, so with a multiroom system a nonworking gantry is not as serious as a nonworking accelerator.

14.10   EXERCISES

1. Calculate the length of a 440 MeV/nucleon in-plane carbon gantry. $L_3 = 0$; $L_2 = 3$ m (room temperature).
2. For the same particle and energy and length constraints as in 1, what would the length and width be of a superconducting in-plane gantry with a 5 T field?
3. What is the optical condition to make a scanned beam with upstream scanning on a gantry with infinite SAD?
4. Plot the bending radius of a room temperature dipole on a gantry as a function of carbon beam energy. Lithium. Helium. Maximum energy is a range of 30 cm.
5. How many magnetic field shifts are required from 230 to 70 MeV in the gantry with a momentum band pass of 3%? The width of the Bragg peak is 5 mm. The target is H_2O.
6. For a roof span of 10 m, what is the necessary thickness just to prevent deflections, under its own weight of less than 1 mm?

15 Safety in Radiotherapy

The processes of particle therapy are multifaceted and multidisciplinary. An important goal is to provide a safe and effective treatment taking into account the entire clinical workflow. One way to do this is to demystify the risk reduction processes and provide self-help tools for the application of the appropriate procedures and mitigations. Different departments may have different workflows and processes, and will therefore need to analyze their specific situations. This chapter provides tools that can help with that.

15.1 PROCESSES TO RAISE SAFETY AWARENESS

It is important to ensure that a particle therapy facility can accomplish the following:

1. Deliver an accurate treatment for effective results
2. Deliver a treatment safely
3. Promote an appropriate culture of safety
4. Provide a safe environment for all personnel

The radiotherapy journey starts with defining the processes and parameters that will be used to treat a patient and continues with those parameters being delivered by the equipment. For the most part, the processes followed are the responsibility of the radiotherapy department. The hardware (HW) functions and procedures are the responsibility of the equipment manufacturer whose parameters are then periodically verified by the establishment of quality assurance (QA) programs at the facility. This had been the status quo for decades in the field of radiotherapy. However, more recently, and motivated in no small part by the New York Times (NYT) articles of 10 years ago as well as incidents in other countries, the questions of who should be responsible for what processes, and what processes in addition to the equipment QA should be included to enhance safety, are continually being revisited. Despite the fact that the incidents reported in the NYT articles are the exception that proves the rule that radiotherapy is and has been very reliable and safe; it is a reminder of what is already known; there can always be room for improvement.

Moreover, after the first publication of the ISO 9000 standard related to quality management systems (QMSs) in 1987 and its widespread use in many different areas of industry and science, the idea of a more general, proactive approach to quality management, including QA and risk management, has been introduced gradually into radiotherapy. Based on the International Organization for Standardization (ISO) 9000 series, an international standard for QMS of medical devices has been established and is called ISO 13485. This can serve as a guideline, when setting up a QMS. Furthermore, in some countries, dedicated standards for QMS in radiotherapy exist, like the German standard "Quality management system in medical radiology – Part 1: Radiotherapy", published in 2009. General guidelines for improving the safety in radiotherapy can also be found in the standards published by the International Atomic Energy Agency (IAEA) like "Safety in Radiation Oncology" (SAFRON), which includes a voluntary reporting and learning system.

At a meeting called by the food and drug administration (FDA) in 2010 to garner input related to the situation defined by the NY T articles, it was clear that the roles of the equipment manufacturer and the facility personnel were effectively being questioned. For example, one panelist said something which indicated that "it is never the fault of the hospital staff member, it is the fault of the equipment manufacturer". Another panelist noted that the equipment manufacturer should do a better job in training the hospital personnel about the safety of the equipment. In 2019, NY Times

DOI: 10.1201/9781003123880-15

articles calling for tighter regulations were published. Although it was not clearly stated at that meeting, it seemed clear to some that a safe environment and a safe system comes about through a process of understanding what is important and managing that safety. This includes understanding what is at risk, how the risk is mitigated and a clear identification of the roles of the equipment and the individuals in the processes of radiotherapy. This is possible with the right attention to details during the design and treatment process.

It is not the case that there is one clear standard to implement safety in radiotherapy. Different facilities follow different workflows; different equipment manufacturers implement checks and/or mitigations in different ways, owing to differences in their equipment (or other reasons). Personnel protocols are as much a part (or have been) of safety management as hardware (HW) and software (SW) interlocks, although this is questioned more and more. One could argue that this is due to a lack of standards, but another viewpoint is that it is due to the diversity in a continuously growing field. Even if it were possible to implement all safety protocols within the equipment, would that be the appropriate thing to do – taking the human out of the equation? It is the case that radiotherapy these days is becoming more and more complex. Self-driving cars still aren't standard. Human intervention may not be possible in time to prevent all possibility of injury during the beam delivery part, but the degree to which this can be designed is still open.

In particle therapy equipment systems, a special problem may be that not many identical systems exist yet and very often additional third-party products are combined with commercial systems. This can result in modifications to standard radiotherapy processes and from a HW perspective, it could result in interface compatibility challenges. Certified components usually have very clearly labeled interface specifications. In the past few years, this has changed, and there are indeed more and more similar systems going into operation. (Remember now there are as many as, or as little as [depending on your perspective] 100 facilities.) It helps to have a clear responsibility for analyzing the risks of these systems, including the interfaces.

This chapter contains some of the information and tools needed to do a self-evaluation of processes and identification of the implementation of safety. The philosophy is that the more engaged the staff is in the understanding of how safety is implemented, or what the potential issues might be, the closer they are to the creation of an appropriate culture of safety and the safer the environment will be. Creating a culture of safety is not simply being careful and watching out for accidents, pointing fingers or retrospectively reporting and discussing specific issues, it encompasses elements of understanding systems and contributing to the proactive quality management methodologies employed in a department and feedback to industry.

Some think that doing the latter is too hard. It is not. To accomplish this goal, it seems useful to not only identify ways in which safety management is implemented generally, but also to provide specific tools that can be used to analyze possible safety concerns in a specific facility and determine whether the mitigations that are necessary have been appropriately implemented. This is not to say that the full hazard analysis (HA) for the equipment performed by an equipment manufacturer must be reproduced by hospital staff, but it is useful, at least, to understand how these can be done, and how it can be extended to other (less technical) aspects of the treatment workflow. For each technique to be described herein, there are multiple ways in which it can be implemented. For the purposes of this book, focus is on methods in common practice and some examples of alternative approaches are mentioned.

15.2   INTRODUCTION TO HAZARDS AND MITIGATIONS

As an introduction, it is helpful to identify a number of concepts to keep in mind when setting out to consider how to think about determining if a process is safe and what to do to help it to result in a safe outcome. Before jumping into any analyses, it's useful to discuss the terms and philosophy. A hazard has various definitions, depending upon the framework of an analysis. A hazard can be a circumstance that sets up the possibility to cause an unplanned or undesirable event (e.g. a live electric wire on the ground). An accident would be someone stepping in a water puddle on a floor on which a live wire is lying. A hazard is something that is potentially dangerous and should be avoided so that an accident

FIGURE 15.1   An accident.

doesn't happen. A live electric wire on the ground is not an accident, just an accident waiting to happen. An accident is typically an unintentional event that might have been prevented. When reading about this topic, take care to understand how the terms are used. There are a number of ways in which to create a hazard. It can be generated by equipment, by humans and by combinations thereof. Also, there are hazards that are a result of a primary action (equipment or human), or a hazard resulting from a reaction to another hazard (e.g. a gas pedal is on the floorboard that is possible to push instead of the brake when trying to stop a car resulting in a car crash). What was the hazard that resulted in the accident in Figure 15.1? Most don't believe that all hazards can be eliminated, but it would be good to have a system that is protected from conceivable hazards. If the system is *simple enough*, a good fraction of the possible hazards can be conceived. Given a hazard, it may be possible to institute mitigations such as:

- Prevention
- Detection
- Reaction

Safety is a goal of the environment in which we live. Sometimes it is believed that an intuitive sense of how to implement safety exists. The photos of Figure 15.2 show the use of prevention, detection and reaction (clockwise). However this doesn't result in 100% success. Systems and workflows that are more complex require a level of analysis beyond our day-to-day instincts. Some of the

FIGURE 15.2   Prevention (left), detection (upper right) and reaction (lower right).

questions posed are the frequency of detection and time for reaction. These are just as necessary in a radiotherapy environment.

One may ask whether we can prevent all errors in radiotherapy. Prevention has a specific meaning but could be interpreted in degrees. From an equipment perspective, take an airplane as an example. Is an airplane prevented from losing altitude or is a loss in altitude detected and reacted to? A plane taxiing on the ground is prevented from losing altitude; however, a plane in flight is monitored and a loss in altitude, when identified, is mitigated so as to not result in an incident. An incident is also an important word with which to take care. It is like shouting fire in a crowded room. In a radiotherapy workflow example: can the wrong image be prevented from being used or is it necessary to detect and react to a mistake before an incident can occur? These are the kinds of questions to be sensitized to while analyzing a given situation.

Another word is risk. In this chapter, risk is the probability that a hazard will end up causing a certain level of accident. Generally, there exist different ways to mitigate risks:

- Risk avoidance by design of the equipment, i.e. a fail-safe design/no hazard
- Risk mitigation by QA, e.g. by a daily check or double check (believing that the time between incidents is greater than the time between daily checks)
- Risk mitigation by notification, e.g. by informing all personnel about specific measures and training them accordingly to avoid hazards

In dealing with risk and/or mitigating a potential hazard, there is the question of whether something should be dealt with by humans or by machine? From the equipment perspective, consider the following: Conventional radiotherapy consisted of ~30 fractions per course of treatment with about 1 minute/fraction. Let's say that human thinking reaction (not instinctual reaction) time is about 1–3 seconds, 3 seconds out of 1 minute is 5%. Perhaps this is what has led to a one-minute treatment; this allows a human to intervene while only a 2–5% error is made. On the other hand, particle beam spot scanning has a time frame of milliseconds per spot and while an error analysis is dependent upon dose rate and specific treatment planning issues, one can see that this is generally faster than human reaction time, for a single spot. Also, in conventional radiotherapy a 1 mm leaf gap results in a 5% dose delivery error, which is also below human detection capability in a clinical treatment scenario. However, all leaves open instead of in a particular pattern is well within a humans' capability to detect. On the other hand, with multi-fraction treatments, there is the chance to make up for some level of errors. Other than beam delivery, most of radiotherapy workflow is human-based. Should all the rest of the tasks become machine-based?

15.3   INTRODUCTION TO RISKS AND CRITICALITY

Two of the key tools that are used as input in safety analyses are risk and criticality. This section covers ways of defining these quantities. According to IEC 601-1-4 and ISO 13489-1, the term risk is defined as the combination of the probability that a hazard will cause an accident and the degree of severity of harm caused by an accident associated with the hazard. The result of such an analysis is related to the importance of the effect and the identification of a mitigation to reduce the severity and/or probability of the effect. While there are many methods and references, there is also an ISO standard ISO 14971:2007. In addition, it is sometimes considered how hard it might be to detect an occurrence of a hazard (if a tree is falling in the forest, and you are home, will you know?). If it wasn't detected that doesn't mean it didn't happen. However, it can be quite serious (if someone was standing under the tree). Detection can lead to mitigation. For example, an overdose may be delivered and in some circumstances (depending on the details), if it is detected and is part of a course of many fractions, it might be compensated. The same is true of a measure of avoidability, which is sometimes explicitly included. This it is obviously related to detectability, but not the same. If you're on a river heading toward a waterfall, the waterfall might be detectable, but can it be avoided? Maybe it depends on when it was detected, or how strong a swimmer you are? But, if it isn't detected, it can't be avoided. While it is not strictly speaking a separate quantity (in some opinions), it is sometimes separated from the probability number.

15.3.1 HIERARCHY OF RISK PARAMETERS

Risk is a combination of the severity of an accident resulting from a hazard and the probability of occurrence of that hazard. There can be several levels of the severity of a hazard. For example, someone touching a bare wire from low voltage lawn lighting is less severe than being subjected to the bare wiring in a 200 amp main circuit breaker panel. Generally, in practice, there are four (or five levels if the no determination category is used) as noted in Table 15.1.

TABLE 15.1
Hierarchy of Severity

Term/Severity	Level	Description
Undetermined	5	No determination has been made yet (assume the worst)
Serious	4	Results in death, permanent impairment or life-threatening injury
Moderate	3	Results in recoverable injury or impairment requiring professional medical intervention
Minor	2	Results in recoverable injury or impairment not requiring professional medical intervention
Negligible	1	Inconvenience or temporary discomfort

These levels may be defined differently for different circumstances and may, for example, be reflective of individual or group risk. For example, in the case of radiotherapy, there may be specific dose constraints associated with each of these levels (e.g. Section 15.6.4).

The levels of frequency of occurrence (or probability) can be defined according to Table 15.2, although different time frames are considered in different circumstances:

TABLE 15.2
Hierarchy of Probability

Term	Level	Description (Occurrences in Installed Base Over Lifetime)
Probability		
Undetermined	U/5	Unknown – i.e. caused by software flaw, misuse or user error. Indeterminate occurrence rates.
Frequent	5	A common/typical occurrence (≥ 1/year).
Probable	4	Likely to occur several times over the lifetime of the product (~1/year to 0.1/year).
Occasional	3	Likely to occur few times over product lifetime (~0.001/year to 0.1/year).
Remote	2	Likely to occur at least once during the lifetime of the product but is highly improbable (<0.001/year).
Improbable	1	Unreasonable to expect occurrence over lifetime of the product but is theoretically possible (<0.00001/year).

In addition sometimes these levels are subdivided further into a total of ten or more depending upon the individual analyzer's tastes or specific implementation. Note that an undetermined probability or severity may be treated as undefined or as the maximum depending upon the degree of conservatism used in the analysis.

The combination of severity and probability determine, by definition, the risk potential number (RPN). For example, in the tables above, the maximum RPN is 25 (= 5 × 5) and the minimum (for those that are defined) is 1. As mentioned in the previous section, sometimes specific qualifiers such as detectability and avoidability are explicitly factored into the RPN, thus modifying the probability number. Sometimes the RPN is referred to as 'criticality' and the RPN numbers may be subdivided into criticality categories. The RPN is then used to inform the analyzer about the necessity of mitigation and type of mitigation (very secure or less secure) to be used to attempt to prevent an accident.

15.4   CATEGORIES AND QUALIFIERS OF RISK

The main goal of this section is to define an RPN that can be used to inform the hazard or failure mode analysis. The RPN can then be used to determine mitigations. Below are various examples of how the RPN can be calculated.

15.4.1   RPN, RISK AND CATEGORIES

The analysis of risk can be simplified, compared to Tables 15.1 and 15.2 (it's not always appropriate to do so) by approximating the risk number contributions as follows:

Severity (S):

Define S1 = Moderate Severity (Less Severe)
Define S2 = Serious Severity (More Severe)

In this example, only S1 and S2 are considered for the severity of the possible accident/injury.

These analyses can be used for all aspects of a system, which includes all the steps leading up to and including the treatment of a patient; from consultation through planning and treatment. Therefore, one consideration is that of the radiation dose that will be or is delivered. For example, in cases where a dose >2 Gy is inadvertently delivered to the bone marrow or optical lens (which could be due to an improperly contoured treatment plan (TP) or a failure of the equipment), these could be considered S2 severity. In addition, situations in which electrocution or heavy objects falling on people, while not directly related to radiation, may also considered S2 levels of severity. Drawing the line is still somewhat subjective and identifying the above considerations is designed to draw a line in the sand and to see if the reader reacts to them and has their own impression of how the rating should be structured.

Probability (P):

Define P1 = Includes Occasional, Remote and Improbable Probability (Lower)
Define P2 = includes Probable and Frequent and Undetermined Probability (Higher)

Probability includes the consideration of a variety of issues.

In the case of a falling object, the probability of a device falling can be considered to be a potential hazard with a certain probability and the probability of an individual being subjected to the hazard (e.g. walking under the falling piano at the right moment) can also be considered, so that these two are considered together. For example, it is more likely that a patient directly receiving therapeutic radiation will be in the treatment room when the beam is on than other individuals and, therefore, the probability may be different for different categories of people. Considering another issue, for example, is it more likely that a computer will transfer data accurately compared to a human transcribing it?

Avoidability (A):

Define A1=High ability to Avoid (and/or Easy to Detect)
Define A2=Very difficult to Avoid (and/or Hard to Detect)

Again, these quantities include a variety of considerations.

As noted earlier, some consider these to be simply modifiers of probability and do not address them separately. (If it's avoidable then it's less probable. Or is it?) Similarly detectability is entwined with avoidability. Indeed there are many aspects of probability and they can't all be addressed separately. This is done here, for the sake of example. For example, if the area under the falling piano is cordoned off, does that make the accident avoidable? When evaluating this quantity for irradiation, consider the quantities associated with beam delivery that are measured online, which are detected, and therefore have a chance of being avoided. Of course, the act of measuring beam quantities on-line is usually a result of a safety mitigation determined from the risk analysis and/or the failure modes and effects analysis (FMEA) (to be discussed later). In this case, the purpose of the mitigation may in fact have been to change the ability to detect or avoid the hazard. Therefore, the mitigation modified the probability in this case. Another method that can be helpful in avoiding accidents includes activities generally supervised by experienced staff. This will again change the relative risk depending upon the category of individual involved. An experienced staff member may notice that a certain TP does not match with the desired patient's field. If a mitigation is introduced, which causes the staff member to check it, there would be a higher probability of detecting it. Or an experienced staff through rote habit may just sign off on a plan without noticing something unusual, so another factor may need to be introduced. A study of skydiving accidents shows a higher incidence of injury among novices and a higher death rate (still very small) among the very experienced. The author used to call the latter group skyg-ds some decades ago. The safest group is that with some experience who still have a healthy respect for the potential hazards.

Given these quantities, one possibility is the matrix of Figure 15.3 that leads to a way of categorizing the level of the risk and determining what type of mitigation could be considered. Given the combination of severity and probability (and avoidance and/or detectability), the value of the RPN to identify a potential risk category criticality category can be generated. It can also be called Risk Assessment Codes (RAC).

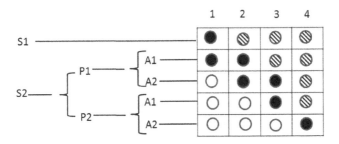

FIGURE 15.3   Risk mitigation category chart.

The letter/number combination along the side are combinations of S and P and A. For example, S1 is low severity and, therefore, other qualifiers (probability and avoidability) are not factored in. (This is how it was done in EN 945-1. In the newer ISO 13489-1, the S1 line now subdivides.) The black circles in the cells are the preferred criticality category(ies) of mitigation (see the numbers at the top of the columns) for the particular combination of parameters, which leads to a given row.

This means that for the given combination of S, P and A, the black circle indicates the level of reaction or mitigation that is recommended. The hatched circles are possible mitigations for additional conservatism or a particular set of concerns (a higher level than otherwise recommended). The empty circles are lower levels of mitigation if there is good reason to do so and the reasons should be recorded. For example, the third row comprises S2 × P1 × A2. In that row, one might perhaps implement a lower category (e.g. category 1 – open circles). The preferred category of risk to evaluation is either 2 or 3 (black filled circles). One can also implement a higher category (hatched circles). Note that the evaluations are essentially multiplicative results of the various components with equal weights. Of course, this will depend upon how the 'numbers' are assigned to the various issues.

In the matrix Figure 15.3, the numbers across the top row refer to the categories of mitigations that are identified in Table 15.3. Based upon this category, certain action(s) or mitigation(s) may be warranted. For example, the following categorization of mitigation from EN 954-1 is sometimes used:

TABLE 15.3
Category Hierarchy Table

Category Number	Implication of Category
1	A safety function implemented should be checked from time to time.
2	Design the system according to well-tested components and safety principles.
3	Design such that one single error does not cause the loss of a redundant safety function and when practical that safety function operation is detected independently or tested, thus minimizing the loss of redundancy.
4	All of the above plus design the system so that the error of a component does not cause the loss of the safety function.

The analyzer has some leeway in determining how to best approach the issues. They are specific to a specific set of circumstances and are depending upon the humans who are evaluating them to understand these circumstances. These can apply equally to a 'system' or a 'process'. Preparing a staff to think in these terms is a very important part of a culture of safety.

15.4.2 BRUTE FORCE CALCULATION

The previous method was particularly rigid and minimized the degrees of risk and probability but is an okay way to do it. Although it was a standard, that standard (as noted earlier) has been updated and there are more gradations. One can perhaps best think of this by allowing a certain degree of flexibility and increasing or decreasing the number of choices of categories and where to choose the threshold for a given RPN/mitigation categories in some situations. Increased choice may also lead to increased difficulty of analysis, or it may enable the individuals doing this work to feel more a part of it and to take more responsibility. A simpler, brute force way is possible.

The result of a straightforward numerical analysis of four categories of severity, five of probability and two of avoidability could be as illustrated in Table 15.4. A simple product of severity, probability and avoidance yields a maximum possible number (called criticality in this table) of 40. Categories are defined based upon the product. The most serious category (IV) can be defined as those items with a score ranging from 31 to 40 and similarly for lower scores from 21 to 30 can be assigned category III. This has a similar format as in the previous figure but is more straightforward to handle numerically. However it may be considered a bit more subjective in that the analyzer must identify that line separating the categories. Mitigations are then assigned based on that category. In Table 15.4, the multiplicative product is placed in the appropriate column. Also the criticality is entered into the fifth column to make it clearer.

TABLE 15.4
Brute Force Matrix

Severity	Probability	Avoidability	Criticality	1–11	12–20	21–25	25–30	>30
				0	I	II	III	IV
1	1	1		1				
		2		2				
	2	1		2				
		2		4				
	3	1		3				
		2		6				
	4	1		4				
		2		8				
	5	1		5				
		2		10				
2	1	1		2				
		2		4				
	2	1		4				
		2		8				
	3	1		6				
		2	I		12			
	4	1		8				
		2	I		16			
	5	1		10				
		2	I		20			
3	1	1		3				
		2		6				
	2	1		6				
		2	I		12			
	3	1		9				
		2	I		18			
	4	1	I		12			
		2	II			24		
	5	1			15			
		2	III				30	
4	1	1		4				
		2		8				
	2	1		8				
		2	I		16			
	3	1	I		12			
		2	II			24		
	4	1	I		16			
		2	IV					32
	5	1	II		20			
		2	IV					40

One important aspect of this analysis is to allow the arithmetic to speak for itself, but then to question the results. If the results are somehow at odds with the intuition or expectation of the analyzers, then it is a learning experience and it is very important to spend the time to understand wherein the difference from intuition is based. This can usually enable a greater understanding of the hazard and allow a better estimate of the parameters. This learning is more probable with a larger table. Fewer numbers may inject more subjectivity in the beginning.

15.4.3   BINARY COMBINATION

The earlier sections outline a systematic means of evaluating the highest risk processes, procedures or technology hazards. Once the highest risk issues are evaluated, it is possible to start reducing risk through safety improvement mitigations. In the simplest form, one might, for example, create a binary structure, basically drawing a line between those risks that require mitigation and those that do not. The first example shown was that in Figure 15.3 and categories were then identified to be associated with the levels of mitigation. As another example, Table 15.5 uses the probability and severity combination of Tables 15.1 and 15.2 (without avoidability) to identify what combinations should be mitigated and which are acceptable risks. This is simply a binary decision without the gradations of the categories in the previous section. One can replace this table with a threshold in a range of other RPN definitions.

TABLE 15.5
Binary Combination of Probability and Severity

	Severity			
Probability	**Negligible**	**Minor**	**Moderate**	**Serious**
Undetermined	Acceptable	Mitigate	Mitigate	Mitigate
Frequent	Acceptable	Mitigate	Mitigate	Mitigate
Probable	Acceptable	Acceptable	Mitigate	Mitigate
Occasional	Acceptable	Acceptable	Mitigate	Mitigate
Remote	Acceptable	Acceptable	Acceptable	Mitigate
Improbable	Acceptable	Acceptable	Acceptable	Mitigate

15.5   MODELS AND METHODOLOGIES

It is possible to analyze the entire workflow in radiation oncology from when the patient arrives for consult to their follow-up visit(s), but, while it's possible to do this, the effort involved for staff that is not experienced in this process may be intimidating. It is often more productive to focus on smaller pieces, for example, the process of beam delivery. (But this particular choice may not be the one with the most benefit to an existing radiotherapy department.) One can build upon this to assemble the entire workflow while being cognizant of interfaces between different pieces of the workflow. It is likely best to start with a high level and then work out individual details. In all cases, it is helpful to identify pieces that are simple enough to understand, which helps to analyze safety more thoroughly. Since this is a book about technology, the focus is placed there.

Over the decades, various methods to analyze potential accidents have evolved.

15.5.1 DOMINOS, SWISS CHEESE, BUT NO WINE

In 1931, Heinrich published the 'Domino Accident Model'. As shown in Figure 15.4, it is based upon the postulate that once you set a chain of events in motion, the result is inevitable and resistance is futile. Identifying the links of that chain is the work. It is a linear chain of events and removing any one domino has the possibility to break the chain. The elements of the chain include social environment, carelessness and actions. This model was used in the early 20th century when factories and people were a main source of accidents.

FIGURE 15.4   Domino model.

FIGURE 15.5   Swiss cheese model.

James Reason proposed the 'Swiss Cheese Model' (Figure 15.5) of system failure in about 1990. It is an evolution from the Domino model in that it incorporates multiple possible sources of error (the holes in the Swiss cheese). Each hole in the Swiss cheese represents an opportunity for a process to fail and these opportunities are to be identified. While it is also a linear progression, the only way for an accident to occur is for the holes in each line of defense to line up. This brings in probability to the identified events. The holes, or the hazards exist, but the probability of encountering it is related to the position and size of the hole. (Note that even these early models incorporate multiple lines of defense or multiple steps to an end.) Which holes are in which position and their sizes may be random.

Coming back to the barn from Section 7.3, the probability of a hole to be in a particular location, or the cross section to hit that hole is the hole area/Swiss cheese slice area. Each of the Swiss cheese slices are independent (assumed), so the probability of going through a line of holes is the product of these probabilities. The holes are of different sizes, so that complicates the evaluation, but the maximum and minimum probabilities can be evaluated. If there are n slices, a binomial distribution can estimate the probabilities that the holes in m slices will line up.

Note that this 20th century model still includes organizational factors such as supervision of workers as prerequisites for accidents. The fact that there are multiple holes may allow for the possibility that various factory processes may come into play. Both of these models incorporate organization, management, social culture and environment. The fact that these are older models does not mean that these factors should be ignored in a modern-day analysis. Some models even incorporate input from government interactions, or in a case such as radiotherapy, the guidelines and reports of such organizations as AAPM and FDA, or IAEA and ISO on an international level, as these may play a role in hazard evaluations, creation of hazards, or failures to do it properly, as well.

15.5.2   FRAM

With modernization comes increasing complexity and the perceived need for more complex analysis models. One such model is the Functional Resonance Accident Model (FRAM), which postulates that even from random noise, circumstances can conspire to cause a coherent-looking signal. In other words, a system can be so complex that it may be impossible to identify all the 'Swiss cheese holes' and there will be some finite possibility that some holes will align at some time because they are always stochastically moving.

15.5.3 STAMP

A more recent development is the System Theoretic Accident Model and Process (STAMP), published by Nancy Levenson. The central assumption is that the nature of technology has evolved to form very complex socio-technological systems. Among the concepts that this model wishes to incorporate are the interactions between subsystems and explicitly identify possible failures of mitigations. According to STAMP, the analyst will first prepare a control diagram of the system in question to identify which control actions exist such as in the diagram of Figure 15.6. This effectively incorporates the processes and mitigations and the criteria for the operation of these process loops. A failure mode in any of the loops (e.g. the boxes or the arrows) is itemized. For any process, there are decisions made based upon data and there are potential errors in this. In this model, not only can a component failure or a control action not being executed cause an accident but there may also be subtleties in the timing of a control action that may be

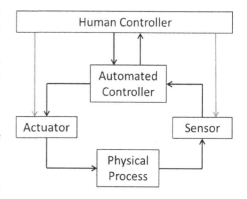

FIGURE 15.6   STAMP.

inappropriate for the desired mitigation. One such example of timing can be so-called SW race conditions. The timing of when a signal arrives to a given controller may not be completely deterministic depending upon what else is happening (e.g. network traffic, high use of CPU) and the SW may not have taken this eventuality into account. For example, the 'everything is okay' signal may precede a 'not ready' signal (which got delayed). Having received the okay signal, the not ready signal may be missed or ignored (from poorly designed SW). While within the original STAMP process no further quantification is recommended, many users have combined the methodology with various risk quantification methods in practice. In this model, some input can even be from management.

15.5.4 TOOLS AND METHODS

Having identified some types of accident models and key elements required for hazard analyses, it is useful to briefly summarize some analysis tools that will be discussed in more detail later. The above models are primarily ways to help think about the problem, with the exception of STAMP that includes analytical methods as well. Typically, the following three 'tools' are used to analyze system safety.

- Hazard/risk analysis
- Fault/event tree analysis
- FMEA/FMECA

Hazard Analysis

A hazard analysis (HA) is used to identify hazardous conditions and risks. It starts by identifying the hazards, evaluating the relevant risks and working out procedures to control or eliminate the risk. In Mesopotamia, around 3200 BC, the Asipu were called upon to predict the future of sick people. They were consulted for risky or difficult decisions. They developed methods of identifying the scope of a problem, identifying alternatives and predicting likely outcomes. This was the first known instance of risk analysis. Early risk analysis and probability theory go hand and hand and developed along the same path. A few thousand years later, the first recorded analysis was made, choosing between two religions based upon whether or not a higher being existed and the probable consequences. Pascal's wager was a true risk analysis with clear mitigations. The concept of causality was included in the writings of Hippocrates who argued a link between disease and the environment. So it has been used in health care for quite some time. The regulation of transportation and the involvement of government in such analyses were already in place during the time of Julius Caesar who prohibited all wheeled vehicles to operate in Rome between sunrise and sunset. In the last 100 years, the analyses of hazard identification and mitigation have evolved, but traffic still hasn't improved.

An HA is a 'top-down approach'. It identifies the hazards of a system and the hazardous situations that these can create. Hazard Analyses are performed to identify hazards, possible high level reasons for the hazards and the significance of these hazards. This leads to identification of safety design decisions. The HA can be conducted during all phases of a design and can become more and more detailed. The hazards are listed and analyzed with the help of some tools. The tools can include fault tree and event tree analyses. Then, mitigations to deal with the causes of, or the hazards can be determined. It could be that the system has or has not included mitigations or a new system is being designed or an old system is being reverse engineered. Typically, a system and/or workflow are separated into many smaller components or steps and each component is analyzed separately for a given hazard or accident. Care must be taken if there are interfaces between components or steps.

Fault and Event Tree Analysis

A fault tree analysis (FTA) is also a 'top-down approach'. A chain of events starting with a specific accident or fault is identified continuing with a list of the factors that may contribute to that accident or ways the hazard was created. It was originally developed in 1962 at Bell Laboratories by H.A. Watson under a U.S. Air Force division contract. The importance of a hazard is classified and the causes of the hazard are listed. The FTA is a deductive process essentially using Boolean algebra logic models to identify how the factors contribute to the hazard (e.g. AND or OR). For example $B \wedge C$ will cause A, or $B \vee C$ will cause A. Next, what will cause B and C? It's called deductive because Boolean logic is used, but there is an inductive component in the identification of the B's and C's because they are being inferred. Perhaps it is called deductive because it uses some mathematical logic. In a particle radiotherapy unit, these components may include aspects of treatment planning, oncology information system functions and building infrastructure. Technical factors may be considered including such details as: ion sources, accelerator, beam extraction system, beam delivery, monitoring system, control system, power supplies, patient positioning devices and imaging devices as all possibly contributing to the same hazard. This method has proven to be so effective that only complex (sometimes referred to as 'nonlinear') interactions are left as sources of error.

An Event Tree Analysis is also a logical model. It starts with, for example, a failure event and a mitigation of that event or a mitigation of the effect of that failure (two different things) and the analysis then walks down the path logically of those issues that can prevent that safety mitigation from working (similar to one of the aspects of the STAMP methodology). At every step the probability of success or failure can be included. It doesn't have to be just success or failure it can be any related event. It can check the effects of functions and error systems. An FTA lays out relationships among events and an ETA explores sequences linked by conditional probabilities.

FMEA

The FMEA (failure modes and effects analysis) was not developed for health care. The use of the FMEA, or as it is sometimes called FMECA (C = Criticality), dates back to the US Department of Defense where it was used by the Navy in the late 1940s. They were interested in finding ways to mitigate malfunctioning munitions. The resulted in MIL-P-1629.

Later, the methodology was taken up by NASA in various forms, and by the 1970s had made its way into manufacturing and other arenas where it is now widely used. Of particular relevance for health care is the fact that since 1990, the FDA requires device manufacturers to conduct a prospective risk assessment as part of the Medical Device Good Manufacturing Practices Regulations. This requirement is often fulfilled with an FMEA. It is a recognized tool by The Joint Commission (TJC) that accredited nearly 22,000 healthcare organizations in the United States. In 2001, the TJC issued rules (Rule LD5.2) requiring organizations to 'At least annually, select at least one high-risk process for proactive risk assessment'. This is often conducted via FMEA.

The use of FMEA in radiation oncology (outside of its use in machine design) is relatively more recent. In 2009, Eric Ford published the first report of the use of an FMEA in a radiation oncology clinic. A number of studies have appeared in the years since including application to stereotactic body radiotherapy, MLC tracking systems and other clinical applications. FMEA is also the subject of a Task Group report within AAPM known as TG-100, which may have some differing methodology as compared to this chapter. Internationally, within radiation oncology, ICRP Publication 112 describes FMEA and advocates its use in risk reduction programs.

The FMEA is a bottom-up method that starts at the component level where a possible failure mode is identified and explores the consequences on the system at a higher level. If the FTA is categorized as deductive, the FMEA is called inductive. Inductive reasoning uses existing information to reach a conclusion that may contain more information than the experience on which it is based. Knowing information about a particular component, it's possible to identify how the functionality of that component can compromise other aspects of a system. Whether that's deductive or inductive may depend on the expertise of the analyzer. The FMEA is used together with, for example, an FTA, ETA and HA to get a full picture.

These tools help to either visualize (if drawn graphically) or at least articulate a clear systematic approach to identifying potential system issues. That is one of the key ingredients. The other is accurate data and information. Feynman wrote in his minority report of the challenger accident "reality must take precedence over public relations, for nature cannot be fooled". The idea of doing these analyses is intimidating sometimes. It may be helpful to remember the saying 'well begun is half done' as quoted by Aristotle, having heard it originally from his nanny Mary Poppins. Using these tools as an aid to clarity can produce good results.

15.5.5 Practical Considerations

In the analyses described earlier, terms like 'identifying a failure is used'. One of the important things that is not so deeply understood is that this is not just a failure of a component (e.g. a resistor), or a step (lost prescription), this could also be a failure of a 'control action', which was performed based upon some input. Some of the newer methods, such as STAMP, explicitly require identification of these types of issues, but it does not find these issues automatically. None of the other schemes preclude inclusion of these types of issues. It may be an obvious statement to say that the input incorporated in any of these tools is generally more complete and accurate if performed by experienced individuals with system knowledge – or a combination of individuals. There is no system that automatically, reliably finds inputs not given by the analyzer, yet. There are, however, attempts to more easily capture the manually entered system requirements and dependencies and this assists in the analysis.

In the end, if a system is too complex, it is difficult to convince oneself, or anyone, that the appropriate set of mitigations has been properly implemented. It is necessary not only to implement risk mitigation strategies but also strategies to verify that these mitigations actually work. This type of verification of risk mitigation is an essential part of the procedure. It is helpful to design a system modular enough to be simple enough so that it can be understood, and documentation should be written that would make reverse engineering possible. One way to do it is to separate functionality, although interfaces then have to be addressed. Finally, it is necessary to define a safe state. Note that the absence of treatment is not necessarily a safe state, since lack of treatment can lead to negative consequences. There should be a safe state to enable (if possible) some time to consider what to do next before cancelling a treatment.

How a given set of processes and/or systems may be analyzed, at this point could be individualized, to some extent. The authors' personal preference is that for any analysis style, taking into account the appropriate possible issues that can arise is the most critical aspect, second only to designing the system in such a way that it's possible to understand it. There may be systems other than radiotherapy where this is not possible, but it is possible for the technology of radiotherapy. No system automatically provides all the inputs, outputs and interrelationships necessary to perform an analysis. It is up to the analyzer to incorporate the necessary data, ask the appropriate questions and go to the appropriate level of detail to optimize the safety of a system. It is also important for the analyzer to do this work, if possible, before the system is fully designed so as to provide feedback to the designers, and then to redo it when the design is completed. In the case of an operating system, the process of incorporating appropriate data can be facilitated by many mechanisms including a safety culture that encourages robust reporting.

In virtually all these processes the following should be identified:

- Requirements for effective system performance
- Processes that are needed to fulfill those requirements
- Hazards that can arise and be present in the execution of these processes
- Methods to mitigate either the cause of or result of the hazard, including failures of the mitigation

However the analysis may be different for a system currently under design and one already in operation depending upon the amount of information available. In these two cases, mitigations may be implemented at different levels. Reverse engineering a system is at best a difficult endeavor, but if there is an issue that staff members experience daily, it should be so noted. In one example, a system was designed so that a device would not start moving unless a component was secured. The staff quickly learned that they could secure the component, start the motion and then remove the component while the motion continued.

15.6 HAZARD ANALYSIS

The HA (hazard analysis) is a comprehensive process that is intended to identify foreseen hazards that can occur as a result of the system workflow and operation and to analyze their potential causes. The goal of an HA is to identify conditions that would give rise to a given identified hazard category and prioritize them for possible safety improvements. A key output of this analysis is the determination of the biggest risks and possible mitigations. It is sometimes done at the start of a design before detailed implementation is accessible, but is repeated as the design evolves. An analysis of this sort requires inputs and generates outputs. Identifying and assembling the inputs is critical to the success of the analysis. Asking the appropriate questions enhances the completeness of the input, which helps to generate the most complete output.

An HA (also sometimes called a risk analysis) (not to be confused with an FMEA)) is a 'top-down' process. Briefly the following steps are to be performed:

1. Pick a hazard (starting with this, means starting at the top) (input)
2. Score the importance (e.g. severity of the potential accident) of that hazard (output)

3. Identify a subsystem or process step (input)
4. Image how that subsystem or process step can create that hazard (output)
5. Identify the probability that the hazard can occur (output)
6. Create a mitigation (output)
7. Identify how that mitigation will be verified (Is a mitigation a real mitigation if you don't know whether or not it is working?) (output)

In the normal flow of the HA, it very quickly gets to the point of identifying sub-steps and how these can create the hazard being considered without going into great underlying detail. In general, the HA will be maintained at a higher level and may go down to a subsystem level but not down to the component level or below a subprocess level which will be dealt with in a more detailed 'bottom-up' FMEA analysis. That means that the mitigations can also be at a higher level.

It is extremely helpful to begin with the inputs to the analysis. These inputs can include:

- A list of hazards
- Device Tree
- A table of risks scores and the RPN system that will be used
- Define a methodology to identify potential causes
- Define a safe state

15.6.1 Pick a Hazard, Any Hazard

Such analyses require inputs and yield outputs. One input is the identification of a hazard. While it is unlikely to identify every single possible hazard or every possible error condition that may arise, it is the expectation that, through a careful and systematic evaluation of the types of hazards identified in ISO 14971 and 21 CFR 820.30(g), and in conjunction with other techniques (e.g. FMEA), it will be possible to achieve an acceptably safe system. One example of a list of hazards is given in appendix C from prEN 1441. Thus, an input to the HA includes a list of the types of hazards to be considered. Hazards that may affect the health and safety of patients, staff and visitors are to be considered. In addition, there may be multiple effects of a hazard. The accident caused by interacting with the hazard should be considered. For each hazard, a score should be assigned appropriate to its severity and probability. Some hazards may seem too simple, but consideration of these creates a wealth of content. For example, consider:

- Deliver (unwanted/any) dose to healthy tissue
- Underdose the target
- Mechanical harm to patient
- Electrical harm to patient
- Other harm to patient

15.6.2 Identify a Subsystem

Another input can be a list of the 'things' that may create the hazard under consideration. One of the subjective elements in the performance of an HA is the selection of which items to include in this analysis including the list of the components of the system (including human or equipment steps) that are to be analyzed. It should be reemphasized that the analyses being discussed can be applied to the entire radiotherapy clinical workflow. Just one element of that is the equipment-related hazards. Thus a 'component' can be a step of the workflow. However, the equipment-related hazards are the primary topic of this book, so it may be helpful to highlight these.

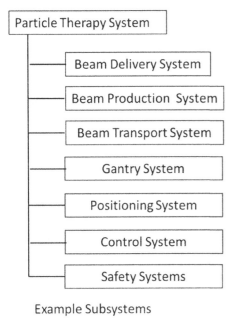

Example Subsystems

FIGURE 15.7 Example of PT subsystems.

Figure 15.7 shows a breakdown, at a high level, of the subsystems that comprise a typical particle therapy system (PTS). When considering a particular hazard, these subsystems should be included. The trick is to identify the 'requirements' of each of these subsystems, or what they're supposed to do, to enable a breakdown of the possible errors that can cause deviations from the requirements that may result from these subsystems. For example, a gantry structure (not including the beamline, which is part of the beam transport system) affects the beam angle but not the beam energy, so hazards associated with the beam range in the patient may not be affected by the gantry. On the other hand, a gantry's motion may also affect alignment of the beam transport elements (sort of an interface), which may, in turn, affect the beam properties such as beam position; thus it can affect more than beam angle. Does this sound familiar? This has been the purpose of the "What can go wrong?" sections. The success of an HA will partly be determined by how well the subsystem requirements and/or functional decomposition (decomposing the functions of that subsystem) are highlighted. It may be found that a requirement needed to avoid or mitigate a hazard, hasn't been implemented. Details of the implementation may not be needed at this level.

This is perhaps a good use of a use case representation, given all the 'actors' involved to analyze this situation. It is a way to break down the steps and identify how different users interact with them, although the results are analogous with an equipment decomposition, it goes a step further and allows various contributors to the analysis to provide input of their own personal expertise. Starting with each actor, at the top, the steps followed by each actor are listed in the actor's column and the interactions between columns are indicated. Once the subsystems and steps and actors and their respective responsibilities and requirements have been accumulated, the cause of the particular hazard being identified may be identified.

15.6.3 Hazard Table

A helpful tool or visual representation to aid in the analysis of hazards is a table. The table helps to ensure that relevant information is captured and the appropriate questions are asked (cells in the table shouldn't be blank). The table does not ensure that all the causes are found, but asking the right questions and having empty cells to fill in, goes a long way toward that goal.

Table 15.6 lists one potential accident associated with a radiation hazard and lists possible causes from previous chapters. The causes may come from a root cause analysis (RCA). Item 1 in the second column is a hazard – the dosimetry system may not be in a state to read dose properly. It doesn't say why or what is the root cause. That is a possible next step. However, sometimes a deep dive to root causes isn't appropriate for a HA. The mitigation can also be at a high level. Item 2 is one step deeper, identifying a readout as a cause of 1, but still at the high level and not going further than that. Mitigations are included as well as an indication of the type of professional or subsystem component (an 'actor' that plays a part) that would be associated with that cause. The subsystems/actors are: (A) dosimetry, (B) accelerator, (C) therapist/physicist and (D) SW.

In addition the risk category approach is used (arising from the RPN divided into categories) to help with identifying the mitigation(s). This table is meant to identify a thought process, not provide all the specific detailed solutions. But it does provide a list of safety decisions that should be followed through in the design.

TABLE 15.6
Irradiation Hazard Analysis Example

Hazard/Accident Description	Possible Causal Factors	Mitigations	Risk Cat	Sub System
Radiation Hazard: Over-irradiation of the entire treatment field	1. The dosimetry system may not be in a state to properly read the dose	Provide a redundant dosimetry system, and/or further analyze the reasons in the FMEA.	3	A
	2. The dosimetry readout may fail to stop irradiation when the dose is reached	Provide a redundant dosimetry system, and/or further analyze the reasons in an FMEA.	3	A
	3. The accelerator may fail to interrupt the beam when instructed to do so	Provide redundant beam shutoffs and provide redundant signals to shut off the beam, and/or further analyze the reasons in an FMEA	2	B
	4. An excessively high irradiation dose is requested by a human error, treatment planning error, TCS error or data transmission error	Introduce strict dose entry procedures. Validate dose entry values, and/or further analyze the reasons in an FMEA	2	C
	5. A patient may be treated while the system is not in treatment mode	Introduce strict system mode procedures. Provide redundant mode checking and/or further analyze the reasons in an FMEA	3	C,D

15.6.4 RADIOTHERAPY RADIATION HAZARD AND RISK EXAMPLES

The following factors arising from the goals of radiotherapy or the highest level requirements in the delivery of radiation are important:

1. The absolute overall dose
2. The relative distribution of a dose
3. The absolute position of the dose

From the point of view of radiation hazards, there are two situations that can arise from any step of the clinical workflow:

1. Absolute Dose
   a. Overdose
   b. Underdose
2. Distribution
   a. Too much dose to the target
   b. Too little dose to the target
   c. Too much dose to healthy tissue
   d. Too little dose to healthy tissue
3. Position
   a. Inside the target
   b. Outside the target

One can consider the effects of this in Table 15.7.

Although this table is no more complicated than the one described earlier choosing between religions, it is still very useful. If the dose outside the target is lower than expected, then the normal tissue receives a lower dose and the severity of the accident arising from this hazard is low (assuming it's not associated with other errors). If the dose inside the target is lower than expected, then

TABLE 15.7
Risk Category

	Overdose	Under Dose
Inside	Hazard severity: low	Hazard severity: medium
Outside	Hazard severity: high	Hazard severity: low

there is a risk of a failure to cure, which could be either a high or medium hazard, depending both on the treatment situation and on the subjectively on the individual doing the analysis. If the dose is too high inside the target, then the main issue may be voiding a protocol, but there may or may not be a big hazard. There may also be other combinations. Not everyone will agree with the assignments in this paragraph. Therein lays some subjectivity in the analysis and results of a risk analysis. But it's important to have a policy and evidence that the policy is enforced. In particle therapy, an incorrect range (e.g. too deep) with an SOBP field can result in an overdose outside the target and a partial underdose in the target. To address the hazards, it is important for the analyzer to determine whether these all have the same severity, or if they are different.

A lack of knowledge could result in a worst case analysis. For example, an incorrect contour could result in the incorrect dose applied, but if the TP and delivery are correct relative to this contour, this type of error might not be determined. It may be likely that it wouldn't be caught early and if full data weren't available, it would not be possible to know what error resulted and the worst case assumption would have to be used. Similarly errors in treatment planning algorithms, imaging simulations and beam delivery could all result in a radiation accident. The situation would change however, if methods of detecting these types of issues were in place.

There could be a certain amount of subjectivity to this analysis. Again, it's useful to have a policy and evidence of its implementation. Ultimately, it's important to perform the analysis including the following:

- Identify the failures that can lead to the hazard and consider the resulting accidents
- Identify the probability of the specific hazard
- Determine the overall risk and from that the mitigation; e.g. frequency of the quality check

Those safety items of highest concern are addressed with safety mitigations. Those issues of treatment precision may possibly be treated at a lower level of concern depending upon the tolerances identified. The remediation strategies are categorized as to their relevance to the various subsystems.

15.7   FAILURE MODE AND EFFECTS ANALYSIS (FMEA)

The goal of an FMEA (e.g. Health Devices 2004; 233) is to identify failure modes that may result in hazards and prioritize them for possible safety improvements. This method points in the bottoms-up direction as highlighted in Figure 15.8. Whereas the HA starts with a generic hazard and asks the analyzer to imagine the types of causes, the FMEA examines, step-by-step, each action of the process, or component of the equipment, and asks the analyzer using inductive logic to imagine what can fail in the step or how a component can fail and from that failure what effects may lead to hazards. The HA and FMEA are sometimes confused. Both are valuable and complementary. Both should normally be done.

In its essence, the FMEA is simple. Briefly the following steps are to be performed:

FIGURE 15.8   Bottoms up.

1. Identify a failure mode from a component or a step (input)
2. Identify the effect of that failure (output)
3. Identify the hazard that can result from this effect (this can be a multistep analysis) (output)

3. Score the importance of that hazard (output)
4. Create a mitigation (output)
5. Identify how that mitigation will be verified (output)
6. Optional – Rescore after each mitigation to determine if further mitigations are needed

The first step mentioned is to identify a list of 'failure modes' – things that can go wrong. In clinical radiation therapy operations, this can be anything from a pure technical failure (e.g. wrong beam energy delivered) to a process-related planning error (e.g. wrong CT scan used for planning) to a mistake in clinical care (e.g. wrong delivered dose due to a miscommunication about prescription). Once the possible failure modes have been identified, the FMEA formalism provides a method for ranking them in priority order. This approach recognizes that time and resources are limited and that it is best to prioritize safety improvement interventions to target the most important hazards. For example, it may not make sense to prioritize a possible error that may occur only once in 10 years even if that error might be moderately serious. Perhaps a higher priority should be given to a failure that occurs once a month even if that failure has less serious consequences. An interesting thought provoking question is when is it necessary to install a traffic signal at an intersection? This is a choice that ought to be made consciously, possibly using the RPN. What could become a semi-random list of failure modes, can be more systematically identified by noting the steps in the treatment workflow and realizing that there can be an error, as noted above in the each of these steps. An FMEA provides a semiquantitative method to balance these factors.

Once the list of failure modes has been developed, some of the outputs include the identification of effects and the hazards arising from these effects. These hazards are then scored in the same way as in an HA. These hazards arising from the failures become the link between the FMEA and the HA. Thus top-down and bottom-up meet and hopefully, fill in any gaps.

As discussed earlier, various scoring scales have been proposed. The AAPM TG-100 report also has one. It is helpful to use what makes sense for the specific environment and not adopt a scale from another place arbitrarily, even one considered a standard if it doesn't make sense for your situation. The actual scale used is probably less important than the need for consistency in the relative scores among different faults using the FMEA tool for different activities, since after all; these scores are only used to generate a relative ranking. A larger range of the scores can enable a finer tuning for determining action priorities.

15.7.1 FMEA Example

The basic approach is the same when applied to either particle radiotherapy equipment or processes. An example of possible failure modes relevant to particle radiotherapy is presented in Table 15.8. Consider a beam delivery subsystem with an ionization chamber, which requires a high-voltage (HV) power supply. The IC can be considered a component. Going one step further, this power supply can also be considered to be a component. One can, of course, go further and consider a particular resistor in that power supply as a component. However, through an understanding of the system functionality, one can identify various failure modes of the power supply without having to look at the failures of

---

TABLE 15.8
Equipment Partial FMEA Table

	Component Failure	Failure Mode	Effects	Hazard	S	P	D	RPN
1	IC power supply failure	Wrong current	It shouldn't draw much current and may invalidate dose.	Unknown	9	2	7	126
2		Wrong voltage	Wrong gain or recombination.	Overdose	9	3	5	135
3		No voltage	Reading zero dose	Overdose	9	2	5	90

each resistor and capacitor. It's also the case that there is little externally that interacts directly with that power supply. Or one can consider the power supply as the component and it either fails or doesn't, not considering what's inside. In more complex systems, further analysis may be necessary, but it's helpful to limit this when possible. In the case of a power supply, there can be a finite number of failures from the perspective of the output of the device, such as wrong voltage or current. In fact, what is being done in this instance is to identify the power supply functions and requirements and a cutoff for the analysis based partially on maintenance and spare parts considerations. It's unlikely to decide to implement safety mitigations at the resistor level but rather detect the error at the output voltage level. That is not to say that the power supply manufacturer might elect to design the power supply with sufficient redundancy so as not to cause a problem if one particular resistor goes bad or to monitor the value of that circuit. (My team has replaced plenty of resistors.) But the radiotherapy department or even equipment vendor is not necessarily the power supply designer and manufacturer.

One example of a wrong output voltage is that it goes to zero. The following *thought process* can then be followed:

- Component: HV on an ionization chamber
- Failure Mode: HV turns off
- Effect: MUs will not be counted
- Hazard: Overdose
- Severity: High
- Probability: Medium
- Detectability: High (if one implements methods for this detection)
- Result: Mitigation required for detectability

It may be instructive to examine some more detail as to the thought process that can go into this situation. Mitigation is not separate from an HA or FMEA. In practice, they should be dealt with simultaneously. The diagram in Figure 15.9 is an example of possible considerations resulting from this seemingly simple failure. In this case, there are two ICs (IC2 and IC3) and HW and SW are used. IS means the ion source of the accelerator. NBD means a 'no-beam-detect' circuit. Among the considerations embedded in this diagrammatic thought tree are the following:

- Can the failure be detected by HW?
- Can the failure be detected by SW?
- What types of redundancy can be implemented to identify that the failure has occurred (e.g. a second ionization chamber)

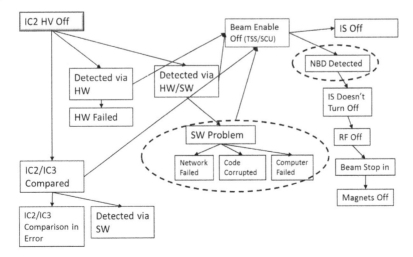

FIGURE 15.9 Fault mitigation thought diagram example.

- What if the detection doesn't work?
- What failure modes are possible if the SW detection method is employed?
- What are the actions to take if this situation is detected?
- What detection is necessary to ensure that the desired action(s) take place?
- What further action is needed if the desired action(s) do not take place?

Okay, a 'thought diagram' may not be found in the literature, but that doesn't mean it couldn't be done, if useful. This is the thought process used by the author and is perhaps most compatible with an 'event tree' type of analysis. It is not a fault tree because it doesn't start with a hazard and work down to the casual factor and it isn't an FMEA because it doesn't factor in the consequence of the failure (the effects). It looks at an error and follows the path of the mitigations, catching the error and reacting to it, but it is not binary and boolean. This thought diagram is an example showing that one can analyze the situation the way one thinks about the problem. It's not always the case that a particular analysis can be made to fit into a standard format. For example, an event tree from this diagram shown in Figure 15.10 does not include the 'diagnostics' boxes that were in Figure 15.9 (e.g. network failed), which could lead to more detailed branches of why a SW error would occur. The event tree follows more of a sequential path rather than multiple parallel paths, which is okay for a simpler probability analysis. It is sometimes the case that a more standard format will help the analysis become clearer and complete. (E.g. the event tree includes some probability estimate of each path.) It would take multiple event trees to analyze the thought diagram. This is a very important consideration when deciding upon the tools for the analysis. Exactly how far one goes in this analysis depends in large part on the risk and on the system complexity. See how this event tree sequential approach starts with detection (first three columns) and ends with mitigation errors (last four columns).

Recognizing that the tables offered are not a replacement for thinking or the thought analysis and interdisciplinary discussion, sometimes the tables can be used as an aid in the thought process. The following table headings represent an evolution over a number of years of working with equipment for PTSs. Table 15.9(A) is actually the left side of the table and Table 15.9(B) is the right side of one full table (which doesn't fit as one in portrait orientation).

It is helpful to see that this table goes in the direction of the considerations of the 'thought diagram'. At the left, it presents the opportunity to drill down to the desired component level and identifies the failure mode and the effect. This effect is further deconstructed to specific issues related to beam properties and, in turn, specific issues associated with treatment hazards. Up to this point,

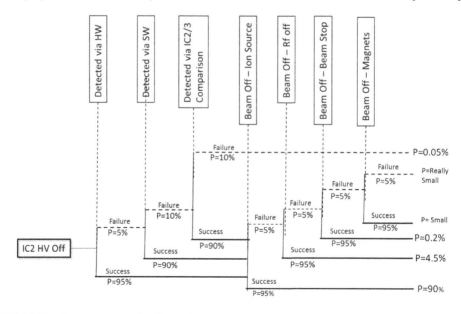

FIGURE 15.10   Event tree analysis of same issue (from thought diagram).

TABLE 15.9

FMEA Example Table

FMEA Example Table Part A (Left Side)

System	SubSystem	SubSubSystem	Failure Mode	Failure Effect	Effect on Beam Properties			Treatment Radiation Hazard		Mechanical Hazard	Equipment Hazard	RPN Analysis			
					Position	Size	Range	Target Dose	Outside Target			Probability	Severity	Category (RPN)	Redundancy Level
Scan Mags	Power Supply	No Volt		Wrong Position							No danger to Patient				

FMEA Example Table Part B (Right Side)

Mitigation A1	Mitigation A2	Mitigation B1	Mitigation B2	Mitigation B3	Even More Mitigations	Mitigation F1 (multi)	Mitigation S1 (multi)	Irradiation State				Machine State				
								Nominal	Pause	Abort	End	Injector On/Off	Accelerator Rf On/Off	Beam Stops In/Out	Beamline Magnets	Scanning Magnets
Equipment Prevention (HUMAN)	*Protocol*	*Hardware Detection*	*Hardware Detection*	*Hardware Detection*	*??*	*Functional Detections*	*Software*									

the system/subsystem columns can be replaced with steps/sub-steps of a process and the beam property columns can be eliminated for non-beam-generating-equipment. (Note that effects on the equipment – e.g. damage to the equipment – is included. If the equipment is damaged, it results in lost treatment, which in itself is a hazard.) Next the RPN analysis is accomplished with mitigation considerations. The RPN results in a level of criticality and a column exists for the level of redundancy required based upon that criticality.

Finally, in addition to mitigations spelled out, the desired irradiation state that the system should be in, if that failure occurs, is identified, as well as the conditions of critical elements of the system to ensure that the system state is appropriately safe. Note that there are several approaches from here. One can use the level of criticality to determine the mitigation levels needed, or one can iterate, as noted earlier, and keep recalculating the RPN after the introduction of each mitigation. Such strategies need to be addressed upfront. Note also that this table does not call out detectability or avoidability explicitly, it is assumed to be in the probability.

15.7.2 FMEA Limitations

Having covered the basics of the FMEA it may be useful to briefly discuss some of the limitations. The first is practical and has to do with effort and usability. An FMEA is sometimes thought to be so labor-intensive as to be almost undoable and/or unusable. This is largely a myth, however, which is the outgrowth of early experience. The original FMEA exercise of Ford et al. required five months of weekly meetings with an 11-member steering committee. However, that study considered the entire radiation oncology workflow, which is very ambitious, and was conducted at a time (2007) when FMEA was essentially unknown in radiation oncology. The learning curve is considerably less steep now. A study by the same group a few years later demonstrated that an FMEA for the full radiation oncology process could be conducted in four one-hour session and showed a measureable risk reduction. A number of other studies have appeared in the radiation oncology literature demonstrating the usability of FMEA. The view of the FMEA as a difficult tool may also be related to AAPM task group 100 on FMEA, which took something like ten years to publish and, when complete, was one of the longest task group reports published. It must be appreciated, though, that a substantial effort of this committee to develop a comprehensive report does not necessarily translate into a tool that is difficult to use. On the other hand, the tools suggested in this chapter are time tested and proven helpful. With the right understanding, preparation and guidance the FMEA has been found to be usable. The lack of understanding of the difference between an HA and an FMEA can add complexity to the process. The HA is a real help in getting one's mind into the process and the FMEA then follows up with details. This author has worked on a score of FMEAs for particle therapy equipment, none of which took much more than the order of a month of integrated time.

Other limitations of FMEA are more fundamental in nature. The first, and most obvious, has to do with the scoring system. S, P and D scores are subjective and variable among people. This translates into an uncertainty in RPN-based ranking of failure modes. As such, the FMEA rankings basically have error bars. The rankings are best suited for identifying the most obvious risk points, the outliers in terms of RPN and for internal comparisons. At an even more fundamental level, the concept of prioritizing risks by a simple multiplicative product of S, P and D is probably oversimplistic. However, it is not a bad approach either. The analyzer group can modify the approach with good cause if needed.

15.8 MITIGATION OF RISK

A key output of the analyses discussed involves the mitigation of potential risks. Some of this has already been mentioned along with risk. The RPNs have been estimated and calculated and what remains is to identify how to reduce the importance of any given risk. Although, mitigation is arguably at least as important as the identification of risk, relatively little literature is dedicated to it, especially for particle therapy. This is certainly true in the radiation oncology literature but is also apparent in the patient safety literature as a whole.

One way to evaluate the utility of mitigation for a particular failure or hazard is to estimate the degree to which this mitigation reduced the probability or severity of the hazard and continue to introduce mitigations until the hazard category is reduced to an acceptable level. While it is easy to see how mitigations may reduce the probability (e.g. through detectability or avoidability), almost by definition, it cannot affect the severity. If an unbreakable barrier is installed in the way of something that may fall, this will very probably prevent the accident unless the thing that may fall is an irresistible force. The hazard is still severe but highly improbable (as long as the unbreakable barrier stays unbreakable).

Another method is to define a high-level strategy of mitigation based upon the category of the hazard. For example, a category 4 from Table 15.3 risk may invoke the introduction of redundant HW sensors or a second look by another individual or the imposition of a non-SW means of mitigation or redundant mechanical supports, whereas a lower category risk may allow for SW mitigation methods or human check alone.

John Grout identifies three ways to address risk or to practice 'mistake proofing'.

1. Make the risk highly unlikely – eliminate the risk
2. Make the error visible – provide the appropriate detection
3. Create a fail-safe design (self-detection and/or self-reacting – e.g. normally open relays)

Unlikely risk (1) corresponds to reducing the probability of occurrence, P, of a failure mode. Making them more visible (2) relates to, D, detectability, and fail-safe design (3) can reduce the probability of a particular event happening to near zero (assuming that the probability of the fail-safe design being fail-safe is near 100%), or it is just a mitigation to reduce the risk.

Often, the categorization of mitigation strategies is reactive instead of proactive. A hazard is uncovered and work is done to make the error visible and mitigate it. Sometimes, a hazard is an inappropriate use of the system, which was not accounted for in the design. If the approved procedures are followed, that hazard may never be observed. One strategy to address these situations is being proactive. It's possible to divide the proactiveness into that of being proactive with an existing system and being proactive during the design phase.

Being proactive with an existing system, to prevent the inappropriate use of the system, can take the form of developing protocols, providing training and ensuring communication. This is part of a proactive quality management program and includes elements of raising a culture of safety and internal review that can lead toward effective mitigation strategies.

Proactive mitigation strategies with an existing system can include:

- Proactive mitigations through training: This category attempts to grow expertise and experience in the staff that will contribute to the improved operation of a system.
- Proactive mitigations through protocol and communication: This is a category that attempts to prevent an accident through human-based information and procedures.
- Proactive mitigations through QA: This is a category that attempts to identify system problems through regular measurements. These measurements occur at various time intervals. The reaction to a finding must be determined and implemented in design and/or protocol.

Proactive mitigation strategies during the design phase can include:

- Mitigation through equipment safety design: This is a category that attempts to either eliminate, or if that is not possible, to minimize the hazard by specific design implementation, or measurement capability together with methods to end the source of the hazard.
- Mitigation through training of the design personal to optimize errors and clarify philosophy.
- Mitigation through communication and protocol.

15.8.1 TRAINING

Training with Respect to an Existing System

The IAEA has a long tradition in collecting information about radiation accidents and analyzing the backgrounds. In a safety report on accidental exposures in radiotherapy (e.g. Safety Report Series 17), it was demonstrated that one of the most frequent source of accidents is connected to the personnel in an institution, namely, a lack of education in radiotherapy physics, a lack of procedures, lack of supervision and communication and training of personnel. This includes proper training by the institutions as well as by the vendor. The institution has to take care that all members of a team receive training in different areas:

- Regular training in internal regulations: These include typically all safety regulations in a facility, as well as radiation protection regulations and regulations about data privacy and include also all existing standard operating procedures (SOPs).
- Regular training on the dedicated systems used for radiotherapy: This may include the vendor but often is done by the institution internally. It includes not only basic instructions on how to work with new system but also information about changes or updates of a system. Also, the users should be informed about the intended use of the products. The train the trainer approach, if kept up too long, or passed down through too many generations, can result in a diluted understanding of the system with a tendency to emphasize shortcuts.
- General continuing training on the job: This includes all kind of teaching activities like teaching courses on different subjects, regular seminars or advanced courses for specialization.

Proper training on a system includes hands-on training and appropriate, understandable documentation material. Such documents are living documents that can be updated if a risk analysis team comes to some new conclusions. Documentation of the participation of the personnel is important for the mitigation measures to have been realized successfully. Too many times, there are excuses for staff not reading documentation. This is on the shoulders of the supervisors.

Training During System Design

To most users and even members of a design team, elements of the design of a system could be a black box. A group of people are working toward the design with little visibility externally. This is not necessarily all bad. However, in a PTS, there are many inputs and there are several subsystems. No one team works on all of these, and yet one of the most important aspects of these systems is that it interfaces both internal and external. Staff must be sufficiently trained in what others are doing and assuming so that any mismatches are minimized.

15.8.2 COMMUNICATION

Communication with Respect to an Existing System

Communication within an institution is crucial. It is important for the personnel to know the safety and quality philosophy of an institution, the background of certain procedures, SOPs or regulations and to be informed about the actual status of the system and the activities around it. Communication should be more than a one-way lecture. All relevant information should reach the personnel who need to know. A robust safety culture facilitates routine reporting of possible hazards and other issues, which eases the desired communication flow, without recrimination. However, formal communications often stifle interaction. It should be possible to instill a culture of informal discussion also.

The vendor of a system may be informed of an accident in another institution and come to the conclusion that a product warning has to be issued. There would be a notification by the vendor to be followed by all users of this system to mitigate the risk of repeating this mistake. Such a product warning is typically sent to a single contact person in an institution (e.g. the head of medical physics, or to a regulatory person). This contact person has to ensure proper distribution of this information to all involved groups in an institution and may also start an internal discussion to get feedback, if changes in the clinical workflow or additions to the internal regulations have to be made. To secure the proper flow of information, it is useful to have a communications path for various groups, involved in different tasks, where the relevant information is distributed. Meetings as needed can be helpful, including informal discussions. It is critical that such information should not give rise to an overreaction. If that happens, the training is insufficient and the safety culture is flawed. Sometimes, the communication will lapse if overreaction is the main reaction.

Communication During System Design

Training may already have occurred but as is typical during a design process, there is an evolution of thought, technique and implementation. It is important that this information is shared, through various forms of communication among the design teams and sometimes the end user in case there are operational ramifications. One example that happened is that one group changed a motor but didn't communicate with the group that was tasked to control it. While it wasn't totally incompatible, eventually weaknesses were exposed. Implied in all this is that the clinical specifications are being followed and explicit analyses are done to ensure that no changes affect the required clinical parameters.

15.8.3 Standard Operating Procedures

To err is human (as well as technical components) and consequently they may introduce risks at any stage of the treatment process. However, it is not necessarily appropriate to offload all responsibility to a machine (a risk analysis would have to be done). One way to limit this human risk is to define SOPs for the most important processes. These may include SOPs for QA and quality control (QC) procedures but also parts of the clinical workflow. One example of a very critical QC procedure is the daily monitor unit calibration, since it directly affects the dose applied to all patients on that day. It usually involves the setup of a phantom, a dose measurement and some kind of analysis. The correct performance of any of these steps requires some kind of expert knowledge and there are numerous ways to introduce errors, many of them too small to be directly visible. Problems arise typically, when a new person joins the team. Sometimes that may be done by medical physics, other times it's done by radiotherapists. To make every detail of the measurement very clear without ambiguity, an SOP is very useful. Such an SOP should always be considered a working document, i.e. it is necessary to collect the feedback of the team working with the SOP and add new aspects or comments to continually improve the SOP. An excellent source for improvements of SOPs are the comments from new people being trained on the system. SOPs should also include appropriate reactions to measurement discrepancies.

It may be daunting to think about writing SOPs, but as soon as they exist, their value becomes immediately clear:

- It makes the procedure more straightforward to do (less error prone)
- It may allow involving more people from a team into procedures that were hitherto thought to be an expert's matter (better use of resources)
- It's invaluable to all new members of a team (shorter learning curve)
- It's important in a risk analysis to have a clear description of a process

Only when a standard is defined can different procedures be analyzed to improve the process. It should also be noted that SOPs are often the result of a risk analysis, since the process helps identify the issues that could contribute to the analysis. It is important, however, not to place all responsibility on the use of SOPs and other regulations. Otherwise people in a team may feel like not carrying any responsibility for their work any longer and not having to think about the process they are performing. This would be counterproductive.

15.8.4 Equipment Design

Design with Respect to an Existing System

Sometimes, a hazard is uncovered for which the only viable mitigation is the redesign of a portion of the system. It could be SW or HW. This is a real possibility and it should not be avoided. Not all staff in a facility do or could understand all of the detailed safety mitigations in a radiotherapy system. However, some limited group of the staff could have some of this knowledge and could be available for questions and training. An analysis will be different for each radiotherapy department and type of equipment. This staff, from the facility, can work with the vendor to identify the best modification and associated testing program.

Design Considerations During Initial System Design

Basically there are two tasks that are relevant. A device in the equipment is set and the value to which the device was set is measured, with a specified accuracy and frequency. Depending upon the level of potential risk of incorrectly setting this device, there may be additional redundant ways to measure the setting of this device. In addition, one can also require a 'functional Redundancy', or a redundant measurement, to ensure that the function desired has been performed. For example, if the function is to deliver the appropriate, accurate dose, a dosimetry monitor chamber would be required, which is independent of the ion source settings. One could, for example, argue that knowing the beam energy and the charge delivered from the accelerator, by the settings that have been applied and measured, is sufficient information. One can, in fact, argue that point possibly successfully; however, the standards require redundant monitor chambers. Taking the above into account, a design implementation can look like the components in Figure 15.11. It shows what must be set to create a certain function and what can be measured to verify

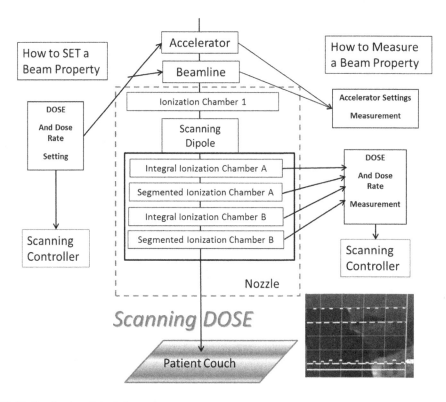

FIGURE 15.11   Set it and don't forget it.

that both the settings are correct and that the function is correctly executed. One of the parameters that determines the dose is the beam current, and components that are related to that include the accelerator and the energy selection system (if the latter exists). These must be set for the right current and must be turned off when the dose is reached, or controlled during the dose delivery. The settings, like voltage and current, are monitored, perhaps redundantly. But to ensure everything is working, the ionization chambers are a functional redundancy since they do not depend on the accelerator settings. They are an independent measure of the beam dose rate and the total dose. Counting the number of redundancies here is an exercise for the reader.

15.9 QUALITY ASSURANCE

QA is part of the clinical workflow and is an important proactive part of ensuring that the particle therapy technology can deliver a safe treatment. The implementation of a QMS includes QA of various processes. As an aside, a QMS is an incredibly valuable set of documentation, but it is avoided at almost all costs in a healthcare environment. What is not avoided is the QA portion of the QMS. Perhaps the most detailed QA is the measurement of the beam and verification of beam measurement devices. All of these aspects of QA ought to be identified in an SOP. Sometimes, without a QMS, it is confusing to determine who is responsible for what or what function should be analyzed. It is helpful, for example, if QA is incorporated into the QMS and that the responsible individuals ensure that the QA checks are implemented throughout the entire workflow. More generally, QA is defined as the set of all planned and systematic actions in a QMS, which is necessary to provide confidence that a product or service will satisfy given requirements. Quality is often limited to measureable performance parameters, like dose accuracy. However, any check that any part of the process is correct could be considered part of QA. For example, a check that a patient's chart has the correct data is QA and verification of a patient specific HW such as a collimator is QA. Having said that, but recognizing that the title of this book is Particle Therapy Technology for Safe Treatment, the focus goes back to technology and beam performance.

Not every aspect of the beam specification and delivery process is possible to verify during the treatment process. Therefore, a clear strategy of measurements with the appropriate frequency is essential. One possibility is to select from the QA options identified by consensus reports such as AAPM TG 142 or the latest 224; however, even such reports indicate that every radiotherapy facility may be a little different and the staff should identify the appropriate procedures to follow for their facility.

15.9.1 BEAM QUALITY ASSURANCE

QA protocols related to beam measurements include the procedures necessary to provide confidence that a radiotherapy machine can produce the required dosimetric parameters for patient treatment. It should also include verification of safety mitigations. Various forms of beam QA includes the acceptance testing of a system, the clinical commissioning and finally an ongoing and living QC process including machine and clinical QA. It may be helpful to remember that QA (especially with respect to safety) is really an error mitigation scenario. There are some possible failure modes of a system and QA is one method to detect and react to these possible errors.

Two key branches of QA can be defined; one is machine QA and the other is clinical QA. These are illustrated in Figure 15.12.

In the case of machine QA (lower left to lower right boxes), it is necessary to 'assure' that the PTS machine produces the beam parameters that a treatment planning system (TPS) uses to determine the dose distribution, including whether the parameters from the TPS were properly transferred to the PTS. If, for example, the TPS requires a particle beam with a range of 20 cm, does the machine produce that particle beam with a range of 20 cm (within acceptable margins)? In the case of clinical

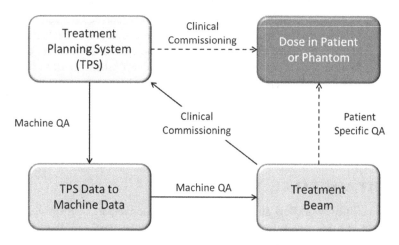

FIGURE 15.12 Aspect of QA.

QA (upper part of figure), it is necessary to assure that, using the beam parameters input in the database of the TPS (measured from the machine originally during clinical commissioning); the dose distribution predicted by the TP is indeed that which will be delivered in the chosen anatomy (assuming the machine produces the correct beam). In other words, are the algorithms and machine data embedded in the TPS correct for a particular case or class of treatment sites? This part does not ask the question of whether the beam produced by the machine is correct, or if the parameters have been transferred correctly. Getting the correct machine beam parameters into the TPS is sometimes called 'Clinical Commissioning' (arrow going from lower right to upper left and upper left to upper right), not to be confused with 'Machine Commissioning', which works toward obtaining the desired machine beam parameters (and not shown in the figure). Both should be considered to be part of a more general commissioning process, which aims at providing all necessary procedures, protocols, instructions and data to start clinical service (which includes the development of QA procedures). The commissioning process could be considered to be the ultimate QA, but routine QA is used only to spot-check some of the commissioning results and look at specific situations. Some beam arrangements, beam ranges and/or scanning patterns may be different than the 'usual' ones for a specific patient and patient-specific QA will have to be done to validate that these can be delivered. Actually this is done most of the time, partly owing to skepticism that the TPS data may not be properly transferred to the PTS. There are efforts to reduce this and to use automated machine measurements to validate that, called log file-based QA. (See, for example, the work of the PTCOG treatment efficiency subcommittee.)

There are many things that might be considered for measurement as part of the QA process. One extreme is the commissioning process (one which can take weeks or months), which cannot be done for every patient treated. Therefore, some method to minimize the routine measurements actually needed needs to be applied. Application of the analysis techniques discussed in this chapter can be used to help determine the necessity and frequency of the measurements on the basis of the risks presented. It is reasonable to measure some components of the equipment to ensure that they are operating correctly, being set correctly and are stable. However, medical physics is most comfortable with assuring that the beam parameters delivered by all the components comprising the equipment is what was requested. These are the dosimetric beam quantities. It is possibly helpful, once again in this book, to separate the dose distribution into three categories:

1. The absolute dose
2. The relative distribution of a dose
3. The absolute position of the dose distribution

In this way, it is easier to separate out those parts of the system that must be verified as part of the dosimetric QA process.

Following from the above logic, machine QA can be defined as the act of checking that the system produces the same parameters as when it was measured for clinical commissioning, and patient QA can be defined as the act of checking that the TP gives the correct output for a specific patient plan. Of course, if the TP implementation is perfect, the two are the same. Patient-specific QA ought to be done when it may be that a specific patient plan could introduce an aspect of the planning that may not have been previously exercised (such as anatomical heterogeneities or artifacts), or it involves a combination of machine parameters, which may be critical. When applied to the machine, the QA verifies that the machine produces the beam parameters and/or the dose distribution that the TP predicted. Determining if the TP predicted the appropriate plan is patient-specific QA. Additionally, do not forget the verification that mitigations are operational, and check that error detection works.

Using the tools and steps that have been discussed in this chapter could be helpful. The subsystems could be identified. The key functions could be listed. In this case, at the highest level, they would be the clinical and machine parameters. Creating an association between the clinical and machine parameters would then give options about what verifications could be done. Some of this follows.

15.9.2 CLINICAL AND MACHINE PARAMETERS

To begin the analysis of what measurements should be done and how often they should be done, it is helpful to collect information that will expedite the work. At the highest level, the tolerances associated with the dose distribution parameters need to be analyzed and the consequences of deviating from them need to be understood (see Chapters 12, 16 and 17). With respect to beam QA, a parameter (such as beam range) can have a tolerance associated (such as ±0.5 mm), but in other aspects of clinical workflow, such as using a CT image, there may only a binary right or wrong image. (Of course, analyzing that image includes converting the CT densities obtained to proton beam stopping power that has in it inherent uncertainties.)

Figure 15.13 brings back the now familiar dosimetric regions of interest for the two key beam delivery methods: double scattering (left) and scanning (right). This is essentially the second item in

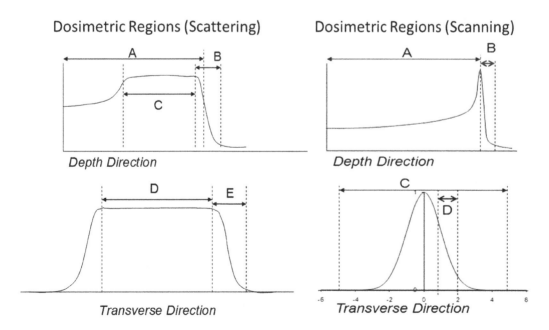

FIGURE 15.13 Dosimetric quantities for scattered and scanned beams.

the dose distribution category list identified in the previous section. There are other beam spreading variations and this highlights the need, for each delivery method, to identify the crucial parameters that characterize the dose distribution.

In the figures for scattering, the following parameters have been identified:

1. The range of the spread out Bragg peak
2. The distal falloff or distal penumbra
3. The width of the spread out Bragg peak
4. The transverse uniformity
5. The lateral penumbra

In the case of scanning, the beam is unmodified and, therefore, the relevant beam parameters include:

1. The range of the pristine Bragg peak
2. The distal falloff or distal penumbra
3. The centroid position of the beam and its stability
4. The lateral penumbra, which is related to the beam size and shape

Another useful quantity for treatment delivery is the relationship between the entrance dose and the dose monitor readout (sometimes referred to as monitor units or proton charge delivered). (This was discussed in Section 11.3.1.2.) The parameters of Figure 15.13 and MU calibration will normally be measured at different locations and different gantry angles if gantries are involved. These parameters are tagged because they can represent the relevant beam parameters created by the beam delivery system and are used by the TPSs to calculate the dose distribution. Also, they represent a further breakdown of the requirements needed to identify the types of potential errors, or fault effects, that may arise in the beam delivery as was used in Table 15.9.

Different institutions or countries may have different definitions of these parameters so they have to be clearly specified. For example, at which position is the range (50% distal to the peak or 90%) defined or the distal falloff measured. In principle, there may also be additional parameters, like the roundness or symmetry of a beam. Appendix D includes these parameters as the key dosimetric beam parameters that may be measured during QA procedures. Measurement of some of these parameters using the functional redundancy mitigation strategy identified earlier is possible. Instruments are used to measure the beam properties created and instruments are used to verify that the other instruments are measuring correctly.

It is the subject of some philosophical debate whether more than a direct beam measurement is needed. It is the opinion of this author that monitoring the equipment used to produce the beam parameters is very important. This can possibly help to predict failures and can be done in real time, possibly in place of some beam measurement QA. There is clearly a difference between measuring clinical beam parameters using external instrumentation (e.g. a water tank) and measuring the status of a piece of equipment that may uniquely determine that beam parameter. While a cable falling in the path of the beam can't be detected by a power supply current stability, the relative probabilities become worth discussing. Determining the relevant measurement is part of the process of designing the QA procedure. Often both measurements may be done consecutively or simultaneously during commissioning to relate them to each other. It is true that in some cases, direct beam measurements of many parameters can be done almost simultaneously (see an example in Section 15.10).

Consider the relationship between clinical and machine parameters in the case of double scattering.

1. The range of the SOBP: In this case, the range is primarily determined by the range of the most distal Bragg peak (although there is an important component of the next closest peak that contributes).

    a.   This range is primarily determined by the beam energy from the accelerator (including the degrader if one is used).

    b.   In addition, the beam energy is modified by materials that the beam traverses including, for example:

       i. Scattering devices

      ii. Range-modulating devices

     iii. Vacuum-to-air transitions

     iv. Instrumentation material

2. The distal falloff: This is determined by the final energy spread in the beam that comprises the following:

    a.   The initial energy spread in the beam from the accelerator.

    b.   The range straggling created when the beam traverses material before entering the target. (See [1.b.] earlier.)

    c.   The range straggling created when the beam penetrates the target.

    d.   The distribution of Bragg peaks in the SOBP

3. The width of the spread-out Bragg peak: The SOBP is generated by combining the appropriate number of Bragg peaks with the correct amplitude by some method.

    a.   The devices used to modulate the Bragg peak such as a range modulator wheel, or binary filters or ridge filters.

4. The transverse uniformity: This results from a variety of devices:

    a.   Scattering material

    b.   Collimator

    c.   Air gap between collimator and target

    d.   Distance between effective source and target

    e.   Beam centering

5. The Lateral Penumbra: This also results from a variety of devices

    a.   Collimator

    b.   Air gap between the collimator and target

    c.   Effective source size

    d.   Multiple scattering in material and the target

The above list is an example. Each system can be so analyzed in more detail. Again some of these topics are addressed in Chapters 16 and 17. It is interesting that in this example for scattering, the only parameters in the original beam (from the accelerator and beam line) that are relevant are the energy and energy spread. That is why this example was chosen instead of scanning. The energy selected by the beam transport system will be determined by the settings of the dipoles. These do not adjust the central beam energy; they only select which energy range transported. Verification of this type of control element or verification of the other elements that actually affect the beam energy are options when considering an error analysis and the QA to be performed to mitigate possible errors. This is normally done for particle therapy accelerator systems. The beam delivery system components affect those other beam parameters mentioned.

Note that in the case of scanning, since the beam is unmodified, the parameters that are to be measured are the same parameters as shown in the right side of Figure 15.13. (Don't forget Figure 4.2 also.) Therefore, measuring the pristine transverse distribution, while not necessary for scattering, is critical for scanning. A key addition to this is the effect of the scanning system including any modification to the beam size and position as the beam is moved across the target. Some scanning systems allow for a variation of beam width, which in combination with variation of energy, may lead to a substantial number of combinations that may have to be checked.

Thus the clinical beam parameters can be related to the beam parameters through the equipment used to spread out the beam. This has been a primary goal of this book. Identification of this equipment can help to start a failure mode analysis.

15.9.3 INSTRUMENT QA

It may be obvious, but still useful, to point out, that some of the measurements identified earlier used instruments to measure some beam properties. The accuracy of the readouts of these instruments and the SW (if any) to analyze these results are components of the system that can fail. Part of the beam QA process is indeed to verify the accurate performance of these instruments. Revisiting the philosophical discussion, it can be argued, either way, that it may be possible to verify these instruments by using the machine setting. For example, a magnetic dipole current can be adjusted to measure that the position readout of an ionization chamber is accurate. This procedure may even be done automatically. However, the alternate side of the discussion is that one should use another instrument, external to the system, to validate the performance of the instrument in the system. It is implied, for some unknown reason, that this instrument cannot be the magnetic dipole. In turn, this external instrument must be sent out to someplace where its performance will be validated. Who will validate the validators? This can also be covered via a risk analysis.

It is important to decide how to verify that the beam is entering the target with the correct parameters. Some will say that only a direct measurement of the beam parameters before the target can definitively tell that the beam is entering the target correctly. However it takes two points to measure a trajectory. Some may desire a second ionization chamber to determine the beam angle. One can argue that downstream ionization chambers that measure the position of the beam before the patient is sufficient; however, this can depend upon the distance of these chambers to the patient. If they are close, then the beam angle will not play a significant role, (assuming the beam angle is not large – think about the distal edge of the target), but if they are far enough away, the angle could be cause for consideration. Introducing a monitor that intercepts the beam will affect the beam profile in a potentially adverse way. Could it be done without those monitors? Consider that the measurement of the current and voltage and magnetic field of a scanning magnet is a triply redundant way to know the effect of that magnet on the input beam trajectory (which can also be measured). The overall system design is important in determining the optimum configuration of instrumentation. Identification of the probability of a hazard and the necessity for mitigation involves these types of considerations.

15.9.4 HOW OFTEN TO MEASURE

From the equipment perspective, it is not enough to identify all the parameters and equipment that can affect the desired beam delivery. It is also useful to ask how often a parameter needs to be checked. This can be determined by analyzing the severity of the failure of this parameter to meet a desired value and the probability that such an event will occur. The RPN (or RAC) number can help. This analysis could be included in an FMEA. All too often, QA procedures transferred from one machine to another follow some standard, which may not be relevant to the specific nature of a given machine. Of course, the sensitivity of the range is clinically the same independent of the machine producing the beam. But different machines may have different equipment and different probabilities of the error, so it is not necessarily the case that the same QA measurement frequency is required. The question of who should identify these unique situations is not addressed here (it should be in the QMS), but the fact that these should be understood to guide the design of a QA process is clear.

Potentially serious failures, such as overdose, are monitored very often independent of the probability of its occurrence. In such a case, it's almost as if the RPN were determined solely by the severity number. In fact, it is monitored in real time during a radiotherapy treatment. If that can be done, why not do it? If it were hard and expensive, would it be designed in the system? In this case, the severity may be far off the charts previously described and the situation could very well be one of $\infty \times \varepsilon$ (where ε is a very small number). This is the type of debate that should be done in creating an FMEA in a QA program. In addition, the instrumentation that performs this monitoring is checked daily during physics QA of the machine. Thus, the question of what can and should be monitored in real time versus less frequently is posed and sometimes the answer is assumed or regulated. But it is good if there can be a systematic approach to ensure safety in the optimal way and justify the workload and personnel needed.

Some time intervals include:

1. Real time or online
2. Online initial check just prior to treatment
3. Daily checks
4. Weekly checks
5. Monthly checks
6. Annual checks

Appendix D has a table that summarizes the clinical beam parameters for scattering and scanning and identifies an example of the frequency of each of these measurements. This is not a standard and only offered to enable a starting point for criticism. The decision for these is system dependent and it is not expected that this example should be implemented without a thorough individual evaluation. In fact, it is hoped that they will be challenged. Once again the particular implementation will depend upon the type of equipment used. For example, in the case of scattering, a ridge filter is a passive device that is unlikely to fail, but in almost all other cases, scattering is not a passive modality and there are various moving parts or equipment that must be verified for proper timing. It's possible that this equipment, or the beam produced by this equipment, should be evaluated more often. Sometimes misuse of the word passive might reduce the awareness of possible errors.

Table 15.10 is an example of a sort of HA that helps to identify the frequency of measurement for some situations. There are many numbers in Table 15.10. This is the result of a subjective analysis, but perhaps the thinking behind some of these numbers can be reflective in the thought process used to determine the relative risks. Some of the clinically relevant beam parameters (specific beam properties, e.g. Figure 15.13) are listed. The effective 'hazard' is that the parameters of the beam delivered for treatment are out of tolerance from what was prescribed. There is no indication of failure mode yet. In the next column, the severity of the result of a failure to obtain the correct parameter is identified. So, for example, if the beam range is incorrect, this can result in an overdose to the normal tissue and/or an underdose to the target, both very severe consequences. In this table, the scoring is 1–3, where 3 is a worst case (very severe or highly probable). A range-spread change is not good, but for most cases of the types of errors, the spread will not cause a problem as severe as a wrong specific range. The SOBP uniformity, also depending upon how severe, is not typically as severe as other errors but can be in extreme circumstances, so perhaps there can be multiple degrees of errors. Regarding the transverse beam properties, the field size and position will play a major role in the incorrect dose distribution, as could the field uniformity, but that will have less of an effect typically. One can, alternatively, take the worst case scenario and score 3 for all of these, but that is left as an exercise to the reader

The next series of columns includes potential causes of the error of the particular beam parameter identified for a double scattering system. It is assumed that each of these contribute independently so that the probability of the error is the product of each of the individual errors. Basically these are all possible failure modes. If this were an FMEA, these would be listed first and the resulting error would be listed second. This table can help to highlight some creative thinking processes since it is a hybrid.

There are many system-dependent considerations factored into the Table 15.10 numbers:

- For a double scattering system, the input beam trajectory has less of an effect on the field position and size due to the use of a collimator but has a strong effect on the transverse field uniformity due to the effect of a second scatterer.
- The effect of the initial beam energy spread has less of an effect on the transverse parameters than on the distal falloff. The latter will have an effect on the SOBP uniformity.

TABLE 15.10

Analysis to Help Identify Measurement Frequency (Scattering)

High Level Beam Property	Specific Beam Property	Severity of Fault	Incident Beam Trajectory	Beam Energy spread	Incident Beam Size	Energy Degrader	Second Scatterer	Mod Wheel and Scatterer	Collimator	Compensator	Product of Probabilities	Normalized Probability	RPN Product	Online Measurements	Mitigation Type Possible?
			\[Probability of Affecting — Intrinsic\]			\[Device Causing Prob.\]									
Beam range	Beam range value	3	1	1	1	3	2	3	1	3	54	0.37	1.13		No
	Range spread	2	2	2	1	2	1	3	1	2	48	0.33	0.67		No
	SOBP uniformity	1	2	2	1	2	2	3	2	3	288	2	2		No
Beam profile	Field size values	3	1	1	2	2	2	3	3	2	144	1	3		Yes
	Field uniformity	2	3	1	2	2	2	3	3	2	432	3	6		Yes
	Position values	3	1	1	1	2	1	1	3	2	12	0.08	0.25		Yes

- The energy degrader will have a major effect on the beam range and the beam range will play a role in the transverse beam parameters, but the latter will depend upon the quantitative amount.
- The modulator wheel, especially a rotating one, can have various failure modes and strongly affects both the beam range and the transverse scattering
- The collimator position and possible edge scattering effects will play a major role on the beam transverse properties and owing to the scattering possibilities, also affect the SOBP uniformity.

There are two columns summarizing the probability result. One is the raw product of all the probabilities in a row, and the other is normalized to scale the maximum probability to be three, so as not to overwhelm the RPN = S × P product. The probability of a particular hazard will be related to the product of all of the probabilities of the individual faults that can occur. This probability, it is important to understand, is before any further mitigation. Note that it is most probable to achieve an error in transverse field uniformity, followed by an error in SOBP uniformity. However, the highest RPN numbers go from transverse uniformity to transverse field size, to SOBP uniformity.

It has been discussed that the RPN numbers should be used to determine the level of criticality and the mitigation. In addition to HW and controls, mitigations to ensure that the scattering and beam-related components behave as desired such as the functional redundancy, can be implemented. The transverse beam field size and uniformity are relatively easy to measure online, but the SOBP uniformity is not easy to measure online. The conclusion may then be that the SOBP uniformity should be measured more often, off-line, than the other parameters that have online mitigations (online measurements) to further reduce the probability. To be clear, higher RPN numbers result in potentially higher measurement frequency.

Alternatively, another strategy is possible; that is, to only look at components with potential failure modes that have a level 3 probability. This would include

- Incident beam trajectory
- Energy degrader
- Second scatterer
- Collimator
- Compensator

These components can be monitored through QA and/or ensure that the QA program includes measurements to confirm that they are behaving properly. This results in a patient-specific QA program to inspect the collimator and compensators before the first use, and beam measurements at various energies to ensure that the beam parameters are correct. The latter can be HW specific (e.g. create a short beam pulse to generate a specific pristine Bragg peak on a moving range modulator wheel) or be more general and measure an output factor. In any case, it is not the usual functional redundant external instrumentation measurements. At the very least, discussion of the numbers applied to the above table can help the analysis team better understand their system, create a safer treatment process and determine how often to do what.

15.10   QUICK QUALITY ASSURANCE

There are important parameters to ensure that a treatment is safely delivered. With the parameters and the tolerances defined, it is possible to create a table of measurements that are appropriate for QA. All of these parameters are measurable and it is convenient to find a quick way to do it. Speed of measurement does encourage the measurement. This just a practical fact of human nature.

The author proposed a QA pattern that could do much of this. Well, it started from an old TV pattern in the 1960s, but no one else in the room was old enough to remember that. So it was fine-tuned by the group assembled by IBA to the patterns in Figure 15.14. The different shades of gray represent different doses that are positioned in a variety of shapes in different locations. This pattern will measure dose, position, beam size and penumbra and the ability of the system to generate patterns. The pattern of Figure 15.14(left) is the ideal sharp-edge beam pattern. The pattern in Figure 15.14(right) is one simulated by a beam with a sigma of 3 mm. Adding a lucite wedge (or two) will also enable measurement of the beam range.

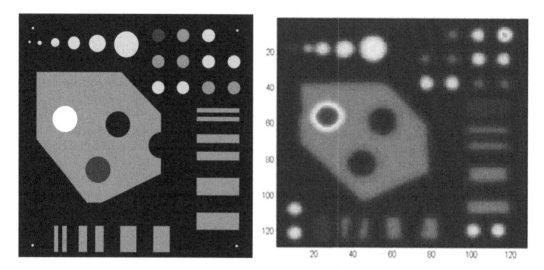

FIGURE 15.14   Ideal QA pattern (left) and pattern with 3 mm beam size (right).

Addressing the question about the timing of the quality testing, this pattern cannot be measured with a patient on the table, so this is a measurement for after treatment hours. However, measurements of beam position, size and dose are possible to measure online during patient treatment. With a scanned beam, since the materials in the beam path are minimized, it is difficult to directly measure the range, online, but by monitoring the degrader and energy selection system (if included) or the synchrotron parameters, the beam energy at the start of the beam transport system can be determined, at least indirectly, but with multiple redundancies.

It may also be a reasonable consideration, now knowing the tolerances of the technical components, to measure them more or less continuously, e.g. a strip chart recording (are you old enough to remember that?) or equivalent. This can include the magnetic fields, power supply currents and voltage. It can perhaps proactively identify degrading performance in advance of anything that will be identified during beam QA. This means that the medical physics QA could be augmented to include some technical knowledge. There are different levels of proactiveness and a quality system can help to define what is useful.

15.11   BEYOND SAFETY

Your first thought now is what could be beyond safety? An old adage says that 'the safest system is one that does not work ☹'. In radiotherapy, that is not true. Failure to treat can be a hazard. Everything cannot always be expected to be perfect and, therefore, safely reducing the sensitivity to errors, or correcting errors in a fast and accurate way that allows treatment to continue, is useful. Stopping on an error may be safer than continuing sometimes, but perhaps addressing that error instantly is beyond safety in a good direction. For example, that error may not even be real (e.g. electronics noise) and it would be unfortunate to stop if that were so.

If there is a position error, it is possible to make vector corrections to adjust the beam position and even do this online. This should allow the beam to be properly located, and the data that is uploaded after a plan is delivered would then include the appropriate beam data, even though the physical magnet settings may not be nominal due to the vector correction made to compensate for machine fluctuations.

It is also important to realize that we do not live in an ideal world. In our world, we have electronics noise, glitches and microphonics, to name a few annoyances. It would be disadvantageous to continually pause the perfectly accurate beam when spurious errors arise. It is not unsafe to account for reality and it is not necessarily safe to continually stop the beam on every perceived error and possibly compromise treatment fidelity. It is possible to include layers of redundancy and include enough that would allow, for example, 2 out of 3 logic before an irradiation is paused. Alternatively, error detection techniques may not need to be simply binary (inside or outside tolerances). Statistical methods can help to allow a certain number of out-of-tolerance measurements below a certain clinically acceptable threshold before taking action. A more statistical approach applies, for example, the Exponentially Weighted Mean Average (EWMA) to account for systematic and random statistical variations separately.

If the beam has been paused or stopped before the irradiation has completed and it is desired to complete that irradiation, this process must account for the entire dose that has been previously delivered. Unlike scattering, where the entire volume is irradiated almost simultaneously, in the case of scanning, there is time dependence to the beam delivery. In the case of resuming a treatment after it has been stopped or paused, the recorded measurement of dose already delivered is crucial.

In all cases, it is necessary to be convinced that the best things are being done to ensure safe and efficient patient treatments.

Be safe out there.

15.12   WHAT COULD GO WRONG – 8?

No new equipment has been added in this section, but errors associated with this subject matter could have some of the most significant consequences. Some things that can go wrong include:

- An appropriate culture of safety is not encouraged.
- The appropriate group of experts is not part of the safety analysis team.
- QA is done in a rote way that does not include an understanding of the equipment operation.
- Treatment is done in a rote way that does not include an understanding of the equipment operation.
- Appropriate communication is not encouraged.
- Measurements are not done frequently enough.
- Analysis tools are not used properly and the results are not correct.
- Insufficient effort is devoted to finding ways to detect that mitigations are working.
- There is an overreaction to unexpected events.

15.13   EXERCISES

1. Create the equivalent of Table 15.10 for scanning. There is no 'right' answer? So no answer key is provided. Also rewrite your own version for scattering.
2. What hazards could have caused Figure 15.1?

3. Swiss Cheese
   a. What is the probability of an accident in the Swiss cheese model if there were four slices of cheese in which the holes took up 10% of the area of the slice? The slices are positioned randomly.
   b. What is the probability of just two or three slices lining up next to each other?
4. Create a version of Table 15.6 with the hazard of incorrect beam position for a beam scanning system with a gantry.
5. Create a version of Table 15.9 for a beam scanning system including failures of the scanning magnets and ionization chambers. Make a list of what technical components affect the parameters of a scanning beam such as the parameters shown on the right side of Figure 15.13. Use that for the FMEA table.
6. Consider Appendix D. Identify any changes that you would make.

16 Sensitivities and Tolerances: Scattering

The author was pleased to have had the chance to work with Russel Wolf to explore the sensitivities and tolerances of the scattering system at the Burr Proton Therapy Center (BPTC) facility. Russel was another talented student loaned to me by Harald Paganetti. The safety of a particle therapy facility is heavily intertwined with an understanding of these parameters.

This chapter is essentially a bigger 'what could go wrong' section, for the scattering mode.

16.1 METHODOLOGY

The geometry of the Ion Beam Applications s.a. (IBA) nozzles used at the Francis H BPTC was modeled using TOPAS (TOol for PArticle Simulation), an extension of Geant4 designed for Monte Carlo simulations in particle therapy. This tool is used to model beam delivery components and examine the beam properties that interact with them. It is then possible to simulate the effect of perturbations to beam-shaping devices within the nozzle on the delivered dose in a water phantom. Similar to what was described in Section 12.2, these components are depicted in Figure 16.1.

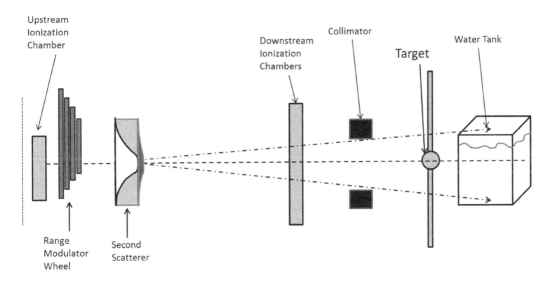

FIGURE 16.1 Scattering system components.

16.1.1 INPUT BEAM PERTURBATIONS

The simulations begin with a beam of given parameters on which the subsequent components act to modify the beam to what is required for treatment. It is useful to consider the impact of modified beam entrance parameters in the absence of any other perturbation. TOPAS by default assumes a beam sigma of 6.5 mm, with an angular spread of 3.2 mrad. (Note that 6.5 mm × 3.2 mrad = 20 mm-mrad, similar to what has been previously used in this book.) Two simple beam-steering

DOI: 10.1201/9781003123880-16

adjustments have been considered here: pure translation of a beam parallel to the central axis and a pure angular kick of a beam starting from the center of the nozzle entrance. In addition, changes in transverse spot size and in the angular spread were examined.

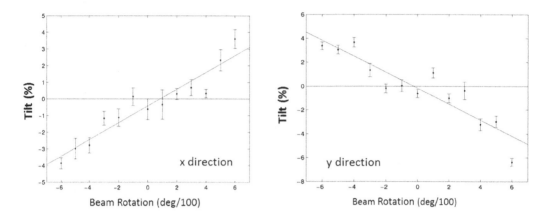

FIGURE 16.2   Effect of beam angle shift on transverse beam tilt.

It has been shown in Section 12.2 that the double scattered beam transverse uniformity is sensitive to its position on the second scatterer. Changing the trajectory angle (rotation) by a few hundredths of a degree leads to measurable changes in the skewness and profile tilt similar to the profile of Figure 12.15. These are shown for more cases in Figure 16.2. The left graph is in the x direction and the right graph is for the y direction. All values in the actual control system are constrained by error bounds. If an error limit is exceeded during irradiation, a warning is issued and the beam is paused. The error bounds on skewness are reached by modifying the input beam parameters somewhere between 0.03 and 0.06 degrees of beam angular kick, with a sensitivity of 0.00438 ± 0.00006 change in skewness per 0.01 degrees of beam angle. The profile tilt has sensitivity $0.63\% \pm 0.05\%$ per 0.01 degrees, with a maximum around $\pm 4\%$.

Translation of the beam entrance position by up to 2 mm affects the same parameters as the beam angle. This is shown in Figure 16.3. Too large a translation will also trigger a beam position

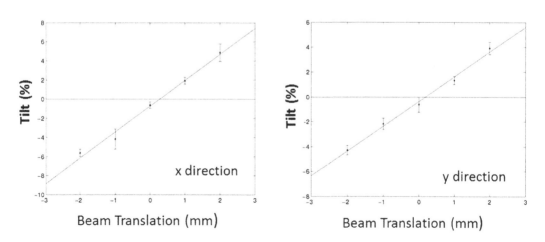

FIGURE 16.3   Effect of beam translation on transverse beam tilt.

error detection in the upstream monitor chamber. The tolerance in one direction is already at 2 mm, although in this case before this is reached, skewness errors from measurements in the downstream ICs are detected first. A translation of 1 mm is already enough to trigger skewness errors in some cases and 2 mm in all cases with a skewness sensitivity of 0.0171 ± 0.0004/mm. The tolerance is somewhat more relaxed for the contour scatterer compared to the post of Section 12.2. Despite somewhat worse skewness over this range than the examined angle changes, the profile tilt is not significantly worsened. A sensitivity of 2.3 ± 0.1%/mm leads to maximum tilt around ±5% at 2 mm translation.

The size and angular spread of the beam were also studied. If the scattering system dominates these quantities relative to the incoming beam, it might be expected that these input parameters would be insensitive. After investigation, the beam angular spread, over the range investigated, appears to have almost no noticeable effect on any measured outputs. This would seem to imply that the scattering in the first scatterer completely dominates the downstream angular distribution.

Deviations in the lateral beam size, on the other hand, create significant changes; but not with respect to the scattering. If the beam spot is too large, then it spills over onto other tracks or steps in the range modulator wheel. If some beam is on the wrong wheel step (Figure 12.7), that portion of the beam will have the wrong energy that could (depending on the field) be detectable in the dose distribution. The TOPAS default 6.5 mm spot is already large enough that a tiny (<1%) bump appears past the distal edge of the spread out Bragg peak (SOBP). Considerable efforts were invested during commissioning to ensure that the beam tunes available resulted in a beam size at the modulator wheel that would not overlap an incorrect step. TOPAS uses a larger beam than measured in the actual system. The modulation width changes by 0.854 ± 0.004 mm for each mm of change in beam spot size between 5 and 8 mm, so that a change in spot size can affect the entrance dose. This result is specific to the range modulator used in the simulation.

16.1.2 COMPONENT PERTURBATIONS: SECOND SCATTERER

The previous section shows the sensitivity of the position of the beam relative to the nozzle components. In Section 12.2, Figure 12.15, the effect of the beam position relative to the second scatterer post was shown for one example. A double scattering system with a post is more extreme than one with a contour second scatterer. It can be shown that the skewness of an intended uniform distribution is related to the flatness or tilt of the flat top. Figure 16.4 shows three distributions on the left side. The solid line is flat, in the uniform region and the dashed lines are tilted. The right side of Figure 16.4 is a plot of the percentage of skewness of these distributions as a function of the

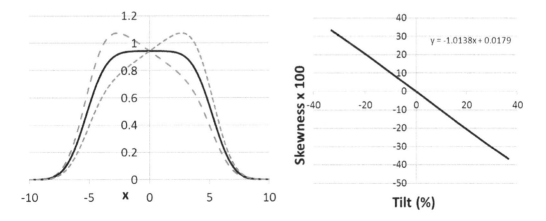

FIGURE 16.4   Tilt vs. skewness.

percentage of tilt. It is very close to linear. The goal is to measure the skewness online, during beam delivery and have a measure of the tilt at the target. The IC is located upstream of the target and while the tilt may not have fully developed upstream, the skewness will tell the story. This IC has strips as shown in Figure 11.8 which can sample the beam at sections across its profile. Skewness is defined as

$$\text{Skewness} = \frac{\sum_i w_i (x_i - \text{mean})^3}{(\text{rms})^3 \sum_i w_i} \tag{16.1}$$

where w_i is the weight or intensity of the ith channel of the ionization chamber strips. The mean and rms are with respect to the data from the i channels. The measurement geometry is also important in properly interpreting the data. This works well when the full beam is available to include in the calculations.

The second scatterer position has a significant effect on the quality of the scattering system's dose shaping. Shifting the second scatterer laterally by a few millimeters affects the tilt of the dose profile (Figure 16.5). The left side shows the x direction and the right side shows the y direction. A linear relationship can be observed between the scatterer position and the measured skewness, which leads to a linear increase in the asymmetry as measured by the tilt in the dose profile. Best-fit lines are plotted to highlight this relationship, and the slope of these lines gives the sensitivity of the skewness and tilt to linear shifts of the scatterer. The skewness for this case has a sensitivity of 0.0079 ± 0.0002 per mm along either axis, and the tilt $2.4\% \pm 0.4\%$ per mm.

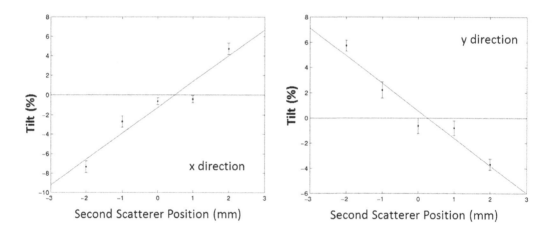

FIGURE 16.5   Tilt vs. second scatterer position.

There are 14 different first and second scatterer options and the same simulations were done for all of them. For most options, the skewness sensitivity lies near 0.005–0.010 per mm, so that generally a 4 mm or larger shift would be needed to trigger skewness errors, although this value can be as small as 1–2 mm in certain cases.

Other second scatterer perturbations seem to have little effect. Tilting the scatterer in x or y does seem to have a linear relation with skewness, but its magnitude is relatively small, and it would take a large deviation to cause an appreciable skewness change. As a result, there is little effect on any other measured beam parameters. A large deviation is unlikely, because a rotation larger than about 1.7 degrees would cause the edge of the scatterer to move past the physical constraints.

16.1.3 CORRECTION CAPABILITY

There are sensitivities to various beam and hardware errors and many of these can be measured during an irradiation. It is very useful to determine if corrections, especially online corrections can be made for any of these errors. The main knob available is steering the beam (which is also the quickest method of correction). Could this be used to correct profile asymmetries, and the use of skewness to measure the required correction?

Simulations (for a different geometry than described earlier, namely for a larger field size and different second scatterer) show a change in skewness 0.0079 ± 0.0008 per millimeter of scatterer translation and a change in tilt, defined as the percent difference between the opposite sides of a best-fit line of the profile plateau, of $4.86\% \pm 0.08\%$ per millimeter (of shift). Steering the beam angle corrects the skewness by 0.0043 ± 0.0002 per hundredth of a degree, and the tilt by $1.0\% \pm 0.1\%$ per hundredths of a degree. Then, it requires only 0.018 degrees per millimeter of scatterer offset to correct the skewness to the nominal value, but 0.049 degrees to fully correct the tilt. Similarly, steering by beam translation corrects the skewness by 0.0167 and the tilt by 4.7% per millimeter, so that 0.47 mm of beam translation will correct the skewness but 1.0 mm is required to correct the profile tilt. The reason for these differences is subtle. The tilt corrections can be understood as steering the beam so that it passes through the middle of the second scatterer and this is what is desired. Then 1.0 mm of steering offset logically corrects 1 mm of scatterer movement, and similarly, 0.049 degrees of beam angle is enough so that the beam center is offset by 1 mm after traveling the 118 cm distance to the second scatterer. However, the corrections to return the skewness to zero are less obvious. Some of these have to do with hardware issues, even those included in the TOPAS simulations. There are statistical differences on the sides of the beam in these data sets, if two are compared that are intended to have a small difference in skewness, the finite size and spacing of the IC strips with the statistical number of particles used may not detect the skewness difference and may have small statistical errors in the calculated results. In real life, the issue is related to how well the instrumentation is calibrated. If the relative gains of the channels are not well enough matched, the skewness calculated from measurements could be affected. A change in the output of one strip can be offset by a change in another strip in an asymmetric way resulting in an unchanged skewness. So while it is possible to calculate corrections, it is important to understand the value that is being calculated to determine which parameter to monitor for the correction. Well, it's also important to calibrate the instruments very well.

16.1.4 IONIZATION CHAMBER PERTURBATIONS

Beam steering correction is based on the measured skewness, so this measurement must be made carefully. As noted earlier, there is sensitivity to the calibration of different strips in the ionization chambers. If an IC set is slightly out of place, it may miss a part of one tail of the beam profile distribution, leading to a measured nonzero skewness even for a symmetric beam. This could have consequential effects, because the system would then steer the beam off-center to correct, and the resulting dose distribution would be incorrect.

The effects of ion chamber movements along both axes are shown in Figure 16.6 for a check on the second scatterer position. The beam position at the second scatterer is inferred by the position of the beam in ICs upstream and downstream of the second scatterer. (This is not a check on the skewness. It is a redundant check on centroid position.) Sensitivities appear to be slightly asymmetric here, and therefore, there are separate numbers for x and y. For the skewness, there is a change of 0.0077 ± 0.0001 per mm of chamber translation in the x direction, and 0.0027 ± 0.0003 per mm in the y direction. The difference here may be that the y-strips are closer to the y-jaws than the x-strips to the x-jaws of an upstream collimator (not shown in the schematic diagram of Figure 16.1). This means that the skewness measurement in x includes more scattering downstream of the relevant collimator, so that there is a larger tail on the edge of the distribution. In the event of an ion chamber

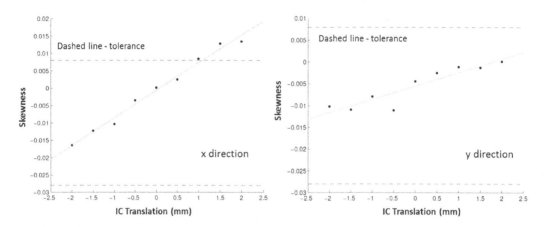

FIGURE 16.6 Centering on the second scatterer vs. IC translation.

displacement, only one side of that tail is seen, which affects the skewness measurement. The lower sensitivity of the y-skewness means that 2 mm of IC translation is not enough to cause an error. In the x direction, a skewness error or warning may be displayed depending on the direction and on whether the beam is slightly off-center on entrance.

From beam measurements, there is some evidence of an actual ion chamber displacement. The simulated skewness measurements were compared with and without an off-center ion chamber. The ion chamber movement was adjusted independently for each gantry to try to minimize the difference between measured and simulated skewness. This process seemed to have worked particularly well for one gantry, except for one option in the y direction. Being able to fit measured skewsness by moving the simulated ion chamber implies that the displacement may actually represent reality. There could be such an ion chamber offset in the physical nozzles and these simulations can help to identify the sensitivities and corrections.

Another parameter that can be investigated is the size of the monitor unit chamber (active area from Chapter 11). The electrodes in this chamber may not be manufactured with perfect precision, leading to a slightly different chamber volume that would increase or decrease the measured charge. If this were so, it would affect various IC ratios, as well as the output factor. The sensitivity of these quantities to the IC chamber monitor radius (see Section 11.3.1.2) was evaluated. These sensitivities are large, and it takes half a millimeter change in the chamber monitor radius to cause an error in the internal IC ratio or change of the output factor by 10% and could affect the simulation results if the parameters input are not exact.

16.1.5 RANGE MODULATOR PERTURBATIONS

The range modulator consists of a large carousal that holds three different wheels as shown in Figure 12.7. Each wheel has three tracks, containing absorbers of varying thickness which produce an SOBP when they rotate while the beam passes through. Consider scenarios in which the range modulator alignment is not nominal. The range modulation system is the most significant asymmetric component of the nozzle, as well as having different tracks for each treatment range option, so there can be variations between options.

A rotational misalignment of the wheel (in the direction of the wheel's normal rotation) is equivalent to a mismatch between the synchronized beam-on and beam-off times by the same amount. It essentially creates a synchronicity error between the wheel and the beam pulse. Because the beam starts on a 'stop block' that blocks protons, there is little effect from changing the starting position. However, the stopping position determines the modulation width, and secondarily the amount of beam that passes through IC, which affects the output factor. Some of these effects are seen in

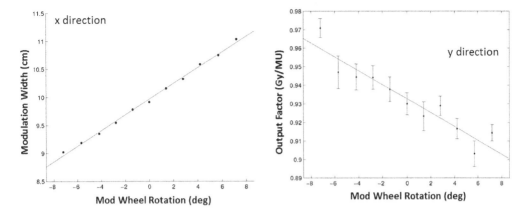

FIGURE 16.7   Modulation width vs. wheel rotation alignment.

Figure 16.7. The modulation width has a sensitivity 0.140 ± 0.001 cm per degree of small wheel misalignment, the output factor 0.0037 ± 0.0002 cGy/MU per degree. The 360 degrees of each small wheel are divided into 256 digits, which define the temporal resolution of the beam. If the on/off time is out-of-sync by one digit, it is equivalent to a rotation of 1.4 degrees, which could change the modulation width by 2 mm. Table 16.1 summarizes some of the results of the simulations.

TABLE 16.1
Summary of Some Sensitivities

Perturbation	Change	Tilt (%)	Skew
Beam angle	0.01 degrees	0.63	0.00438
Beam trans	1 mm	2.3	0.0171
SSC trans	1 mm	2.4	0.0079
SSC tilt	0.01 degrees	Small	0.0079
IC x	1 mm	–	0.0077
IC y	1 mm	–	0.0027
Tolerance	–	4	0.02

16.2 DISCUSSION

The author would like to acknowledge the TOPAS-nBio Collaboration for the development of TOPAS and making it available for these simulations. The TOPAS model of the gantry nozzles imparts a considerable amount of information about nozzle sensitivities, suggesting ways that misaligned components might affect the final dose distribution. The sensitivities calculated can be used as a basis for diagnostics and identification of the probability of hazards and faults that can lead toward quality assurance priorities.

A few general trends are worth emphasizing. The beam range is quite stable under every perturbation thus far examined. This is obviously a very important aspect of particle beam therapy. Were the range tolerances less robust, it might necessitate an additional distal margin in the size of the treatment field. This is relevant to the probability that ought to be used for possible range errors. On the other hand, the modulation width can be sensitive to beam size or the range modulator geometry, so this should be accounted for in beam steering to avoid millimeter-order errors in the proximal edge of the treatment field. The beam profile is sensitive to asymmetric perturbations of the scatterers, and it may be that care should be taken when using skewness measurement to correct the tilt unless the ICs calibrations are checked and optimized prior to measurements.

17 From Clinical to Technical Tolerances: Scanning

Even today, the philosophical dilemma of whether the chicken or the egg was first is argued. Aristotle claimed that both had always existed. Plutarch thought otherwise. In the field of particle therapy, the author argues that simulations and calculations should precede system design to identify the required sensitivities and tolerances for the machine designers who would be tasked to ensure that these could be realized. The opposite side of the argument is framed by the question, tell me what the machine capabilities are, and the available parameters can then be evaluated. This causality dilemma maintained the status quo of stalemate until this author had the pleasure to meet Jonathan Hubeau. Then, with the support of Ion Beam Applications, we were able to investigate the sensitivities of scanning beam delivery. This investigation is about the clinical sensitivities of beam parameters, independent of how those beam parameters are implemented. Only after that is understood, can the technical implementation be derived. The author still firmly believes that the safety of a particle therapy facility, and therefore its initial design, is heavily intertwined with an understanding of these parameter tolerances. Proactive mitigation in design is the optimal way to implement safety. This chapter is essentially the biggest 'what could go wrong' section of this book, done for the scanning mode.

17.1 THE CHICKEN OR THE EGG

Equipment for particle therapy is manufactured on the basis of many requirements including safety and clinical specifications. The requirements, which the system must satisfy, may have a significant effect on the difficulty of the design, fabrication, overall cost and commissioning of the equipment. Conversely, the requirements, that the system can satisfy, will fully determine the type of treatment and the particular clinical sites that are possible to treat. While this should be a generally true statement, sometimes the link between the clinical requirements and the machine specifications is not that tight. It is not always clear whether the clinical use is primarily based upon what the machine capabilities are or whether the machine has been built on the basis of some desired clinical application requirements. Of course, in practice, the former can be limiting, and to some extent, the latter is usually, simpler to use. Furthermore, even when 'clinical' requirements are identified; it is not always clear where that came from. Therefore, it seems useful to create a procedure to more rigorously link the two ends of the development process.

While the link can be devised theoretically, there are a number of practical limitations that must be accounted for. There must be a cost-benefit analysis comparing the clinical gains and safety features of a particular beam or machine specification with the resulting complexity of physical implementation, commissioning and operation. Also, some hardware limitation may give rise to an upper or lower limit on obtainable parameters or the stability attainable for these parameters.

Discussion in Chapter 15 of safety mitigations suggested that it is important to understand the magnitude of the effect that an error can have on a treatment. In that way, the severity and probability of the effect can be analyzed. This chapter is intended to help to provide some of that input for scanning.

DOI: 10.1201/9781003123880-17

17.2   THE ACCEPTANCE CRITERIA

The key parameters of a scanning beam, that have been summarized in Figure 12.48, are reproduced here in Figure 17.1. Some have referred to scanning as a more complicated, active method of beam delivery. This is perhaps because of its great flexibility. This flexibility comes from the use of the four simple parameters of Figure 17.1, A, B, C and D which were discussed in Chapter 12. These beam parameters are used to create a dose distribution which matches the prescription. The two may deviate from each other if the beam parameters are not able to create the desired distribution. The questions are, how much does it deviate, what is the acceptable difference and why is it different? The connection between the technical and the clinical is how well the prescribed beam distribution can be matched. The efficacy of the clinical prescription is related to the medicine and biology. Therefore, in this book, the criteria are based on the fidelity between the prescription and the delivered dose, without questioning the prescription.

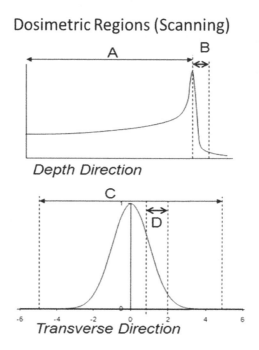

FIGURE 17.1   Scanning parameters.

   Remembering the important cornerstone, put the right dose in the right place, simplifies the goal of the comparison. The right dose and the right place can be given by a treatment plan. The dose distribution that is measured or otherwise simulated is to be compared with that. There are two key criteria: is the dose within tolerance, and/or within an acceptable distance from the right place? Figure 17.2 helps with this. For the former, define the dose difference criteria δD_{gc}, and for the latter, define the distance-to-agreement (DTA) criteria δd_{gc}. The dose distribution (call it the simulated distribution) that is in question is to be compared with the reference dose distribution. If the difference between the reference dose, D_0, and the simulated dose, D_{Sim}, at a point r_0, $(D_0(r_0)-D_{sim}(r_{sim}))$ is within $\pm\delta D_{gc}$ (vertical axis) and the distance (r_0-r_{sim}) is within $\pm\delta d_{gc}$ (horizontal axis) of the reference distribution, or inside the rectangle of Figure 17.2, things are okay. The two comparisons were separate. Doing this is possible and indeed it used to be done, but it did not lend itself to convenient visualization. Instead consider the ellipse inside that rectangle. The ellipse of Figure 17.2 is

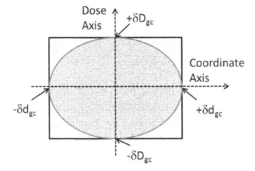

FIGURE 17.2   Dose and distance error.

a two-dimensional shape with one axis indicating the dose and the other axis indicating the coordinate. The equation of an ellipse is

$$\frac{D^2}{\delta D^2} + \frac{d^2}{\delta d^2} = 1 \tag{17.1}$$

Any coordinate (d,D) such that

$$\frac{D^2}{\delta D^2} + \frac{d^2}{\delta d^2} \leq 1 \tag{17.2}$$

is inside that ellipse.

17.3   GAMMA INDEX

Let r be a distance to a two-dimensional coordinate, from an origin. A particular voxel in the reference dose is located at r_0. This is the center of the coordinate system in Figure 17.3. The axes x and y are in the plane defined by the shaded region surrounded by the black dashed curve. The dose is plotted orthogonal to that plane. The actual reference location is indicated by the black filled circle.

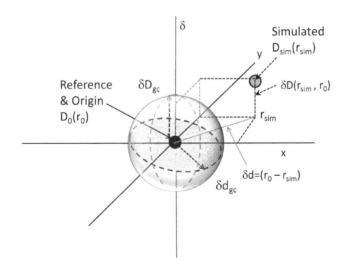

FIGURE 17.3   Coordinates for simulated vs. reference comparison.

The simulated voxel, to compare with the reference voxel, is indicated by the gray filled circle. It is at a coordinate indicated by r_{sim} and dose given by $D_{sim}(r_{sim})$. The difference between that dose and the reference dose, $D_0(r_0)$ is δD. The three-dimensional (3D) ellipsoid (looking like a sphere) has dimensions given by the criteria δD_{gc} and δd_{gc}. If the gray filled circle were inside the ellipsoid, things would be okay.

The gamma function is a measure of the goodness of the simulated dose distribution taking into account the possible deviation of dose due to mispositioning as well as the dose deposited at a point. Like the ellipse of Equation 17.2, it would be good to identify an expression which would numerically indicate if the dose and position met the desired criteria; that value would be less than or equal to 1. The γ-index $\gamma(r_0)$ at each point $\mathbf{r}_0$ is defined as

$$\gamma(r_0) = min \sqrt{\frac{\left(\left|r_0 - r_{sim}\right|\right)^2}{\delta d_{gc}^2} + \frac{\left(D_0 - D_{sim}\right)^2}{\delta D_{gc}^2}} \tag{17.3}$$

Now it's clear that the subscript gc stands for gamma criteria. This form allows the simplified criteria to be if $\gamma(\mathbf{r}_0) \leq 1$. In this case, the comparison test is passed. Be careful of the units, whether dose difference is in fractions or percentages. This calculation must be done for every point in the treatment volume. For every reference point, a simulated point must be found that can be compared and that minimizes the gamma function. This can essentially be done in a brute force way in that each reference point is compared with every simulated point in the entire volume to find the minimum value (min in Equation 17.3). Since there are thousands of voxels in a treatment plan, this is a huge number of pairs, so the calculation can take some significant time. In fact the number of comparisons can be N^{2k} where N is the number of voxels in one dimension and k is the number of dimensions. For example, if there are 10×10 voxels per plane or 100 voxels per plane, with two dimensions there would be 10,000 comparisons, with three dimensions 1,000,000 comparisons. With 900 voxels per plane, in a 3D volume 10^9 calculations are made. That time is shortened using tricks like looking only in close-by regions, or ignoring points with very small dose, etc. One example of such an approximation is to look at the gamma index in three planes that cut through the volume as in Figure 17.4. This is just a sampling but may provide a decent representation, depending on the dose distribution.

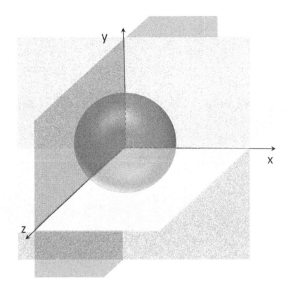

FIGURE 17.4 Approximation to evaluate gamma index in three planes.

To get a better feeling for the calculation, consider a plane with four voxels as in Figure 17.5. The reference voxels are indicated by the black filled circles surrounded by the black dashed rectangle. The simulated voxels are indicated by the gray open circles. The circles are the positions of the voxels. The dose within these voxels is given in the upper cells. There is no orientation change that will

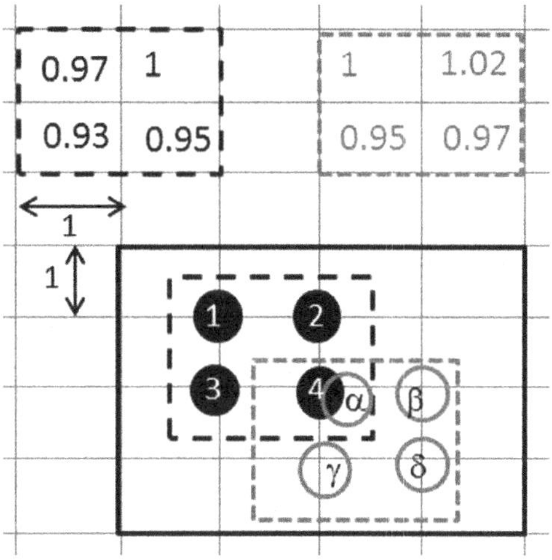

FIGURE 17.5   Gamma index calculation.

make the two regions match, so it is necessary to evaluate the gamma index. The black circles are labeled with white numbers and the gray circles are labeled with gray Greek characters. The grids upon which the circles lie are 1 unit per side. Now $(r_0 - r_{sim})$ and $(D_0 - D_{sim})$ can be evaluated for every combination. Starting with (and finishing with) number 1, Table 17.1 is filled out.

TABLE 17.1

Gamma Index Example

Ref	Sim	δd	δD	$\gamma(r_0)$
1	α	$\sqrt{2}$	0.03	1.134
1	β	$\sqrt{5}$	0.05	1.873
1	γ	$\sqrt{5}$	0.02	1.014
1	δ	$\sqrt{8}$	0.0	0.943

The unexpected result is that the best pair to minimize the gamma index is 1 and δ. This is because the dose matched exactly. The remaining comparisons can be done, but hopefully the point is made. Just to be clear, the comparison is made between one reference point and one simulated point, not the two closest points in dose for the dose comparison and another point which may be closer in position for the position comparison. The goal is to find one point that fits best within the criteria ellipsoid. Be careful to check if units are % or absolute dose.

In the case three planes are explored, to obtain a sampling, the yz plane will be called x and is from the front to the back of that target through the middle of that target (the sphere in the center of Figure 17.4) at $x = 0$. The darker section of the ball is on the left side of the yz plane. (Stare long enough at the ball and you'll see it.) The xz plane at $y = 0$ will be called y and is from the front to the back of the target through that midplane. The xy plane is at $z =$ mid target, is called z and is the full transverse dimension of the target, but at $z =$ mid target. Therefore, each calculation is a 3D exploration in two position coordinates and one dose value at each coordinate. This reduces the number of voxels checked from a lot to about less than a lot.

The overall criteria for acceptance needs to be defined. Normally the dose distribution is accepted if at least 95% of all the points in the target volume are within $\gamma(\mathbf{r}_0) \leq 1$. The choice of the (DTA) and dose difference parameter is dictated by a clinical specification which needs to be defined. This criteria is represented by pairs of tolerances such as (1 mm, 1%), (2 mm, 2%), (3 mm, 3%) or others. Normally the (3 mm, 3%) criteria is used, but others can be explored.

Now comes the visualization part. The distribution of these γ values can be represented on a γ index volume histogram (gamma volume histogram [GVH]). This plots the percentage of the voxels in the volume of the dose distribution that have a gamma index greater than a given value. It's sort of the complement of the complementary error function, if the distribution of gamma indices were Gaussian distributed. Figure 17.6 shows an example of this kind of plot. On this plot, two key points are identified; γ_{95} is the gamma index value such that 95% of the target volume has a gamma index below γ_{95}, or 5% are higher than that. That's why the curve value is 5% at that point. That's just how it's done. P1 is the percentage of the target volume that has a gamma index above 1. On this curve, the 95% point is at a gamma of about 1.3 so less than 95% of the voxels have a gamma index below 1 and the comparison wouldn't have passed. The gamma index is a criterion of clinical acceptance of a delivered plan. More practically, it is an important criterion which will contribute to the determination of whether a dose distribution is appropriate to be delivered to a patient. In this chapter, it will be the determination of whether a simulated dose distribution is clinically acceptable. So poetic license is being taken with the word clinical. There is more to the visualization capabilities of this formalism, so stay tuned.

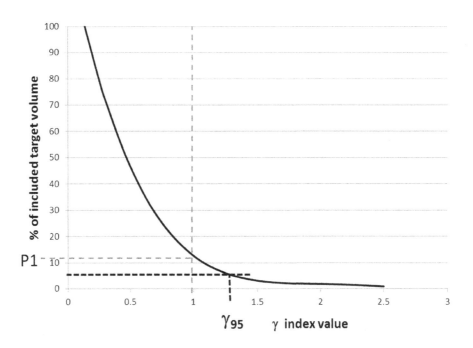

FIGURE 17.6   Gamma volume histogram example.

17.4   TOLERANCES

The goal is to identify the effect of errors in the key parameters of the scanned beam on the delivered dose distribution, when compared to an 'ideal' (reference) distribution. It is important to note that different treatment sites may be more or less sensitive to some of the beam parameters when compared with another site. For example, a treatment site which has a critical structure right next to the distal edge will be more sensitive to the range of the beam. Just a word about a treatment plan; a treatment plan is essentially a map of beam parameters which have been optimized to deliver a prescribed dose distribution. This map contains beam position and dose at that position, for example. To evaluate a tolerance, the following steps are followed:

- Create a treatment plan for a specific treatment site.
- That plan is optimized with the beam parameters available. This is the reference dose distribution.
- Use the fluence map that was calculated in the earlier treatment plan and in addition to evaluating the dose in the patient, use that same map to generate a dose distribution in water.
- Modify the beam parameters in question (introduce errors), and regenerate the dose distribution in water.
- Calculate the difference between the original and modified dose distribution in water with a gamma index.
- Analyze the results for beam parameters perturbations.
- Determine the level of perturbations acceptable relative to the criteria established.
- Use that to determine the tolerances to beam parameter errors for different clinical sites.
- There may be other effects if the patient anatomy was used, but this is not addressed here.
- Translate the tolerances to technical parameters.

Evaluating all possible combinations of errors is neither possible nor necessarily desirable. In some cases, there are clear links among these parameters, and where possible they will be identified. As the error increases, there will likely be fewer and fewer points that will satisfy the gamma index criteria. Figure 17.7 is an example of how this data can be analyzed. Many GVHs are created, one for each error simulation, clinical site and set of gamma criteria etc. The value of γ_{95}, or the value of gamma which contains 95% of the points is plotted for each group of GVHs as a function of the magnitude of the error introduced. Since the criteria for acceptability is for γ_{95} to be ≤ 1, any error which results in that value is acceptable. By identifying the range of errors inside of which $\gamma_{95} \leq 1$ on this new plot, the maximum and minimum error tolerances can be determined. The black arrows in Figure 17.7 cross the γ_{95} value of 1 and point to the max and min errors allowed.

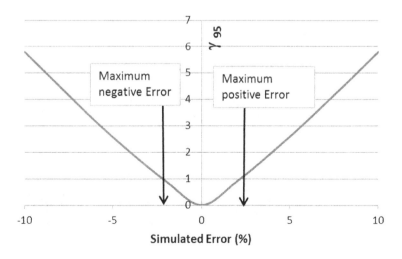

FIGURE 17.7 γ_{95} vs. error plot example.

17.4.1 Clinical Cases Used for the Study

It can be very instructive to provide a useful link between clinically relevant situations and the beam and machine parameters needed. Some clinical cases are more sensitive to certain beam parameters than other cases. Therefore, a variety of sites are chosen hoping to emphasize not only the possibility that different sites will highlight different sensitivities, but to determine if some sites can be treated with relaxed specifications, allowing for a choice in the machine to be used or a patient to be chosen commensurate with the skill of the staff and capabilities of the machine. This former point is important in that it is the opinion of some that not all machines need treat all possible sites, but that it is clear what sites a machine can treat, and it should be a conscious decision to develop a machine with more stringent specifications, capable of treating more difficult targets. In these simulations, it will be attempted to find tolerances that are applicable for all cases. But this was also used to identify a first treatment case as will be seen toward the end of the section.

The following sites were chosen:

1. Head and neck: nasopharyngeal carcinoma
2. Prostate carcinoma
3. Retroperitoneal sarcoma
4. Base of skull ependymoma

Figure 17.8 shows examples of the two-dimensional reference dose maps for some layers in the clinical sites.

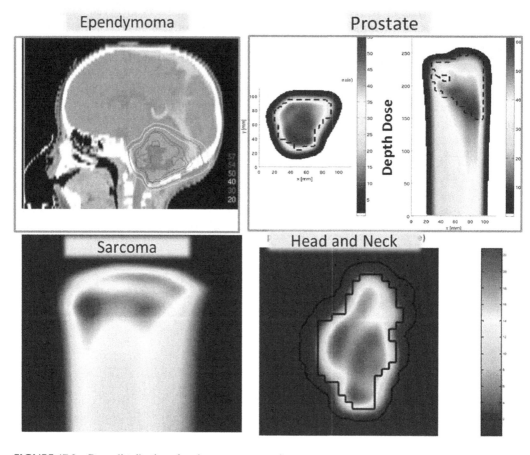

FIGURE 17.8 Dose distributions for chosen treatment sites.

The head and neck case is expected to have increased sensitivity to the beam size and position, owing to the proximity of critical structures. The prostate example also should be sensitive to beam size, since the treatment plan uses fields that rely on the penumbra of the beam to avoid the bowels, as opposed to the range; however, due to the different clinical tolerances, each case should emphasize different parameters. The results might be different if the treatment fields were differently planned.

17.4.2 ERROR SIMULATIONS

Beam errors are simulated independently of each other so as to correctly assess the impact of each individual error on the dose distribution. Errors in beam delivery are then combined so as to realistically include all the errors in the system. Each of the reference fields (a total of 12 in all) were optimized for Gaussian beam spot sizes of 3, 5 and 7 mm. Both random and systematic errors were added to the beam parameters. The parameters explored include

1. Range
2. Transverse beam size (or σ)
3. Transverse beam position
4. Dose

17.4.2.1 Accuracy vs. Reproducibility

Accuracy is a measure of the average absolute value of something with respect to a reference. Reproducibility is the tendency of a measurement to repeat with the same value. Nothing is said about how that value compares to the reference.

The targets with projectile groupings in Figure 17.9 illustrate those points. A systematic error of a beam parameter would mean that it is not accurate, but it could be reproducible (e.g. the leftmost target). A random error of a beam could imply that it is accurate (if it is randomly distributed about the reference value), but it may not be reproducible (e.g. the center target). Reproducibility requires tight groups. Therefore, systematic errors are associated with a lack of accuracy and random errors are associated with a lack of reproducibility. The right target displays the ideal case.

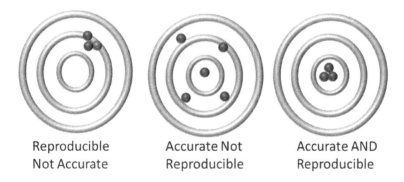

Reproducible   Accurate Not   Accurate AND
Not Accurate   Reproducible   Reproducible

FIGURE 17.9 Accuracy and reproducibility.

Each of these errors can come from different sources. Systematic errors can come from a wrong number copied at the start of the process or a miscalibration of a beam delivery component. It is repeated over and over again, until caught. Random errors arise from something which changes from time to time and could result from an unstable power supply, or different people positioning a patient from day to day.

17.4.2.2 Error Introduction

Systematic and random errors are introduced in the simulations for each of the parameters. In general, three random number seeds were used for each case. The first seed is pseudorandom in that it forces the data to alternate between minimum and maximum errors.

Random Range Errors

The depth penetration (R) of each pencil beam was modified randomly. For each beam

$$R_{\text{beam}} = R_{\text{beam},0} + \delta R_{\text{beam}} \qquad (17.4)$$

with $\delta R_{\text{beam}} \in [-2, +2]$ mm. The subscript 0 indicates the reference range. Three random number seeds were used.

Systematic Range Errors

The depth penetration of each layer was modified by a constant value. For all identical range beams,

$$R_{\text{layer}} = R_{\text{layer},0} + \delta R_{\text{layer}} \qquad (17.5)$$

with $\delta R_{\text{layer}} \in [-2, +2]$ mm. The modified beam was calculated by scaling the database Bragg peaks depth axis. Three random seeds were run.

Random Beam Size Errors

The transverse beam σ of each beam was modified randomly. For each beam,

$$\sigma_{\text{beam}} = \sigma_{\text{beam},0} + \delta\sigma_{\text{beam}} \qquad (17.6)$$

with $\delta\sigma_{\text{beam}} \in [-20, +20]\%$ of the nominal beam σ at patient entrance. The σ deviation is, of course, taken into account at every depth of the pencil. Three random seeds were run.

Systematic Beam Size Errors

The beam σ throughout the whole irradiation was modified by a constant value

$$\sigma = \sigma_0 + \delta\sigma \qquad (17.7)$$

with $\delta\sigma \in [-20, +20]\%$ of the nominal beam σ at patient entrance. The σ deviation is taken into account at every depth of the pencil.

Random Beam Position Errors

The position of each beam was modified randomly. For each beam,

$$x_{\text{beam}} = x_{\text{beam},0} + \delta x_{\text{beam}} \qquad (17.8)$$

$$y_{\text{beam}} = y_{\text{beam},0} + \delta y_{\text{beam}} \qquad (17.9)$$

with $\delta r_{\text{beam}} = \sqrt{\delta x^2_{\text{beam}} + \delta y^2_{\text{beam}}} \in [-1, +1]$ mm. The dose contribution of the beam was calculated with a δr_{beam} shift in the transverse plane. Three random seeds were run.

Systematic Beam Position Errors

The beam positions of all the spots are shifted in the transverse plane by a constant vector

$$x_{\text{beam}} = x_{\text{beam},0} + \delta x \qquad (17.10)$$

$$y_{\text{beam}} = y_{\text{beam},0} + \delta y \qquad (17.11)$$

with $\delta r^2 = \delta x^2 + \delta y^2 \in [-1, +1]$ mm. The dose contribution of each beam was calculated with a δr shift in the transverse plane.

Random Beam Weight Errors

The weight (related to number of protons) of each pencil beam was modified randomly. For each beam,

$$\omega_{beam} = \omega_{beam,0} + \delta\omega_{beam} \tag{17.12}$$

with $\delta\omega_{beam} \in [-10, +10]\%$ of the nominal spot weight.

Systematic Beam Weight Errors

The weight of all the beam spots is shifted by a constant percentage

$$\omega = \omega_0 + \delta\omega \tag{17.13}$$

with $\delta\omega \in [-10, +10]\%$ of the nominal spot weight.

Errors in the weight of the spots have many possible ramifications. It could be as simple as the wrong dose at particular location in a uniform field resulting in a difference with respect to the uniform dose. It could also be an incorrect dose at a very specific spot where it is desired to achieve a sharp dose gradient and the error of a spot in that gradient could cause a proportionally much larger relative error.

17.4.3 EVALUATION OF TOLERANCES

The acceptance criteria for the gamma index are chosen to be compared among (1 mm, 2.5%), (1.5 mm, 2.5%), (2 mm, 2.5%) and (2.5 mm, 2.5%). To demonstrate how the dose distribution is affected by the simulated errors, a few representative results are plotted. Figure 17.10 shows the point-by-point dose difference resulting from a 0.8 mm pseudorandom range error in the middle range transverse cut and in the $y = 0$ plane. Figure 17.11 presents corresponding γ-index values in the middle range cut and in the $y = 0$ plane. Since the pseudo random error means constraining to max and min errors, the gaps in the depth direction are clear. This is also another example of the visualization afforded by the gamma index formalism. It is much prettier in color. In Figure 17.11, the upper plots use a gamma criteria of (1 mm, 2.5%) which is more stringent than the criteria in the

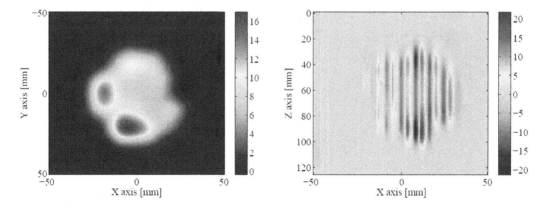

FIGURE 17.10 Ependymoma – 0.8 mm pseudorandom systematic range error.

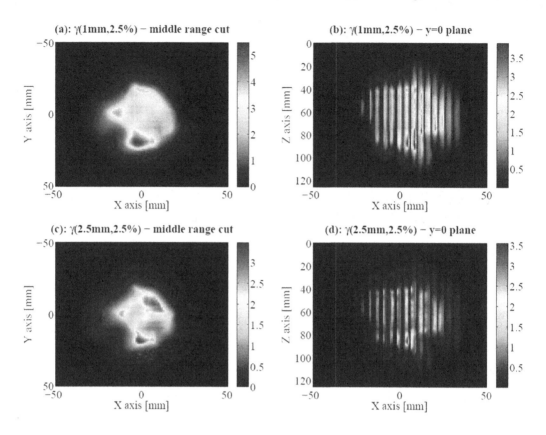

FIGURE 17.11   Ependymoma – Gamma index plots for different criteria.

lower plots that use (2.5 mm, 2.5%). It's somewhat hard to see in grayscale, but the looser criteria results in a lower error in the plot (in this plot it is darker overall, where darker in this case is better).

In the case of the prostate, right lateral field, the results of beam position errors are represented in Figure 17.12. It shows the dose difference between the nominal dose and the dose that would be delivered with the errors that have been simulated. The grayscale is in percent of dose deviation, with darker grays being outside clinical tolerances. The visual representation shown can provide a

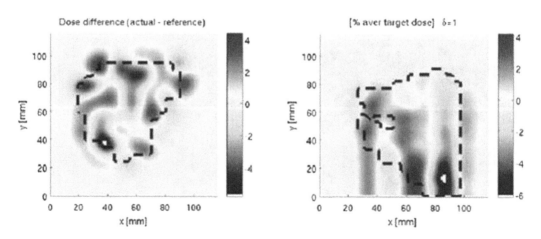

FIGURE 17.12   Prostate error dose difference.

feeling for the results of the simulations in certain cases, and that they are working; however, it is also useful to provide a method which will allow a quantitative evaluation and comparison of the various simulated dose error maps, perhaps even automated. Some examples of this analysis are given for the clinical cases studied. Inspection of those figures yields the desired tolerances.

Figure 17.13 shows two GVHs for some of the cases and parameters investigated. These show results analyzed with gamma parameters of (2.5 mm, 2.5%) and a beam size of 7 mm. The graph to the left is for a 0.7 mm systematic position error applied to the sarcoma case, and the head and neck case, on the right, was simulated with a random range error of 1 mm. If it is desired to have 95% of the points less than a value of 1, it is necessary to start at 5% on the vertical scale and move horizontally to the right until the gamma index value is crossed (see the dashed horizontal line on the right plot). If the horizontal line crosses the curve above a gamma index of 1, the comparison failed, if it is below that point, the comparison passed. These simulations were done for a wide range of errors. In the case of the H&N simulation, the data in the x and z direction pass, but the data in the y direction did not pass for this error. This is also done with different tolerances to quantify that dependence. All pass for the sarcoma.

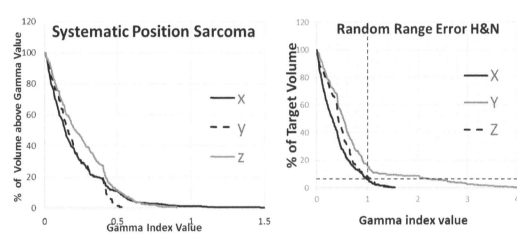

FIGURE 17.13 Two GVH examples.

Each of the GVH curves were analyzed to determine the gamma index value which included 95% of the target volume (whether it was larger or smaller than 1). The gamma index (GI) value data were assembled and plotted as a function of the error. Some examples are shown in Figures 17.14–17.16. Figure 17.14 is a summary of the results for systematic position error of a 5 mm sigma beam treating prostate carcinoma. There are twelve curves (use your magnifying glass). The solid curves are for the most stringent gamma index criteria of (1 mm, 2.5%). The black dashed curves are of (1.5 mm, 2.5%). The result for dimensions x, y and z are individually plotted. The slope of the curve is an indication of the sensitivity to the error introduced. A steeper curve indicates more sensitivity. Since this is a systematic position error, it reflects the situation that would arise when Gaussian spots are mispositioned systematically. Two breaks in the curves are noted. The first break does not introduce much error and is well within acceptable limits for γ_{95} below 1. This is likely the penumbra moving and the target seeing some non-uniform dose. First the margins help, but then the uniform region moves off the target. The fact that the prostate carcinoma is a deep seated target, resulting in considerable multiple scattering, the error at the target could soften. The vertical solid black lines at about ±0.65 mm are the $\gamma_{95} = 1$ limit for very tight criteria of (1 mm, 2.5%). The curves are less steep, and the second set of vertical lines is spaced further out, for less stringent gamma criteria of (1.5 mm, 2.5%). More standard gamma criteria of (2.5 mm, 2.5%) result in tolerances beyond ±1 mm.

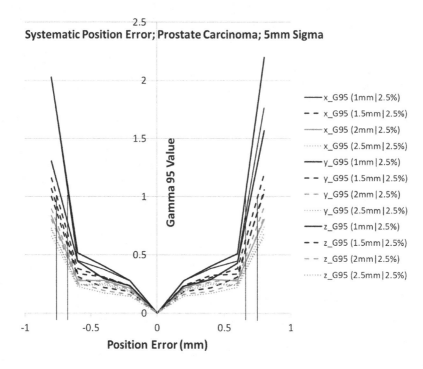

FIGURE 17.14   Prostate – Systematic position error.

Figure 17.15 shows a summary of the results for a random beam size error in the case of a head and neck target with a 3 mm beam spot. No vertical lines are drawn, since for all the errors simulated, none exceeded the $\gamma_{95} = 1$ limit for any gamma index constraints checked. Thus, in this case, the treatment is very insensitive to the beam spot size in the range explored and perhaps the beam does not need to be as small as 3mm in this case. Remember, random errors can average out.

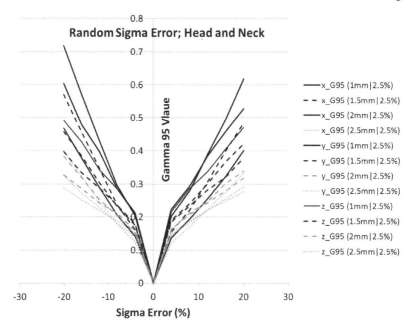

FIGURE 17.15   Head and neck – Random sigma error.

Figure 17.16(left) and 17.16(right) show a summary of the results for random range errors in the case of a base of skull target, with a 3 mm sigma beam size used. These results are a bit more complex. Each of the two figures is for a different random seed. Figure 17.16(right) (seed 2) is asymmetric, showing that the treatment is more sensitive to a shallower range, requiring an accuracy of less than 0.5 mm. The deeper range is less sensitive in that simulation. Figure 17.16(left), with a different random seed also shows an asymmetry, but not as strong and in the opposite direction. Thus, to assemble the worst case tolerances, all seeds and all dimensions are included.

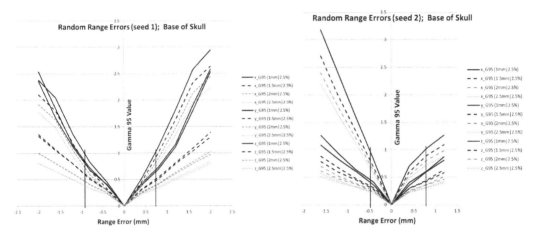

FIGURE 17.16 Random range error 1, BOS (left) and random range error 2, BOS (right).

The analysis shows that the sensitivity of the 12 analyzed fields to the errors introduced could be extremely different. For example, a base of skull field has a much greater sensitivity to systematic range errors as compared to the prostate fields. One reason for this is that the base of skull field is more shallow and, therefore, without any materials in the beam path (e.g. ridge filter or range shifter), the Bragg peaks are sharper and are more sensitive to smaller changes in the range. See Figure 12.50, which is actually from this case. The prostate fields, however, are more sensitive to random weight errors. Table 17.2 summarizes the worst case tolerances. These are not from one treatment site or one field, they are the worst of all the 12 fields, i.e. one parameter from one field and another may be from another field.

TABLE 17.2

Worst Case Tolerances for Scanning Beam Parameters

Beam Size	γ	Range	Range	Sigma	Sigma	Position	Position	Weight	Weight
σ	Criteria	Accuracy	Stability	Accuracy	Stability	Accuracy	Stability	Accuracy	Stability
mm		mm	mm	%	%	mm	mm	%	%
3	$\gamma(1.0, 2.5)$	±0.5	±0.4	>±16	>±10	[−0.8, 0.9]	±0.6	±4	±2
	$\gamma(2.5, 2.5)$	[−1, 1.5]	±0.6	>±16	>±10	>±1	±0.6	±4	±2
5	$\gamma(1.0, 2.5)$	[−0.5, 0.7]	±0.3	>±16	>±10	±0.6	±0.6	±4	±2
	$\gamma(2.5, 2.5)$	[−0.8, 1]	±0.5	>±16	>±10	>±1	[−0.7, 0.8]	±4	±2
7	$\gamma(1.0, 2.5)$	±0.4	±0.3	>±16	>±10	±0.6	±0.6	±4	±2
	$\gamma(2.5, 2.5)$	[−0.6, 0.9]	[−0.4, 0.5]	>±16	>±10	>±1	±0.8	±6	±3

When all errors are introduced at the same time, their individual impact on the dose distribution tends to add. This is illustrated in Figure 17.17 where all the worst case errors were introduced on all the 5 mm σ maps. The GI for all 12 fields are plotted for four different gamma criteria. The GI for all cases exceeds one. Therefore, the worst case individual values are not acceptable overall tolerances, but still may be okay for particular cases. Actual beam requirements must, therefore, ensure that the combination of all the errors remains acceptable. The challenge is to further tighten the error budget.

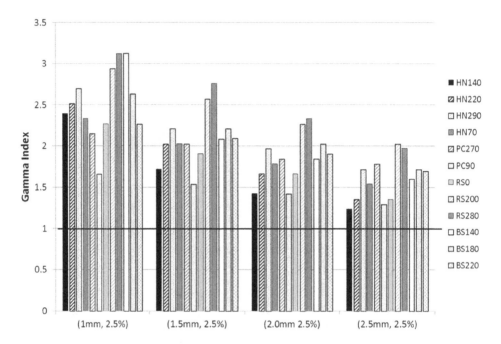

FIGURE 17.17   Results of inclusion of all errors; 12 sites; 4 criteria.

In the case when a given field can be repeatedly delivered, as in the case of fractionated treatment, or repainting of a given fluence map, the results of random errors can be reduced by 1/(square root of the number of deliveries). Assuming that several such deliveries can be accommodated, this will significantly reduce the size of the effect of random errors, and given the results of the calculations described earlier, it would then be sufficient to consider only the effect of the systematic errors. This was not done hererin.

Combining all the errors, it is possible to find a set of values which will lead to acceptable dose distribution within tolerances of the chosen acceptance criteria. This is illustrated in Figure 17.18. All values analyzed with gamma criteria of (2.5 mm, 2.5%) are within tolerance, as is most of the tighter tolerances greater than (1 mm, 2.5%). The values found are summarized in Table 17.3.

The first patient treated with a scanned beam at Massachusetts General Hospital (MGH) was treated for a large sarcoma. Analysis of the sensitivities of this treatment results in the values in Table 17.4. This shows that the tolerances were fairly relaxed when compared to the general set of tolerances earlier.

Now the goal of translating clinical (as defined herein) tolerances to beam parameter tolerances has been achieved. Take a breath.

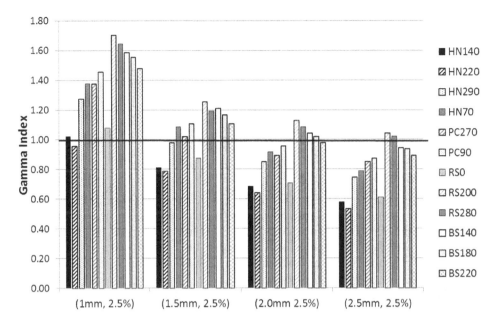

FIGURE 17.18 Reduced error set; inclusion of all errors.

TABLE 17.3
Final Beam Scanning Tolerances That Satisfy (2.5 mm, 2.5%)

Range (mm)		Sigma (%)		Position (mm)		Weight (%)	
Stability	Accuracy	Stability	Accuracy	Stability	Accuracy	Stability	Accuracy
±0.1	±0.5	±5	±2.5	±0.5	±0.25	±2.5	±1

TABLE 17.4
Tolerance for Large Sarcoma First Patient Treatment

Range (mm)		Sigma (%)		Position (mm)		Weight (%)	
Stability	Accuracy	Stability	Accuracy	Stability	Accuracy	Stability	Accuracy
±1.5	±1.5	±25	±25	±2	±2	±10	±5

17.5 FLOW TO TECHNICAL TOLERANCES

The next step is to translate the beam parameter tolerances to technical tolerances. Recall the flow of requirements summarized in Figure 4.1 and reproduced here in Figure 17.19. Having started the book with this and placing it here, close to the end is somewhat satisfying. It indicates its importance and pervasiveness throughout the process. In fact more or less all of what this represents has ultimately been covered.

This chapter has been an example of tying together many of these concepts. The clinical requirements have been identified as the conformity to the treatment prescription and a quantitative measure

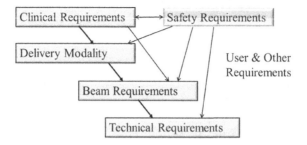

FIGURE 17.19 Flow of requirements.

of this has been the gamma index using the gamma index criteria. The beam delivery modality chosen was that of scanning. Once the beam modality was chosen, the beam parameters (requirements) could be defined (with help from information in Chapter 12). Next, simulations to determine the effect of errors of the identified beam parameters were conducted, and the results give tolerances for these beam parameters. What remains now is to translate these beam parameter tolerances into accuracy and reproducibility technical performance criteria and ensure that the equipment can perform within these specifications.

It is very important to note that many of the procedures of Chapter 15 with respect to building a risk analysis and a failure mode and effects analysis (FMEA) have been followed in this chapter leading to a determination of how to specify technology for a safe treatment. The parameters that are important have been identified. There are additional levels which may be included. For example, this chapter identified the tolerance to the weight of a spot, or basically the dose tolerance. In Chapters 11 and 12, several ways that the dose could be incorrectly deposited were identified and quantitative estimates of parameters including ionization chamber (IC) behavior and timing issues were covered. The beam parameter tolerances can now be used to give guidance to the technical design.

17.5.1 BEAM RANGE

Recalling Equation 9.39, the range momentum relationship of the particle can be written as

$$\frac{\Delta P}{P} = \frac{1}{3.5} \frac{\Delta R}{R} \tag{17.14}$$

where R is the beam range and P is the momentum. According to the beam parameter, tolerance results the beam range must be accurate to ± 0.5 mm and stable to ± 0.1 mm. This is an absolute number which was relevant to the specific treatment site that the most stringent tolerance came from. However, a worst case scenario would be to assume a range of 30 cm. The device that controls the energy in a cyclotron-based system is the energy selection system or just the degrader in a gantry mounted cyclotron. The energy control of a synchrotron is just the synchrotron. In the case of just a degrader, its mechanical position must be set with a accuracy that is based on the type of mechanical device that it is. No approximations are made here. For the energy selection system and the synchrotron, the magnet settings determine the energy that is transported to the treatment site (also the momentum bandwidth plays a role). In these cases, the accuracy requirement leads to a momentum accuracy of 0.05% setting accuracy. This is a reasonable technical power supply parameter. The reproducibility or stability requirement is closer to 2×10^{-4} which is within the capability of the magnetic field regulation loop. Therefore, the technical tolerances are achievable. Experience from proton therapy sites have shown that a range accuracy of ± 0.5 mm is routinely achieved.

17.5.2 BEAM SIZE

Recalling Figure 12.44, sometimes a quadrupole doublet is included in the scanning nozzle equipment to fine tune the beam size at the target. A quadrupole doublet can be the final device that sets the beam size. While all the focusing elements in the beam transport system will affect the beam size at the target, it's useful to get a feel for the scale by considering just this doublet. Figure 17.20 shows three beam envelopes. The beam is moving from the left to the right and is focused near the target that is at the right. The two quadrupoles are represented by the outer partial rectangles to the left. The beam envelope curves shown are for different quadrupole settings. The black solid line increases the vertical focusing at the expense of the horizontal focusing. The beam size, at the target, as a function of the magnetic quadrupole gradients of each of the quadrupoles is shown in Figure 17.21. The accuracy requirement is ±2.5% and the stability requirement is ±5%. This is actually quite a worst case scenario, since many treatment sites can be more tolerant. Using these tolerances and the quadrupole doublet, the sensitivity with respect to the first quadrupole for accuracy in x is 1.5%, and for y, it is 0.26%. The tolerance is larger for x since, as seen in the graph, the beam is near a waist, and near its minimum size, so the quadrupole setting is less sensitive than for

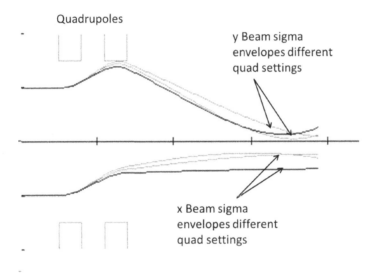

FIGURE 17.20 Quadrupole focusing sensitivity.

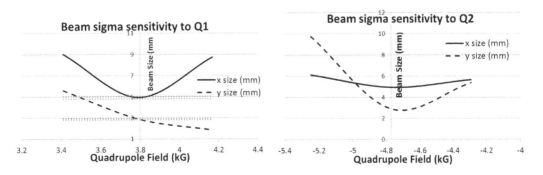

FIGURE 17.21 Sensitivity to Q1 (left) and Sensitivity to Q2 (right).

the y size. The requirement for reproducibility in x is 2.1%, and for y, it is 0.5%. All these values are well within the stability and accuracy capabilities of quadrupole power supplies. Measurements of the beam stability at the nozzle entrance showed the beam σ stability to have a root-mean-square distribution close to 1%.

17.5.3 DOSE WEIGHT

In Chapter 12, the determination of the maximum beam current which would enable a safe delivery of a spot scanned beam within 2% was determined. The tolerance for the beam weight according to this simulations referred to here is ±1%. Therefore, if the current stability is within this 1%, it could be possible to deliver the dose with up to 0.2 nA (0.5 times the limit calculated in Chapter 12). Beam current tests conducted at MGH have shown that a beam current stability of ±5% is achieved, so continuous scanning may not be possible. For dose-driven scanning, the accuracy of the dose delivery mainly relies on the dose monitoring accuracy in the nozzle. The currently achieved accuracy on the dose accuracy is ±1.0%. Most sites have a much less stringent tolerance.

17.5.4 BEAM POSITION

The accuracy requirement for beam position is ±0.25 mm and the stability requirement is ±0.5 mm. These were the worst case values among various simulations which included beam spot sizes of 3, 5 and 7 mm. Therefore, the accuracy requirement translates to 8.4%, 5% and 3.5% of the three spot size. In Section 12.3.2, it was shown that a shift of 0.2σ (20%) results in a 4% fluctuation in dose uniformity. The initial, independent tolerances calculated has the beam position accuracy between 0.6 and 1 mm for all beam sizes. In any case, the technical tolerances can be estimated.

There are three devices that are close to the target area that can cause a beam position change (well maybe five – but focus on the first three). These are the last gantry dipole and the scanning dipoles. (The other two are the scanning quadrupoles. If they are misaligned with respect to the beam trajectory, magnetic field fluctuations could cause beam missteering.) The gantry dipole is at some distance from the target, for the purposes of this estimate 3 m is assumed. The closest scanning dipole can be assumed to be 2.25 m from the target. It is assumed that the field size in the x direction (bend plane of the gantry dipole) is 30 cm. This will change if, for example, upstream scanning magnet(s) are used and will depend on the subsequent optical settings.

Therefore, in this example, the scanning dipole must bend ±3.82 degree (66.67 mrad) to achieve ±15 cm at the target. The accuracy requirement compared to half the field size is (0.25 mm/15 cm) = 0.001667. This corresponds to a tolerance of 0.11 mrad for the scanning dipole. This is an absolute value as was the tolerance number. The magnetic field of the scanning magnet is continually changing. It could be energized (1) to place 235 MeV proton beam at +15 cm or (2) energized to place a 235 MeV proton beam at +1.5 cm. Also, it could be set to (3) place a 100 MeV proton beam at +1.5 cm (and many other configurations). In case (1), the tolerance relative to the full bend angle translates to 0.11 mrad/66.67 mrad = 0.167%. In case (2), the tolerance translates to 0.11 mrad/6.67 mrad = 1.67% relative to the present magnet current setting (not the maximum angle). In case (3), the tolerance is also 1.67% of the current setting (at 1.5 cm). These ratios are with respect to the current setting of the scanning dipole. However, the power supply specifications are normally listed with respect to the maximum current capability of the power supply (the author learned this the hard way). In case (1), the maximum energy beam is bent through the maximum angle, so the magnet current is at its maximum. In case (2), the magnet current is at 10% of maximum and in case (3) the magnet current is at 6.3% of maximum. In cases (1) and (2), at the same energy, it takes the same change in the dipole current to move the beam an amount equal to the tolerance. So the magnet current tolerance, relative to the maximum power supply current, will be 0.167%. However, the absolute amount of power supply current change for the tolerance at 100 MeV compared to the maximum current of the power supply will be 0.63 × 0.167% or 0.105%. Therefore, it's important to identify the lowest setting of the power supply at which the tolerance is desired. The tolerance of the power supply current will be something close to 1×10^{-3}, which is achievable.

This tolerance is the value that the magnetic field has to reach before the beam can be turned on, to ensure the beam is at the correct location when it is turned on. Given the inductance in the magnet system, the tighter the tolerance, the longer the wait. Regarding position tolerances, the word hysteresis has not been mentioned previously, but it does play a role in the specifications of the system including how it will be specified to operate.

The gantry dipole is nominally set to be constant such that the beam is bent 135 degree or 2,356 mrad. A 0.25 mm movement of the beam requires the dipole field to change the bend angle by 0.083 mrad or 0.0035% (3.5×10^{-5}) of the full field at the particular energy it is set to bend. If the momentum is half the maximum momentum (e.g. about a 65 MeV proton beam), the amount of current required to move the beam within tolerance is half what it was at the maximum energy. Therefore, relative to the maximum power supply current, the tolerance at this energy will be 1.75×10^{-5}. This sets the scale for the requirements of the gantry dipole.

The stability requirement is satisfied with a current stability a factor of two higher than the accuracy. Measurements of the beam stability at the nozzle entrance showed the position stability to have a root-mean-square distribution close to 0.1 mm.

Congratulations, clinical (physics-centric) requirements have been identified for scanning, many of the beam parameters have been identified, tolerances have been evaluated and some were translated to equipment technical parameters to enable a safe particle therapy treatment. Hopefully this journey was rewarding and it provided insight into the basic knowledge and connections between the technology and safe clinical utility. Thank you for joining me.

17.6   EXERCISES

1. Finish the analysis of Table 17.1.
2. What are the power supply stability requirements for the scanning dipoles which are designed for 440 MeV/nucleon carbon, in the case of 220 MeV/nucleon carbon beam at a position of 5 cm out of a maximum of 30 cm designed maximum.

18 Afterword

Hundreds of thousands (over 260,000) of patients have been treated with particles. Millions of fields have been delivered. The enumeration of potential error modes with each new set of components discussed is not meant to illicit concern, but rather to show what it took and what it can take to identify potential failure modes and that there are very robust processes to identify mitigations to address these. Understanding the basic principles of the operation of such equipment can help to provide more assurance that the system can perform safely. It can also provide the basis for improvements in the next generation. The particle therapy systems in operation are extremely safe. This safety came about as a result of experienced professionals from multiple disciplines identifying requirements and designs that take safety into account and doing this in a systematic and regulated way. You can help improve that even further. By clearing the way for an appropriate culture of safety, constructive awareness can lead toward open eyes and new ideas. There is always a balance between cost and performance and safety. This is not too difficult to navigate with the methodologies described in this book. Vigilance is needed. Communication is important. Knowledge is important. Teamwork is important. Passing down lessons learned to the next generation is also important, since this knowledge should not be lost. People fear what they don't understand. That is one of the reasons for having written this book. It is hoped that someone who is interested can find in these pages useful information about the multidisciplinary connections between technology and clinical efficacy that make up the field of particle therapy. This can pave the way to follow up with the next generation of faster, more accurate, adaptive charged particle radiotherapy systems with the appropriate context and translational filters necessary for safe implementation.

DOI: 10.1201/9781003123880-18

Acknowledgments

I would like to try to acknowledge all those with whom I talked and interacted, who in no small way led to the topics covered in this book. This list is just too large so I apologize in advance for all the unintended omissions. The long walks with Alejandro Mazal, and the short walks and long talks with Hsaio-Ming Lu and Niek Schreuder were crucial for my understanding of the clinical perspective. The support of Jay Loeffler enabled me to keep on the path of optimizing treatments for our patients. There were a number of individuals whose work with me has already been included and acknowledged elsewhere in this book. I do also want to express my gratitude to Ernie Ihloff, who was a big help in all phases of my work. Thanks also to Harald Paganetti for all his help and support throughout the years.

So much of the work we do depends upon the tools created by our colleagues and that also deserves acknowledgment. For example, some of the optics calculations rely on the PSI graphic framework by Urs Rohrer, based on a version of TRANSPORT by Karl Brown. I remember when I derived the second-order path length terms for TRANSPORT (necessary for my work on the recirculator system), Karl, and Roger Servranckx were very helpful. Urs Rohrer, for that matter, helped me on several occasions throughout the decades. I do also have to say that I wouldn't have believed the capabilities of Microsoft Office until I started working on this book. Finally, the software package of TOPAS has grown to be a power tool. This was published by Shin, Schumann, Faddegon and Paganetti in Med Phys 39, 2012.

It's also important for me to recognize Paul Boisseau, Bill Nett and Chris Pendleton et al. of Pyramid Technical Consultants. From the first day I made my shopping trip at their offices, to the celebration when scanning was complete and the spirited ongoing discussions; these have all been highlights of my professional and personal life.

I also acknowledge the kind sharing of photos from the institutions mentioned in the book. And I would like to thank Taylor and Francis Group for the opportunity to share my work with the rest of the community. Most of all I thank the Particle Radiotherapy community, including the 4,500 members of PTCOG and the 100+ particle facilities in the world for the continued motivation and enthusiasm while I have participated in the growth of this exciting field. It is so rewarding to have had the opportunity to combine many different disciplines and to work with the teams I've been part of to help improve the lives of patients, staff and families.

DOI: 10.1201/9781003123880-19

Appendix A: Particle Therapy Facilities (as of June 2021)*

Country	Who, Where	Particle	S/C/SC* Max. Energy (MeV)	Gantries	Fixed Beams	Scanning	Start of Treatment
Austria	MedAustron, Wiener Neustadt	p	S 253	1	3	Y	2016
Austria	MedAustron, Wiener Neustadt	C	S 403/u	0	3	Y	2019
Belgium	UZ Leuven Particle Proton Centre, Leuven	p	SC 235	1	1	Y	2020
China	WPTC, Wanjie, Zi-Bo	p	C 230	2	1	N	2004
China	SPHIC, Shanghai	p	S 250		3	Y	2014
China	SPHIC, Shanghai	C	S 430/u		3	Y	2014
China	Heavy Ion Cancer Treatment Center, Wuwei, Gansu	C	S 400/u		4	Y	2019
Czech Republic	PTC Czech r.s.o., Prague	p	C 230	3	1	Y	2012
Denmark	Dansk Center for Partikelterapi, Aarhus	p	C 250	3	1	Y	2019
England	Clatterbridge	p	C 62	0	1		1989
England	Rutherford Cancer Centre South Wales, Newport, Wales	p	C 230	1	0	Y	2018
England	The Christie Proton Therapy Center, Manchester	p	C 250	3	0	Y	2018
England	Rutherford Cancer Centre Thames Valley, Berkshire	p	C 230	1	0	Y	2019
England	Rutherford Cancer Centre North East, Northumberland	p	C 230	1	0	Y	2019
France	CAL/IMPT, Nice	p	SC 235	1	1	Y	1991, 2016
France	CPO, Orsay	p	SC 230	1	2	Y	1991, 2014
France	CYCLHAD, Caen	p	SC 230	1	0	Y	2018
Germany	HZB, Berlin	p	C 250	0	1	Y	1998
Germany	HIT, Heidelberg	p	S 250	1	2	Y	2009, 2012
Germany	HIT, Heidelberg	C	S 430/u	1	2	Y	2009, 2012
Germany	WPE, Essen	p	C 230	4	1	Y	2013
Germany	UPTD, Dresden	p	C 230	1	0	Y	2014
Germany	MIT, Marburg	p	S 250	0	4	Y	2015
Germany	MIT, Marburg	C	S 430/u	0	4	Y	2015

(Continued)

Country	Who, Where	Particle	S/C/SC* Max. Energy (MeV)	Gantries	Fixed Beams	Scanning	Start of Treatment
India	Apollo Hospitals PTC, Chennai	p	C 230	2	1	Y	2019
Italy	INFN-LNS, Catania	p	C 60	0	1	N	2002
Italy	CNAO, Pavia	p	S 250	0	4	Y	2011
Italy	CNAO, Pavia	C	S 480/u	0	4	Y	2012
Italy	APSS, Trento	p	C 230	2	1	Y	2014
Japan	HIMAC, Chiba	C	S 800/u	1	3	Y	1994, 2017
Japan	NCC, Kashiwa	p	C 235	2	0	Y	1998
Japan	HIBMC, Hyogo	p	S 230	1	0	N	2001
Japan	HIBMC, Hyogo	C	S 320/u	0	2	N	2002
Japan	PMRC 2, Tsukuba	p	S 250	2	0	Y	2001
Japan	Shizuoka Cancer Center	p	S 235	3	1	N	2003
Japan	STPTC, Koriyama-City	p	S 235	2	1	Y	2008
Japan	GHMC, Gunma	C	S 400/u	0	4	N	2010
Japan	MPTRC, Ibusuki	p	S 250	3	0	Y	2011
Japan	Fukui Prefectural Hospital PTC, Fukui City	p	S 235	2	1	Y	2011
Japan	Nagoya PTC, Nagoya City, Aichi	p	S 250	2	1	Y	2013
Japan	SAGA-HIMAT, Tosu	C	S 400/u	0	4	N	2013
Japan	Hokkaido Univ. Hospital PBTC, Hokkaido	p	S 220	1	0	N	2014
Japan	Aizawa Hospital PTC, Nagano	p	C 235	1	0	N	2014
Japan	i-Rock Kanagawa Cancer Center, Yokohama	C	S 430/u	0	6	N	2015
Japan	Tsuyama Chuo Hospital, Okayama	p	S 235	1	0	N	2016
Japan	PTC Teishinkai Hospital, Sapporo, Hokkaido	p	S 230	1	0	N	2016
Japan	Hakuhokai Group Osaka PT Clinic, Osaka	p	S 235	1	0	N	2017
Japan	Kobe Proton Center, Kobe	p	S 235	1	0	N	2017
Japan	Narita Memorial Proton Center, Toyohashi	p	C 230	1	0	Y	2018
Japan	Osaka Heavy Ion Therapy Center, Osaka	C	S 430/u	0	6	Y	2018
Japan	Hokkaido Ohno Memorial Hospital, Sapporo	p	C 235	1	0	Y	2018

(Continued)

Country	Who, Where	Particle	S/C/SC* Max. Energy (MeV)	Gantries	Fixed Beams	Scanning	Start of Treatment
Japan	Takai Hospital, Tenri City	p	C 230	1	0	Y	2018
Japan	Nagamori Memorial Center of Innovative Cancer Therapy and Research, Kyoto	p	S 220	2	0	Y	2019
Poland	IFJ PAN, Krakow	p	C 230	2	1	N	2011, 2016
Russia	ITEP, Moscow	p	S 250	0	1	N	1969
Russia	JINR 2, Dubna	p	C 200	0	1	N	1999
Russia	MIBS, Saint-Petersburg	p	C 250	2	0	Y	2018
Russia	MRRC, Obninsk	p	S 250	0	1	N	2016
Russia	Federal HighTech Center of FMBA, Dimitrovgrad	p	C 230	4	0	Y	2019
South Korea	KNCC, IIsan	p	C 230	2	1	N	2007
South Korea	Samsung PTC, Seoul	p	C 230	2	0	N	2015
Spain	Quironsalud PTC, Madrid	p	SC 230	1	0	Y	2019
Spain	CUN, Madrid	p	S 220	1	0	Y	2020
Sweden	The Skandion Clinic, Uppsala	p	C 230	2	0	Y	2015
Switzerland	CPT, PSI, Villigen	p	C 250	3	1	Y	1984, 1996, 2013, 2018
Taiwan	Chang Gung Memorial Hospital, Taipei	p	C 230	4	1	Y	2015
Taiwan	Chang Gung Memorial Hospital, Kaohsiung	p	C 230	3	0	Y	2018
Netherlands	UMC PTC, Groningen	p	C 230	2	0	Y	2018
Netherlands	Holland PTC, Delft	p	C 250	2	1	Y	2018
Netherlands	ZON PTC, Maastricht	p	SC 250	1	0	Y	2019
USA, CA	J. Slater PTC, Loma Linda	p	S 250	3	1	N	1990
USA, CA	UCSF-CNL, San Francisco	p	C 60	0	1	N	1994
USA, MA	MGH Francis H. Burr PTC, Boston	p	C 235	2	1	Y	2001
USA, TX	MD Anderson Cancer Center, Houston	p	S 250	3	1	Y	2006
USA, FL	UFHPTI, Jacksonville	p	C 230	3	1	Y	2006
USA, OK	Oklahoma Proton Center, Oklahoma City	p	C 230	1	3	Y	2009
USA, PA	Roberts PTC, UPenn, Philadelphia	p	C 230	4	1	Y	2010

(Continued)

Country	Who, Where	Particle	S/C/SC* Max. Energy (MeV)	Gantries	Fixed Beams	Scanning	Start of Treatment
USA, IL	Chicago Proton Center, Warrenville	p	C 230	1	3	Y	2010
USA, VA	HUPTI, Hampton	p	C 230	4	1	N	2010
USA, NJ	ProCure Proton Therapy Center, Somerset	p	C 230	1	3	Y	2012
USA, WA	SCCA ProCure Proton Therapy Center, Seattle	p	C 230	1	3	Y	2013
USA, MO	S. Lee Kling PTC, Barnes Jewish Hospital, St. Louis	p	SC 250	1	0	N	2013
USA, TN	ProVision Cancer Cares Proton Therapy Center, Knoxville	p	C 230	3	0	Y	2014
USA, CA	California Protons Cancer Therapy Center, San Diego	p	C 250	3	2	Y	2014
USA, LA	Willis Knighton Proton Therapy Cancer Center, Shreveport	p	C 230	1	0	Y	2014
USA, FL	Ackerman Cancer Center, Jacksonville	p	SC 250	1	0	Y	2015
USA, MN	Mayo Clinic Proton Beam Therapy Center, Rochester	p	S 220	4	0	Y	2015
USA, NJ	Laurie Proton Center of Robert Wood Johnson Univ. Hospital, New Brunswick	p	SC 250	1	0	N	2015
USA, TX	Texas Center for Proton Therapy, Irving	p	C 230	2	1	Y	2015
USA, TN	St. Jude Red Frog Events Proton Therapy Center, Memphis	p	S 220	2	1	Y	2015
USA, AZ	Mayo Clinic Proton Therapy Center, Phoenix	p	S 220	4	0	Y	2016
USA, MD	Maryland Proton Treatment Center, Baltimore	p	C 250	4	1	Y	2016
USA, FL	Orlando Health PTC, Orlando	p	SC 250	1	0	N	2016
USA, OH	UH Sideman CC, Cleveland	p	SC 250	1	0	N	2016

(Continued)

Country	Who, Where	Particle	S/C/SC* Max. Energy (MeV)	Gantries	Fixed Beams	Scanning	Start of Treatment
USA, OH	Cincinnati Children's Proton Therapy Center, Cincinnati	p	C 250	3	0	Y	2016
USA, MI	Beaumont Health Proton Therapy Center, Detroit	p	C 230	1	0	Y	2017
USA, FL	Baptist Hospital's Cancer Institute PTC, Miami	p	C 230	3	0	Y	2017
USA, DC	MedStar Georgetown University Hospital PTC, Washington, DC	p	SC 250	1	0	Y	2018
USA, TN	Provision CARES Proton Therapy Center, Nashville	p	C 230	2	0	Y	2018
USA, GA	Emory Proton Therapy Center, Atlanta	p	C 250	3	2	Y	2018
USA, OK	Stephenson Cancer Center, Oklahoma	p	SC 250	1	0	Y	2018
USA, MI	McLaren PTC, Flint	p	S 250/330	3	0	Y	2019
USA, NY	The New York Proton Center, East Harlem, New York	p	C 250	3	1	Y	2019
USA, DC	Johns Hopkins National Proton Center, Washington	p	S 250	3	1	Y	2019
USA, FL	South Florida Proton Institute, SFPTI, Delray Beach	p	C 250	1	0	Y	2019
USA, FL	UFHPTI, Jacksonville	p	C 230	1	0	Y	2019
USA, VA	Inova Schar Cancer Institute PTC, Fairfax	p	C 230	2	0	Y	2020
USA, FL	UM Sylvester Dwoskin Proton Therapy Center, Miami	p	C 250	1	0	Y	2020
USA, AL	University of Alabama PTC, Birmingham	p	C 250	1	0	Y	2020
USA, MA	Gordon Browne Proton Center, MGH, Boston	p	S 250	1	0	Y	2020

* Table courtesy of PTCOG.

* S = Synchrotron

* C = Cyclotron

* SC = SynchroCyclotron

Appendix B: Some Useful Constants

Parameter	Sym	Value	Units	Comment
Constants of Nature				
Speed of light	c	2.9979×10^8	Meters/second (m/s)	
Avogadro's number	N_A	6.02214×10^{23}	Atoms/mole (mole^{-1})	
Planck's constant	h	6.62607×10^{-34}	Joules second (J s)	
Coulomb force constant	k	8.98755×10^9	(Newton meter^2)/Coulombs^2 ((N m^2)/C^2)	$1/4\pi\varepsilon_0$
Vacuum permittivity	ε_0	8.8542×10^{-12}	Farads/meter (F/m)	
Air permeability	μ_A	$4\pi \times 10^{-7}$	Henry/m	
Iron permeability	μ_{Iron}	6.3×10^{-3}	Henry/m	
Energy/Ion pair – air	W	34.3	Joules/Coulomb (J/C)	eV/Ion pair
Particles				
Elementary charge	e	1.60218×10^{-19}	Coulombs (C)	
Electron mass	m_e	9.10938×10^{-31}	Kilograms (kg)	
		0.511	MeV/c^2	
Proton mass	m_p	1.6726×10^{-27}	Kilograms (kg)	
		938.272	MeV/c^2	
Atomic mass units	u	1.66×10^{-27}	kg	Aka – Dalton (Da)
		931.494	MeV/c^2	
Density of air	ρ_{air}	1.225	Kilograms/meter^3 (kg/m^3)	@Standard T and P
Conversions				
Ampere	A	1	Coulomb/s (C/s)	
Electron volt	eV	1.60218×10^{-19}	Joules (J)	
GeV/c	GeV/c	5.34×10^{-19}	(Kilograms meter)/second ((kg m)/s)	(momentum)
Calorie	Cal	4.18	Joules (J)	
Inductance	Henry		(kg m^2)/(s^2 Amp^2)	
			(T m^2)/Amp	
Values				
e	e	2.7183		
π	π	3.14159		

UNITS
International System of Units (SI)

Time	s	Second	
Length	m	Meter	
Mass	kg	Kilogram	
Electric current	A	Ampere	
Magnetic field	T	Tesla	kg/(C s)

Appendix C: Hazard Topics

Help identifying Hazards: prEN 1441

- **ENERGY HAZARDS**
  - Electricity
  - Heat
  - Mechanical force
  - Ionizing radiation
  - Nonionizing radiation
  - Electromagnetic fields
  - Moving parts
  - Suspended masses
  - Patient support device failure
  - Pressure (vessel rupture)
  - Acoustic pressure
  - Vibration
- **BIOLOGICAL HAZARDS**
  - Bio-burden/bio-contamination
  - Bio-incompatibility
  - Incorrect output
  - Incorrect formulation
  - Toxicity
  - (Cross-)infection
  - Pyrogenicity
  - Inability to maintain hygienic safety
- **ENVIRONMENTAL HAZARDS**
  - Electromagnetic interference
  - Inadequate supply of power or coolant
  - Restriction of cooling
  - Likelihood of operation outside prescribed environmental conditions
  - Incompatibility with other devices
  - Accidental mechanical damage
  - Contamination due to waste products and/or device disposal
- **HAZARDS RELATED TO THE USE OF THE DEVICE**
  - Inadequate labeling
  - Inadequate operating instructions
  - Inadequate functional specification for operation of the computerized control system
  - Overcomplicated operating instructions
  - Unavailable or separated operating instructions
  - Use by unskilled personnel
  - Use by untrained personnel
  - Human error
  - Insufficient warning of side effects
  - Inadequate warning of hazards likely with reuse of single use devices
  - Incorrect measurement and other metrological aspects
  - Incorrect diagnosis
  - Erroneous data transfer
  - Misrepresentation of results

- **HAZARDS ARISING FROM FUNCTIONAL FAILURE, MAINTENANCE AND AGEING**
  - Inadequacy of performance characteristics for the intended use
  - Lack of specification for maintenance
  - Inadequate maintenance
  - Lack of adequate determination of end of device life
  - Loss of mechanical integrity
  - Inadequate packaging (contamination and/or deterioration of the device)

Appendix D: Beam QA Frequency Possibility

TABLE D.1
Straw Person QA Table

	SOBP						PBS												Range of Values Used for the Testing
	Measure Online Values	Measure Online Settings	Measure Daily QA Values	Measure Daily QA Settings	Measure Less Frequently	Check the Checks	Measure Online Values	Measure Online Initial Check	Measure Online Settings	Measure Daily QA Values	Measure Daily QA Settings	Measure Less Frequently	Check the Checks	Gantry Angles	PPS Positions	Ranges Tested	Patterns Tested	Currents Tested	Safety Check
Depth Dose Characteristics																			
Energy (range)																			
Range values	Y	Y	Y	N	N	Y	n/a	?	Y	Y	N	Y	Y						M
Range accuracy	n/a	Y	Y	N	N	Y	n/a	?	Y	Y	N	Y	Y						M
Energy spread (distal falloff)																			
Energy spread values	n/a	Y	N	N	Y	n/a	n/a	n/a	Y	N	N	Y	n/a						Yr
Energy spread accuracy	n/a	Y	N	N	Y	n/a	n/a	n/a	Y	N	N	Y	n/a						Yr
Mod width	Y	Y	N	N	N	Y	n/a	n/a	n/a	n/a	N	Y	Y						M
Cross Field Characteristics																			
Beam profile																			
Size values	Y	Y	N	N	N	Y	Y		Y	N	N	Y	Y						M
Setting accuracy	Y	Y	N	N	N	Y	Y		Y	N	N	Y	Y						M
Gaussianess	N	n/a	N	n/a	Y	n/a	Y		Y	N	N	Y	Y						Yr
Beam position																			
Position values	Y	Y	N	N	N	Y	Y		Y	N	N	Y	Y						M
Position accuracy	Y	Y	N	N	N	Y	Y		Y	N	N	Y	Y						M
Time to get at position	n/a	n/a	N	N	Y	n/a	n/a		n/a	Y	N	Y	Y						Yr
Beam velocity																			
Velocity values	n/a	n/a	n/a	n/a	n/a	n/a	n/a		Y	N	N	Y							Yr

(Continued)

TABLE D.1 (Continued)
Straw Person QA Table

	SOBP — Measure Online Values	SOBP — Measure Online Settings	SOBP — Measure Daily QA Values	SOBP — Measure Daily QA Settings	SOBP — Measure Less Frequently	SOBP — Check the Checks	PBS — Measure Online Values	PBS — Measure Online Initial Check	PBS — Measure Online Settings	PBS — Measure Daily QA Values	PBS — Measure Daily QA Settings	PBS — Measure Less Frequently	PBS — Check the Checks	Gantry Angles	PPS Positions	Ranges Tested	Patterns Tested	Currents Tested	Safety Check
Velocity accuracy	n/a	n/a	n/a	n/a	n/a	n/a	n/a	Y	N			N	Y						Yr
Field size	Y	Y	N	N	Y	Y	n/a	Y	N			N	Y						?
Field uniformity	Y	Y	N	N	Y	Y	n/a	Y	N			N	Y						M
Field conformity	n/a	n/a	N	N	Y	Y	Y	Y	N			N	Y						M
SAD	N	N	N	N	Y	Y	n/a	n/a	N			N	Y						M
Dosimetry																			
Number of protons																			
Intensity values	Y	Y	N	N	Y	Y	Y	Y	N			Y	Y						Yr
Intensity accuracy	Y	Y	N	N	Y	Y	Y	Y	N			Y	Y						Yr
Integrated fluence values	Y	Y	Y	N	Y	Y	Y	Y	Y			N	Y						Yr
Integrated fluence accuracy	Y	Y	Y	N	Y	Y	Y	Y	Y			N	Y						Yr
Dark beam current	N	N	N	N		n/a	Y	Y	N			N	Y						M
dI/dt	n/a	n/a	n/a	n/a		n/a	Y	Y	N			N	Y						Yr
Irradiation time																			
Set range time	N	N	N	N	N	n/a	Y	Y	N				Y						Yr
Layer change time	n/a	n/a	n/a	n/a		n/a	Y	Y	N				Y						Yr
Time per layer	n/a	n/a	n/a	n/a		n/a	Y	Y	N				Y						Yr
Number of paintings	n/a	n/a	n/a	n/a		n/a	Y	Y	N				Y						Yr
Gamma index	N	N	N	N	Y	n/a	N	N	?				Y						Yr

Y = Yes, N =No, n/a = Not applicable, ? = Maybe.

OL = Online, D = daily, W = weekly, M = monthly, Yr = yearly.

Check the checks = verify tolerance interlocks and appropriate reactions.

Appendix E: Some Element and Compound Parameters

Element Name	Symbol	Z	A	Z/A	Density (g/cm³)	$\rho/\sqrt{A}$ (g/cm³)	L_R (g/cm²)
Hydrogen	H	1	1.01	0.9921	8.38×10^{-5}	8.34×10^{-5}	63.04
Helium	He	2	4.00	0.4998	1.66×10^{-4}	8.31×10^{-5}	94.32
Lithium	Li	3	6.94	0.4323	0.534	0.203	82.77
Beryllium	Be	4	9.01	0.4439	1.848	0.616	65.19
Boron	B	5	10.81	0.4625	2.37	0.721	52.68
Carbon	C	6	12.01	0.4995	2.21	0.638	42.70
Nitrogen	N	7	14.01	0.4998	1.17×10^{-3}	3.11×10^{-4}	37.99
Water	H₂O	7.5	18	0.417	0.997	0.3525	36.08
Air		7.6	15.2	0.499	$1.22 \times 10^{-3*}$	3.1×10^{-4}	36.62
Oxygen	O	8	16.00	0.5000	1.33×10^{-3}	3.33×10^{-4}	34.24
Fluorine	F	9	19.00	0.4737	1.58×10^{-3}	3.62×10^{-4}	32.93
Neon	Ne	10	20.18	0.4955	8.39×10^{-4}	1.87×10^{-4}	28.93
Sodium	Na	11	22.99	0.4785	0.971	0.203	27.74
Magnesium	Mg	12	24.31	0.4937	1.74	0.353	25.03
Aluminum	Al	13	26.98	0.4818	2.699	0.520	24.01
Silicon	Si	14	28.09	0.4985	2.329	0.439	21.82
Phosphorus	P	15	30.97	0.4843	2.2	0.395	21.21
Sulfur	S	16	32.06	0.4991	2.0	0.353	19.50
Chlorine	Cl	17	35.45	0.4795	2.98×10^{-3}	5.01×10^{-4}	19.28
Argon	Ar	18	39.95	0.4506	1.66×10^{-3}	2.63×10^{-4}	19.55
Potassium	Kr	19	39.10	0.4860	0.862	0.138	17.32
Calcium	Ca	20	40.08	0.4990	1.55	0.245	16.14
Scandium	Sc	21	44.96	0.4671	2.989	0.446	16.55
Titanium	Ti	22	47.87	0.4596	4.54	0.656	16.16
Vanadium	V	23	50.94	0.4515	6.11	0.856	15.84
Chromium	Cr	24	52.00	0.4616	7.18	0.996	14.94
Manganese	Mn	25	54.94	0.4551	7.44	1.004	14.64
Iron	Fe	26	55.85	0.4656	7.874	1.054	13.84
Cobalt	Co	27	58.93	0.4581	8.9	1.159	13.62
Nickel	Ni	28	58.69	0.4771	8.902	1.162	12.68
Copper	Cu	29	63.55	0.4564	8.96	1.124	12.86
Zinc	Zn	30	65.38	0.4589	7.133	0.882	12.43
Gallium	Ga	31	69.72	0.4446	5.904	0.707	12.47
Germanium	Ge	32	72.63	0.4406	5.323	0.625	12.25
Arsenic	As	33	74.92	0.4405	5.73	0.662	11.94
Selenium	Se	34	78.97	0.4305	4.5	0.506	11.91
Bromine	Br	35	79.90	0.4380	7.07×10^{-3}	7.91×10^{-4}	11.42
Krypton	Kr	36	83.80	0.4296	3.49×10^{-3}	3.81×10^{-4}	11.37
Rubidium	Rb	37	85.47	0.4329	1.532	0.166	11.03
Strontium	Sr	38	87.62	0.4337	2.54	0.271	10.76
Yttrium	Y	39	88.91	0.4387	4.469	0.474	10.41

(Continued)

Element Name	Symbol	Z	A	Z/A	Density (g/cm³)	$\rho/\sqrt{A}$ (g/cm³)	L_R (g/cm²)
Zirconium	Zr	40	91.22	0.4385	6.506	0.681	10.2
Niobium	Nb	41	92.91	0.4413	8.57	0.889	9.92
Molybdenum	Mo	42	95.95	0.4377	10.22	1.043	9.8
Technetium	Tc	43	97.00	0.4433	11.5	1.168	9.58
Ruthenium	Ru	44	101.07	0.4353	12.41	1.234	9.48
Rhodium	Rh	45	102.91	0.4373	12.41	1.223	9.27
Palladium	Pd	46	106.42	0.4322	12.02	1.165	9.2
Silver	Ag	47	107.87	0.4357	10.5	1.011	8.97
Cadmium	Cd	48	112.41	0.4270	8.65	0.816	9.00
Indium	In	49	114.82	0.4268	7.31	0.682	8.85
Tin	Sn	50	118.71	0.4212	7.31	0.671	8.82
Antimony	Sb	51	121.76	0.4189	6.691	0.606	8.73
Tellurium	Te	52	127.60	0.4075	6.24	0.552	8.83
Iodine	I	53	126.90	0.4176	4.93	0.438	8.48
Xenon	Xe	54	131.29	0.4113	5.48×10^{-3}	4.79×10^{-4}	8.48
Cesium	Cs	55	132.91	0.4138	1.873	0.162	8.31
Barium	Ba	56	137.33	0.4078	3.5	0.299	8.31
Lanthanum	La	57	138.91	0.4104	6.145	0.521	8.14
Cerium	Ce	58	140.12	0.4139	6.77	0.572	7.96
Praseodymium	Pr	59	140.91	0.4187	6.773	0.571	7.76
Neodymium	Nd	60	144.24	0.4160	7.008	0.584	7.71
Promethium	Pm	61	145.00	0.4207	7.264	0.603	7.51
Samarium	Sm	62	150.36	0.4123	7.52	0.613	7.57
Europium	Eu	63	151.96	0.4146	5.244	0.425	7.44
Gadolinium	Gd	64	157.25	0.4070	7.901	0.630	7.48
Terbium	Tb	65	158.93	0.4090	8.23	0.653	7.36
Dysprosium	Dy	66	162.50	0.4062	8.551	0.671	7.32
Holmium	Ho	67	164.93	0.4062	8.795	0.685	7.23
Erbium	Er	68	167.26	0.4066	9.026	0.698	7.14
Thulium	Tm	69	168.93	0.4084	9.321	0.717	7.03
Ytterbium	Yb	70	173.05	0.4045	6.903	0.525	7.02
Lutetium	Lu	71	174.97	0.4058	9.841	0.744	6.92
Hafnium	Hf	72	178.49	0.4034	13.31	0.996	6.89
Tantalum	Ta	73	180.95	0.4034	16.65	1.238	6.82
Tungsten	W	74	183.84	0.4025	19.3	1.423	6.76
Rhenium	Re	75	186.21	0.4028	21.02	1.540	6.69
Osmium	Os	76	190.23	0.3995	22.57	1.636	6.68
Iridium	Ir	77	192.22	0.4006	22.42	1.617	6.59
Platinum	Pt	78	195.08	0.3998	21.45	1.536	6.54
Gold	Au	79	196.97	0.4011	19.32	1.377	6.46
Mercury	Hg	80	200.59	0.3988	13.55	0.957	6.44
Thallium	Tl	81	204.38	0.3963	11.72	0.820	6.42
Lead	Pb	82	207.20	0.3958	11.35	0.788	6.37
Bismuth	Bi	83	208.98	0.3972	9.747	0.674	6.29
Polonium	Po	84	209.00	0.4019	9.32	0.645	6.16
Astatine	At	85	210.00	0.4048	–	–	6.07
Radon	Rn	86	222.00	0.3874	9.07×10^{-3}	6.08×10^{-4}	6.28
Francium	Fr	87	223.00	0.3901	1.87	0.125	6.19
Radium	Ra	88	226.00	0.3894	5.00	0.333	6.15

(Continued)

Element Name	Symbol	Z	A	Z/A	Density (g/cm³)	$\rho/\sqrt{A}$ (g/cm³)	L_R (g/cm²)
Actinium	Ac	89	227.00	0.3921	10.07	0.668	6.06
Thorium	Th	90	232.04	0.3879	11.72	0.769	6.07
Protactinium	Pa	91	231.04	0.3939	15.37	1.011	5.93
Uranium	U	92	238.03	0.3865	18.95	1.228	6.00
Neptunium	Np	93	237.00	0.3924	20.25	1.315	5.87
Plutonium	Pu	94	244.00	0.3852	19.84	1.270	5.93
Americium	Am	95	243.00	0.3909	13.67	0.877	5.80
Curium	Cm	96	247.00	0.3887	13.51	0.860	5.79
Berkelium	Bk	97	247.00	0.3927	9.86	0.627	5.69
Californium	Cf	98	251.00	0.3904	15.1	0.953	5.68
Einsteinium	Es	99	252.00	0.3929	–	–	5.61

* Like all gasses, this depends on temperature and pressure.
These numbers are for approximations only.
Some numbers from Lawrence Berkeley tables.

Index

This index is divided into subject sections with the following titles: **Acceleration, Attenuation, Beam Dynamics, Beam Parameters, Beam Spreading, Distributions, Exponentials, Gantry, Interactions, People, Radiobiology, Safety Processes,** and **Target Volume.** Some of the same words may appear in multiple sections, but have different meanings.

Printed in the United States
by Baker & Taylor Publisher Services